Advanced Mechanics of Solids and Structures

Advanced Mechanics of Solids and Structures

Dimitrios G Pavlou, PhD

ELSEVIER

Academic Press is an imprint of Elsevier
125 London Wall, London EC2Y 5AS, United Kingdom
525 B Street, Suite 1650, San Diego, CA 92101, United States
50 Hampshire Street, 5th Floor, Cambridge, MA 02139, United States

ISBN: 978-0-443-23778-2

For Information on all Academic Press publications
visit our website at https://www.elsevier.com/books-and-journals

Publisher: Peter Linsley
Acquisitions Editor: Stephen Merken
Editorial Project Manager: Koul Bhaviya
Production Project Manager: Erragounta Saibabu Rao
Cover Designer: Harris Greg

Typeset by MPS Limited, Chennai, India

Printed in the United States of America

Last digit is the print number: 9 8 7 6 5 4 3 2 1

Contents

Preface

Mechanics of Solids and Structures (or Strength of Materials, or Mechanics of Deformable Solids, or Engineering Mechanics) is a crucial field of study in Mechanical, Civil, Aerospace, and Naval Engineering. It deals with materials and structures response to external forces and provides tools for structural design. It is a prerequisite course for Finite Elements, Machine Elements, Concrete Structures, Soil Mechanics, Structural Analysis, and Structural Dynamics. Aim of Mechanics of Solids and Structures is the determination of the stresses and strain field and the estimation of the strength of components and engineering structures.

BSc, MSc, PhD students and professional engineers can be benefited by the analytical description of the physical principles and theorems, and the step-by-step implementation of the theory in real-engineering problems. The book provides the theoretical tools for structural analysis and design, material characterization and selection, failure prediction, and structural optimization.

The book contains a historical overview of the milestones in Mechanics, 14 chapters and an appendix. The theory starts with fundamental concepts and progressively more complicated principles are exhibited, allowing a smooth transition of the reader to deeper knowledge of the methods. The mathematical formulations are analytic and detailed methodologies are explained and implemented in integrated examples.

The **History and Milestones of Mechanics of Solids and Structures** starts from 400 BC where the first scientific foundation of Mechanics was published. All the important works and achievements in the period of the Classical Greece (500-400 BC), the Hellenistic and Roman period (323 BC-476 AD), the Medieval Europe (500-1500), the Early Modern European History (1500-1700), the first Industrial Revolution (1700-1900), the second Industrial Revolution (1900-1950), and the Digital Revolution (1950-today) are summarised and commented. Short CV of the life and the contribution of the pioneers of Mechanics of Solids and Structures is provided.

The **Chapter 1** deals with the fundamental concepts of stress analysis to estimate the strength of materials and structures. The stress vector and tensor in a material element is described and their transformation in a rotated coordinate system is provided. Analytic derivation of the formulae for the principal stresses and the corresponding planes in material elements under triaxial and plane stress conditions is presented and applications in real problems are demonstrated. Differential equations of motion in Cartesian and Polar coordinate system are formulated for analysing the behaviour of solids under various types of stress conditions. This chapter provides the necessary background for understanding the failure criteria of materials and estimate the safety margins of complex structures.

The **Chapter 2** provides the fundamental concepts of strain analysis to estimate the deformation of materials and structures. The chapter starts with the definition of the normal and shear strains and their correlation with displacements. The strain vector and tensor in a material element are described and their transformation in a rotated coordinate system is provided. The strain compatibility equations are formulated for analysing the behaviour of solids under various types of strain conditions. This chapter aims to provide the necessary background for understanding the terms of stiffness, ductility, strain hardening, strain energy density, etc and to investigate the mechanics of material deformation.

The **Chapter 3** describes how the stresses and strains are correlated and how elastic materials respond under multiaxial stress state. The constitutive equations of elasticity for isotropic and anisotropic solids are provided for understanding the effects of the mechanical properties of materials in the mechanical behaviour of structural components. The chapter 3 provides the necessary background for stress and strain analysis of isotropic and composite materials.

Chapter 4 deals with the main mechanical properties of the engineering materials. The basic mechanical tests for quality assessment and characterization of materials are described and the effect of the stress, temperature, strain rate, and the surface quality on the material properties is discussed. The chapter focuses on the methods for measuring the elastic modulus, yield stress, ultimate stress, ductility, fracture toughness, and fatigue endurance limit from the tensile test, impact test, and fatigue test. The chapter aims to provide the necessary background for understanding the nature of the mechanical properties and the material selection criteria for structural design.

Chapter 5 includes the yield criteria of materials under multiaxial stresses and investigates the capacity of beams to carry bending moments beyond the elastic limit. It provides the theoretical tools for the structural design under multiaxial loading conditions and the estimation of the safety margin of structures against yielding. The methodology for the residual stress distribution on beam cross-sections after unloading from the fully plastic state is described and demonstrated on typical examples.

Chapter 6 describes the analysis of beams under bending. Bending induced normal stresses, bending induced shear stresses, and bending induced displacements of beams of compact and thin-wall open cross sections as well as the location of the shear centre are included in the chapter. Special focus is paid on the analysis of beams on elastic foundation where a new technique based on the Fourier Cosine Integral Transform is presented along with implementation on representative examples.

Chapter 7 aims to present the analysis of beams under torsion. The general torsion theory of beams of arbitrary cross-section is presented first, and implementation on beams of elliptical and circular cross section is provided. The method of the Prandtl membrane analogy is presented and torsion of slender and non-slender rectangular, composite, hollow thin-wall single or multicompartment cross-sections is analysed. Analytic formulae for shear stresses and angle of twist versus the torque are derived, and demonstration of the method in characteristic examples is given.

Chapter 8 provides the fundamental energy principles for analysing mechanical and structural engineering problems. It starts with the law of conservation of energy

(1st law of thermodynamics) and the correlation of the stress and displacement components to the elastic energy density of solids. The principles of the Virtual Work, Minimum Potential Energy and the Castigliano's theorem are explained in details and formulae for analysing statically determinate and indeterminate structures are derived. Implementation of Energy methods in force and displacement analysis of typical structures is provided. The Energy methods is the necessary background for understanding the fundamentals of numerical methods like Finite and Boundary Elements in Mechanics.

Chapter 9 deals with the analysis of flat plates. Cylindrical pure bending, bending in two directions, displacement and stress analysis of rectangular plates under general loading, boundary conditions of plate problems, circular plates under axisymmetric loading, and plates on elastic foundation are included in Chapter 9. A new technique based on Hankel's integral transform is used to analyse plates on elastic foundation under axisymmetric loads, and implementation of the method on representative examples is provided.

Chapter 10 presents a powerful theoretical tool for analytical stress analysis of components and structures. Airy's Stress Function is based on the equilibrium equations of stresses and the compatibility equations of strains. It is used for obtaining exact solutions in complex engineering problems. With the aid of this method the difference between exact and approximate solutions for bending stresses in beams can be assessed. The investigation of the effect of holes in the structural integrity and the calculation of the hole-induced stress concentration in structural components is provided.

Chapter 11 includes the fomulation of the equations for the stress distribution within the wall of thick-wall cylinders under internal and external pressure. With the Airy's Stress Function stress and displacement analysis is performed and the governing biharmonic equation is derived. The obtained formulae are applied to analyse compount cylinders after shrink-fitting. Implementation of the method on characteristic problems is provided.

Chapter 12 deals with local stresses associated with the force transfer from one body to another. Stress and deformation analysis on the contact surface between two bodies and the condition assessment of safe force transfer between solids is provided. The chapter aims to investigate the reasons of material failure in the contact surface and to describe the failure mechanism between rolling bodies in contact. The provided formulae are the necessary tools for designing joints, supports, and foundations. Contact mechanics is the background of the principles of engineering tribology.

Chapter 13 is devoted to the understanding of the nature and the consequences of elastic instability of structural members. Classification of elastic bifurcation buckling is provided firstly. With the aid of Euler's theory, analytical formulation of the governing equations and the boundary conditions for calculating the critical loads causing elastic instability in long, intermediate, and short columns and beams is given. The validity range of Euler's formula for buckling is discussed and the Johnson's formula is presented. Formulae for calculating the critical buckling load for beams with geometric imperfection (eccentric loading) and flexural-torsional buckling of beams are derived.

Chapter 14 provides an introduction of the Mechanics of composite materials. With the aid of Kirchhoff's assumption, the Classical Lamination Theory for fiber reinforced laminates is presented analytically. Formulae for stiffness and compliance matrices of multilayered composite materials and the steps for calculating the principal stresses on each layer of a laminate is described. The Tsai –Wu failure criterion is explained, and examples are provided.

The **Appendix** offers a collection of representative Solid Mechanics problems with solutions. The solution of the provided problems require combination of the theory of several chapters of the book and are suitable for the preparation of students for the exam.

Stavanger, spring 2025

Dimitrios G. Pavlou, Ph.D.,
Professor
Department of Mechanical and Structural Engineering
and Materials Science,
University of Stavanger, Norway

History and Milestones of Mechanics of Solids and Structures

The earliest structures may have commenced between 4000 and 2000 BC in ancient Egypt when humans started building shelters for their protection. However, Mechanics of Solids and Structures as an engineering science started in ancient Greece where the first books were found in the Corpus of Aristotle.

Until 200 BC most of the principles of statics of unreformable solids were known and used for making large structures. In the period between 200 BC and 1600 there is a gap in scientific development in Mechanics. This historical period is mainly identified with the period of the European Middle Ages. After 1600, in the early modern European history, the foundation of Mechanics of deformable solids take place and considerable research achievements were published. All the physical principles we use up today were developed with the aid of Mathematics and Mathematical Physics in the period 1660–1920 when the first industrial revolution took place. In 1940–1970, during the period of the 2nd industrial revolution, the theoretical foundations of the numerical methods Finite Elements and Boundary Elements for analysing complex structures were developed and provided the necessary background for the development of the computational mechanics. The digital revolution era starts in the decade of 50s when the first digital computers were developed. After 1980 the Mechanics of Solids and Structures was benefited by the high-speed computers and the development of software packages for analysing engineering structures is the new trend. Nowadays, the Artificial Intelligence (AI) and Machine Learning (ML) have multiplied the human capabilities in designing and health monitoring of smart structures and infrastructures. The milestones in the history of the Mechanics of Solids and Structures are summarized schematically in Figure 1.

MILESTONES IN MECHANICS OF SOLIDS AND STRUCTURES

The first scientific foundation of Mechanics is attributed to the ancient Greek mathematician and engineer **Archytas** in 400 BC. His book was originally entitled in Greek "*Μηχανικά*" (in Latin: Mechanica, in English: Mechanics) and provided rigorous definition and description of the equilibrium condition of the lever, the pulley, and the balance, and expound with considerable success some of the main principles of statics, the law of virtual velocities, the parallelogram of forces, and the law of inertia." (Ross, Sir David, Aristotle, Methuen, London, 1923, p. 12).

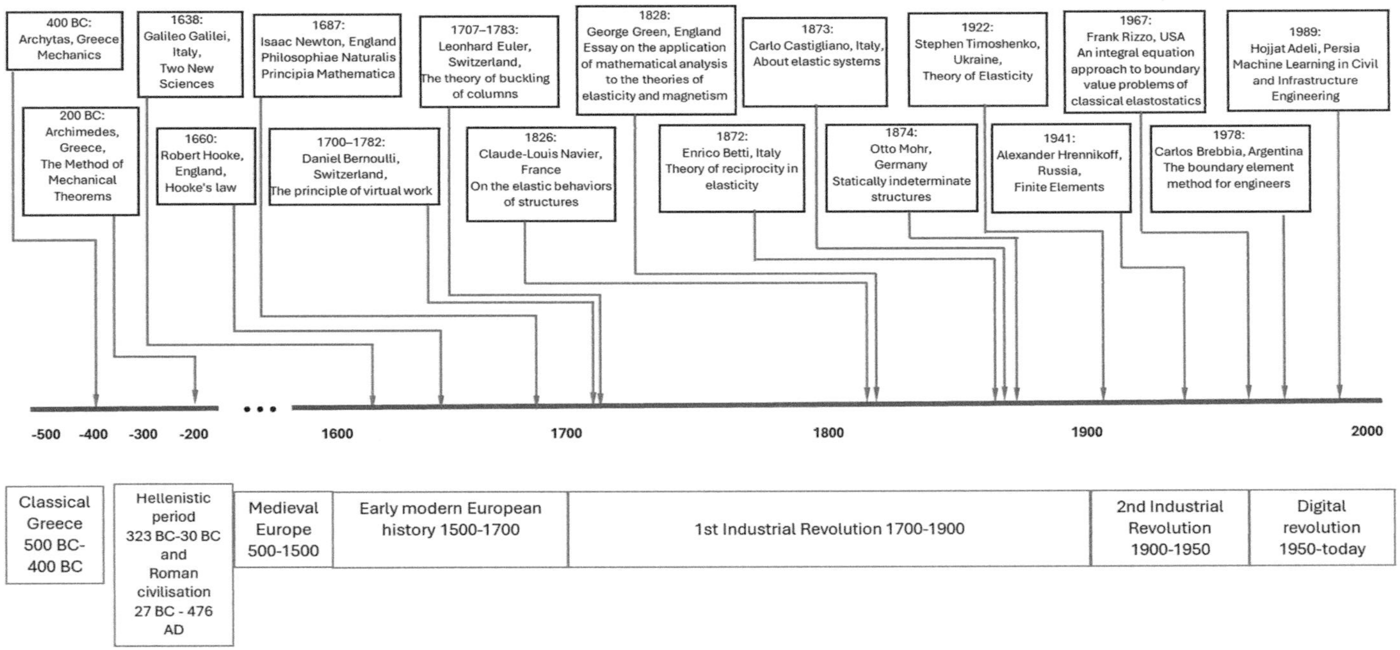

FIG. 1

Milestones in the history of the Mechanics of Solids and Structures.

In 287–212 BC, the ancient Greek mathematician and engineer **Archimedes** published the books "On the Equilibrium of Planes", "The Method of Mechanical Theorems", and "On Floating Bodies" where included a proof of the law of the lever, the widespread use of the concept of center of gravity, and fundamental principles of hydrostatics (Archimedes's principle).

In 1564–1642, the Italian physicist and engineer **Galileo Galilei** made several contributions to what is now known as engineering, as distinct from pure physics. His works "Dynamics" and "Mechanics" were based on Aristotelian–Archimedean fluid dynamics and mechanics and studied velocity, gravity, inertia, and described the properties of the pendulum and hydrostatic balances. His book "Two New Sciences" in 1638, deals with kinematics and the strength of materials.

In 1660, the English architect and engineer **Robert Hooke** discovered the law of elasticity, known as "Hooke's law". He described the linear variation of extension versus force in an elastic spring that was a base for inventing the first mechanical watch to keep time with reasonable accuracy.

In 1687, the English mathematician, physicist, and engineer **Isaac Newton** published the pioneering book Philosophiæ Naturalis Principia Mathematica (Mathematical Principles of Natural Philosophy). This book is considered as the first unification in physics and establishes the classical mechanics. Newton formulated the laws of motion and universal gravitation that dominates the classical mechanics until now.

In 1700–1782, the Swiss mathematician and physicist **Daniel Bernoulli** worked (with Euler) on elasticity and the development of the Euler–Bernoulli beam equation. He proposed the principle of superposition and a key concept in fluid dynamics that relates pressure, density, speed and height which is known as Bernoulli's principle.

In 1707–1783, the Swiss mathematician, physicist and engineer **Leonhard Euler**, contributed to the development of the Euler–Bernoulli beam equation, which became a milestone in Mechanics of Solids and Structures. He achieved to derive a formula for calculating the critical buckling load of a straight bar, which depends only on its length and flexural stiffness. Euler is regarded as one of the most prolific mathematicians in history and the greatest of the 18th century.

In 1821, the French civil engineer **Claude-Louis Navier** formulated the mathematical theory of elasticity. His theory provides good accuracy and is still used for designing engineering structures. In 1826 in the work "On elastic behaviours of structures" he established the elastic modulus as a material property independent of the second moment of area.

In 1828 the British mathematical physicist **George Green** published *An Essay on the application of mathematical analysis to the theories of elasticity and magnetism.* His work in elasticity theory introduced the idea of potential functions and the concept of the "*Green's functions*" that provided the foundation for the development of the Boundary Element Method.

In 1872, **Enrico Betti**, an Italian mathematician proposed the theory of reciprocity in elasticity that is known as "Betti's theorem". This theorem is used to define influence lines in elastic structures and provided a basic tool for the development of the Boundary Element Method.

In 1873 the Italian mathematician and engineer **Carlo Castigliano** published an energy method for determining displacements in an elastic system that is called “Castigliano’s theorem”. This theorem is an effective tool for analysing statically indeterminate structures and is a standard topic in textbooks.

In 1874, **Otto Mohr**, a German structural engineer, formalised the term of a statically indeterminate structure and provided a graphical method for representing stress in two and three dimensions. His method is known as “Mohr’s circle” and used for approximating the stress components on rotated planes. Mohr’s circle is useful for estimating the maximum shear stress and the corresponding plane on a material element.

In 1922 the Ukrainian engineer **Stephen Timoshenko** published seminal monographs in mechanics of solids and structures. His works “Theory of Elasticity”, “Theory of Plates and Shells” and “Theory of Elastic Stability” are cornerstones in the modern Mechanics of Solids and Structures. Timoshenko is considered as the father of modern engineering mechanics.

In 1941, **Alexander Hrennikoff**, a Russian-Canadian structural engineer, published the work “Solution of Problems of Elasticity by the Frame-Work Method, ASME J. Appl. Mech. 8, A619–A715”. In this archival work he developed the lattice analogy for simulating membrane and plate bending of structures and shell buckling problems as a lattice framework. He has also considerable contribution to methods for plastic analysis of metal structures. His research on lattice analogy for modelling the bending of structural elements as a lattice framework provided the foundation of the Finite Element Method and Hrennikoff is considered the originator of this method.

In 1967, the American engineer **Frank Rizzo** published the archival work “An integral equation approach to boundary value problems of classical elastostatics”. This work is a considerable contribution in the development of the Boundary Integral Equation Method.

In 1977, the Argentinian civil engineer **Carlos Brebbia** published (together with J. Dominguez) the work Boundary Element Methods for Potential Problems, Appl Math Model; 1:372–8., and one year later the monograph The Boundary Element Method for Engineers, Pentech Press. His contribution in computational mechanics is enormous and he is considered as the father of the Boundary Element Method.

Hojjat Adeli (1950-), a Persian-American engineer, applied mathematician, and neuroscientist, was the first to introduce Artificial Intelligence (AI) and Machine Learning (ML) in analysing and designing engineering structures and systems. With his 16 ground-breaking books and 650 research papers, he is considered the pioneer on computer-aided structural engineering and mechanics. His impactful work (h-index=141 in November 2024) leaves an indelible footprint in the era of digital revolution.

PIONEERS OF MECHANICS OF SOLIDS AND STRUCTURES

Archytas 435–360 BC was an ancient Greek mathematician, engineer, philosopher and music theorist affiliated with the Pythagorean school. He is the father of mathematical mechanics. He developed the most famous argument for the infinity of the universe in antiquity and he was the first to solve the problem of doubling the cube (the so-called Delian problem) with an ingenious solution. According to Favorinus, Archytas developed a wooden model of a flying dove. His invention was characterized by mechanical ingenuity, and it was perfectly balanced with weights. Archytas was using lines generated by moving objects to derive proportions between magnitudes and was the first to introduce mechanical motions in geometry. Among his achievements is the proof that supernumerary ratios cannot be divided by a mean proportional. Archytas is considered as the most sophisticated music theorist who introduced a mathematical simulation of the musical scales.

Archimedes (287–212 BC) is regarded as the greatest mathematician and engineer of ancient history, and one of the greatest of all time (John M. Henshaw (2014), An Equation for Every Occasion: Fifty-Two Formulas and Why They Matter, Johns Hopkins University Press. p. 68. ISBN 978-1-4214-1492-8). He was one of the first to apply mathematics to statics and hydrostatics. His achievements include a proof of the law of the lever, and the widespread use of the concept of center of gravity (Berggren, J. L. (1976), Spurious Theorems in Archimedes' Equilibrium of Planes: Book I, Archive for History of Exact Sciences, 16 (2): 87–103). He anticipated modern calculus and analysis by applying the concept of the infinitely small and the method of exhaustion to derive and rigorously prove a range of geometrical theorems (Kirfel, Christoph (2013), A generalisation of Archimedes' method, The Mathematical Gazette. 97 (538): 43–52). Archimedes authored many seminal books that have been survived until today. Archimedes's surviving works are "Measurement of a Circle", "The Sand Reckoner", "On the Equilibrium of Planes", "Quadrature of the Parabola", "On the Sphere and Cylinder", "On Spirals", "On Conoids and Spheroids", "On Floating Bodies", "Ostomachion", "The cattle problem", "The Method of Mechanical Theorems", "Apocryphal works", "Archimedes Palimpsest".

Galileo Galilei (1564–1642), was an Italian astronomer, physicist and engineer. He is considered as the father of observational astronomy and the early classical physics. In 1580 Galileo was admitted at the medical school of University of Pisa. One year later, he investigated the motion of a swinging chandelier by air currents and created two pendulums for studying the parameters of pendulum vibration. One hundred years later, his findings were utilized for the invention of an accurate timepiece. After attending a lecture on geometry, Galileo decided to study mathematics and natural philosophy instead of medicine. In 1589, he was appointed professor of mathematics in Pisa but three years later he moved to the University of Padua where he appointed as professor of geometry, mechanics, and astronomy. The thermoscope, and the hydrostatic balance are among his inventions. Apart from mathematics and physics, he was also inspired by art. Galileo made significant discoveries in astronomy and strength of materials.

Robert Hooke (1635–1703) was an English physicist, engineer, astronomer, and architect. He discovered the law of elasticity (Hooke's law) i.e. the linear variation of force with elongation in an elastic spring. His work on elasticity was utilised later for the development of the first accurate, spring-powered watch. Apart from Mechanics, Hook made considerable contribution in astronomy by using a refracting telescope to observe the Great Red Spot of Jupiter, he investigated the formation of the lunar craters and illustrated them in the Micrographia, and observer of the rings of Saturn. He invented a sophisticated joint for torque transfer of shafts whose axes are inclined to each other, the micrometer screw for chieving high precision micro-measurements, and used a silvered glass mirror to develop the first practical telescope. Hooke proposed a model of gravitation. Though his model was not yet universal, however, his hypothesis was more realistic than previous hypotheses.

Isaac Newton (1642–1727) was an English mathematician, physicist, and astronomer. The most significant achievements of Newton were the laws of motion and universal gravitation that were published in Principia and dominated the classical mechanics for centuries until the theory of relativity. He used his model of gravity to derive laws of planetary trajectories. Newton invented the first reflecting telescope, and in 1704 published the archival book Opticks. He proposed a law of heat transfer and calculated the speed of sound. With his proposal of the notion of a Newtonian fluid he offered a significant contribution in hydrodynamics. Apart from his works in mechanics and hydrodynamics, he was among the first researchers who investigated the electricity and provided the idea of the field theory of the electric force. As a mathematician, Newton provided an enormous contribution on calculus with his studies on power series and the method for approximating the roots of a function among other achievements.

Daniel Bernoulli (1700–1782) was a Swiss mathematician and physicist. Bernoulli was born in Groningen, Netherlands, and he was a son of Johann Bernoulli who is recognized as one of the developers of calculus. Daniel Bernoulli is among the pioneers in probability, statistics, and fluid mechanics. He is renowned for his contribution in mechanics of solids, especially in vibration of strings. Together with Euler, Bernoulli proposed the classical beam theory that is a cornerstone in structural mechanics. Apart from the beam theory he also developed mathematical models for plates. In 1753 he proposed for the first time the principle of superposition for linear systems where stated that the response caused by two or more actions is the superposition of the responses that would have been caused by each action individually.

Leonhard Euler (1707–1783) was a Swiss mathematician, physicist, and engineer. He is considered as the most influential scientific mind in mathematics of the 18th century. He proposed important theories on infinitesimal calculus and the theory of functions of a complex variable. Apart from mathematics, he made influential discoveries in mechanics, fluid dynamics, and astronomy. Among his publications are the two volumes of "*Mechanica*" where explained the elastic deformations of deformable solids and the motion of rigid bodies. The Euler–Bernoulli beam equation is regarded as a cornerstone of structural engineering. He also derived the formula for calculating the critical buckling load of an ideal rod which depends only on its elastic modulus, length and moment of inertia. Euler applied analytic mathematics in classical mechanics and determined accurately the orbits of comets and other celestial bodies. In fluid dynamics, Euler predicted the phenomenon of cavitation, and the Euler number was derived in his work on turbines. Apart from the studies on cavitation and turbines, he proposed equations for inviscid flow in fluid dynamics (Euler equations).

Claude-Louis Navier (1785–1836) was a French engineer and physicist. He is the father of the general theory of elasticity (1821) that provided accurate tools for structural design for the first time, and he is considered as the founder of modern structural engineering. Apart from the theory of elasticity, Navier made important contribution in fluid mechanics. The set of the Navier–Stokes equations (1822) that describing the motion of viscous fluids and express momentum balance and conservation of mass is the most important mathematical model in fluid mechanics. This work offered an important tool for studying many physical phenomena (e.g. the air and water pollution, weather forecast, oceanography, etc) and designing aerodynamic efficient aircrafts and cars, study of fluid flow, etc.

George Green (1793–1841) was a British mathematical physicist. He introduced the idea of potential functions in physics, and the concept of what are currently called "*Green's functions*" that is the foundation of the boundary element method in Mechanics. Apart from mathematics, he had enormous contribution in electrical engineering as he created a mathematical theory of electricity and magnetism in his famous work "An *Essay on the Application of Mathematical Analysis to the Theories of Electricity and Magnetism*" that was published in 1828. Green contributed also in the in research of the motion of waves and proposed the Cauchy-Green tensor. According to Albert Einstein (comment during a visit to Nottingham in 1930) Green had been 20 years ahead of his time. Green published important research monographs in Physics and Applied Mechanics: "Mathematical investigations concerning the laws of the equilibrium of fluids analogous to the electric fluid…", "Researches on the vibration of pendulums in fluid media", "On the reflexion and refraction of sound", "On the motion of waves in a variable canal of small depth and width", "Note on the motion of waves in canals", etc.

Enrico Betti (1823–1892) was an Italian mathematician. He published important work on the theory of equations and on topology. Betti (together with Maxwell) discovered the reciprocal work theorem stating that for an elastic structure subjected to two sets of forces, the work done by the first set through the displacements produced by the second set is equal to the work done by the second set through the displacements produced by first set. This theorem together with the "*Green's functions*" provided the foundation of boundary element method in Mechanics of solids and structures.

Carlo Castigliano (1847–1884) was an Italian mathematician and physicist. He proposed a theorem (that currently has his name) for determining displacements in a linear-elastic system based on the partial derivatives of elastic strain energy. His theorem states: "If the strain energy of a linearly elastic structure can be expressed as a function of generalised force, then the partial derivative of the strain energy with respect to generalised force gives the generalised displacement". Although this concept is simple since the change in energy is equal to the causing force times the resulting displacement, the Castigliano's theorem offered an effective tool for analysing statically indetermined structures.

Otto Mohr (1835–1918) was a German civil engineer. At the early period of his life he was working as a railway engineer and designed some famous bridges in Hanover. Later, he was appointed as professor of mechanics at Stuttgart and at Dresden Polytechnic. Mohr was among the first structural engineers who used steel trusses for his constructions. He had significant contribution in structural engineering, and he studied extensively the mechanical stresses and the strength of materials. He is renowned for the "*Mohr's circle*" that provides graphical representation of the stress tensor. Mohr formalised for the first time the concept of the statically indeterminate structures.

Stephen Timoshenko (1878–1972) was a Ukrainian engineer and academician, and a founding member of the Ukrainian Academy of Sciences. In 1905, he moved for one year to the University of Göttingen where he worked with Ludwig Prandtl. In 1906 he was appointed to the Chair of Strengths of Materials at the Kyiv Polytechnic Institute and in 1911 was fired from this position because he signed a protest letter against the Minister of Education. In 1911–1917 he was appointed Professor in the Electrotechnical Institute and the St Petersburg Institute of the Railways. During this period, he worked on the theory of elasticity and elastic stability. In 1922, Timoshenko moved to the United States where started working as an engineer at the Westinghouse Electric Corporation. In 1927 he became professor of engineering mechanics at the University of Michigan and in 1936 he was appointed professor at Stanford University. He was elected fellow of the American Philosophical Society and the United States National Academy of Sciences. His seminal research monographs "Theory of Elasticity", "Theory of Plates and Shells" and "Theory of Elastic Stability" are still widely used and have been translated in 36 languages. Timoshenko is the father of modern engineering mechanics.

Alexander Hrennikoff (1896–1984) was a Russian structural engineer. He was born in Russia but later he moved to Canada. He received his D.Sc degree from the Massachusetts Institute of Technology and became professor of Civil Engineering at the University of British Columbia. With his work in lattice analogy where he modelled the plate bending as a lattice framework, Hrennikoff made the theoretical foundation of the Finite Element Method. However, his work did not attract significant attention at this time because of the lack of powerful computers. Later he extended the lattice models in buckling of shells. He also contributed considerably to the plastic theory of metals.

Frank Rizzo (1938-) was an American Engineer. In 1964 he received his Ph.D. degree from University of Illinois at Urbana-Champaign. At the same year he became Assistant Professor at the University of Washington. In 1987 he was Head of the School of Engineering Sciences and Mechanics at Iowa State University, and in 1989 he returned to the University of Illinois at Urbana-Champaign where he served as Head of the Department of Theoretical and Applied Mechanics. In 1966 he published the work "*An integral equation approach to boundary value problems of classical elastostatics*". This work offered a significant contribution to the boundary integral equation method.

Carlos Brebbia (1948–2018) was an Argentinian civil engineer and Professor. After obtaining his PhD at Southampton University he worked for the Central Electricity Research Laboratories in the UK. He left the Laboratories to take up an academic position at the University of Southampton. During his time at Southampton, he established collaborations and visited as a Visiting Professor many important universities, including an academic year at Princeton. After his appointment as a Full Professor at the University of California, Irvine, he decided to resign his position and return to Southampton to set up the Wessex Institute of Technology. Brebbia is renowned worldwide as the originator of the Boundary Element Method. He has published many important scientific papers, been author of 14 books, co-author of numerous volumes and editor or co-editor of over 500. Professor Brebbia held many special honours, including the Medaille de la Ville de Paris, Echelon Argent; Medaille of the Masonnet Foundation, University of Liege, Belgium; Fellow of the Institution of Mechanical Engineers in the UK; Fellow, and Founding President of the American Society of Civil Engineers

UK Chapter; Honorary PhD at the University of Bucharest; Fellow of the Royal Society of Arts; and Member of the European Academy of Sciences and Arts. He was the founding Editor-in-Chief of the journal "*Engineering Analysis with Boundary Elements*" that is among the main channels for publishing novel works in Boundary Elements.

Hojjat Adeli (1950-) is Emeritus Academy Professor at The Ohio State University where he held the Abba G. Lichtenstein Chaired Professorship for ten years. He received his Ph.D. from Stanford University in 1976 at the age of 26. He has authored over 650 research and scientific publications including 16 ground-breaking and trend setting high technology books in various fields of computer science, engineering, applied mathematics, and medicine, with a Google Scholar h-Index of 141 (Nov. 2024). He has been a Web of Science Highly Cited Researcher and among the World's Most Influential Scientific Minds eight times since 2014 in three categories of Computer Science, Engineering, and Cross-field. He has received 80 awards and honors including eight Honorary Doctorates, several Honorary Professorships, and membership in five different European Academies. He has presented 125 Keynote Lectures at conferences in 47 different countries. He has served as Honorary Editor, Consulting Editor, Associate Editor, or member of Editorial Board of 145 journals. He is the Founder and Editor-in-Chief of the high-impact and prestigious international journals *Computer-Aided Civil and Infrastructure Engineering*, now in 40th year of publication, and *Integrated Computer-Aided Engineering*, now in 33rd year of publication. He has also been the Editor-in-Chief of *International Journal of Neural Systems* since 2005. He is a Distinguished Member of ASCE, and a Fellow of AAAS, IEEE, AIMBE, and American Neurological Association. Two different major publishers have created annual awards in his name: *Hojjat Adeli Award for Innovation in Computing* established by Wiley in 2010 and *Hojjat Adeli Award for Outstanding Contributions in Neural Systems* established by World Scientific in 2011. In 2010, he was profiled as an Engineering Legend in the American Society of Civil Engineers journal of *Leadership and Management in Engineering* authored by a noted biographer of legendary engineers including Leonardo da Vinci, Hardy Cross, George Ferris of the Ferris Wheel fame, and the father of modern engineering mechanics, American Academician of Ukrainian decent Stephen P. Timoshenko. He is a polymath in the tradition of the 9th century Persian polymath Al-Khwarizmi and the 18th century Swiss engineer, mathematician, and physicist Leonard Euler with immense, penetrating, and impactful contributions to many fields including structural engineering and mechanics, intelligent transportation systems, construction engineering and management, design optimization, system engineering, biomedical engineering, signal processing, computer science, neuroscience, and neurology. He holds a U.S. patent in the area large-scale design optimization. With his disciples, recently he was awarded a provisional U.S. patent for an algorithm that can predict seizure in patients with intractable epilepsy twenty minutes in advance of its occurrence. His pioneering research in the area of computational neurology is intended to transform the practice of neurology.

CHAPTER

Stress Theory

1

Chapter Outline

1.1 DEFINITION OF STRESS

Fig. 1.1 shows a solid in equilibrium under external forces. Let's assume an arbitrary cross-section AA′ that divides the solid into two parts I and II. Part II is in equilibrium under the external forces and the internal forces distributed over the cross-section AA′. The intensity of the internal forces is defined by the amount of force per unit area. If the force amount ΔF acts on the elementary surface ΔA, the intensity

$$stress = \lim_{\Delta A \to 0} \frac{\Delta \vec{F}}{\Delta A} \tag{1.1}$$

is called stress. The force $\Delta \vec{F}$ can be analyzed into two components (Fig. 1.1); a component $\Delta \vec{F_N}$ oriented in a direction perpendicular to the surface ΔA, and a component $\Delta \vec{F_S}$ lying on the surface ΔA. With the aid of the components $\Delta \vec{F_N}$ and $\Delta \vec{F_S}$,

Advanced Mechanics of Solids and Structures. DOI: http://doi.org/10.1016/B978-0-443-23778-2.00001-9

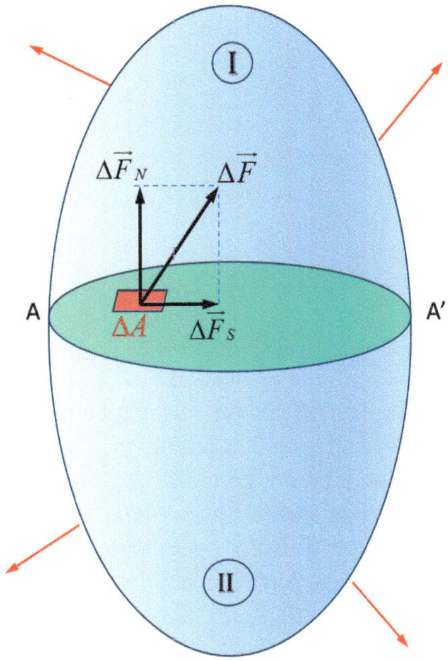

FIG. 1.1

Definition of stress.

the stress can be analyzed to a normal and a shear stress component, σ_N and σ_S, respectively:

$$\sigma_N = \lim_{\Delta A \to 0} \frac{\Delta \overrightarrow{F_N}}{\Delta A} \tag{1.2}$$

$$\sigma_S = \lim_{\Delta A \to 0} \frac{\Delta \overrightarrow{F_S}}{\Delta A} \tag{1.3}$$

The shear stress σ_S can also be symbolized by the symbol τ.

1.2 NOMENCLATURE OF STRESSES—STRESS COMPONENTS OF A POINT

With the aid of a material element in a Cartesian coordinate system, the stress components in the 3D medium can be identified in Fig. 1.2. The dimensions of the three edges of the element are infinitesimal. On each plane of this material, three stress components exist; one normal stress and two shear stresses. On the plane x (i.e., the plane oriented perpendicular to the axis x), the normal stress component

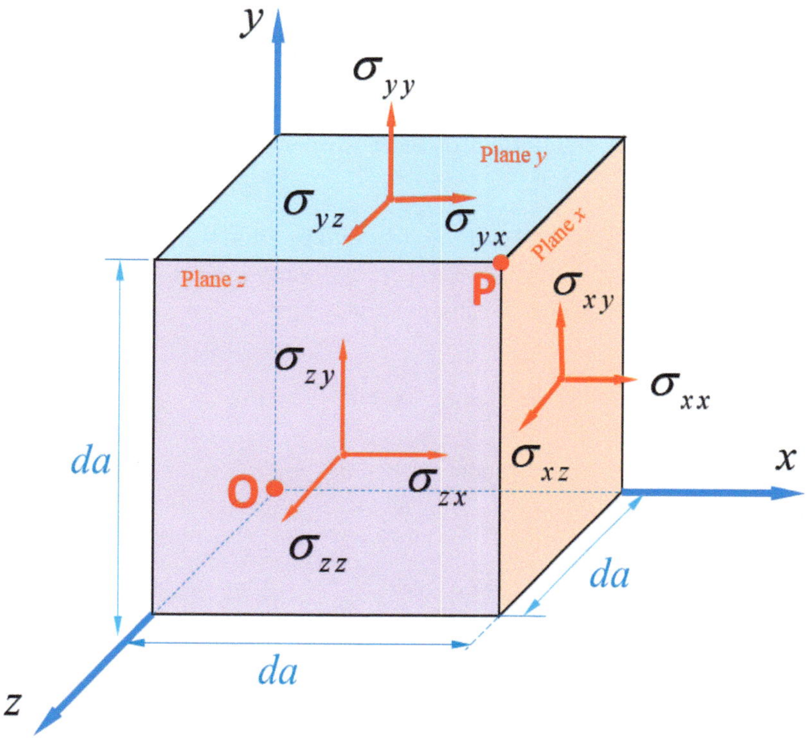

FIG. 1.2

Nomenclature of the stress components on a material element.

is symbolized by σ_{xx} and the shear stress components are symbolized by σ_{xy} and σ_{xz} (Fig. 1.2). The first index on the subscript of the symbol of each stress component indicates the plane where the stress acts, and the second index indicates the direction of the stress component.

We have a total of nine stress components that compose the stress tensor of the point *P*.

1.3 STRESS TENSOR—SYMMETRY OF STRESSES

The stress components of Fig. 1.2 can be tabulated in an array as follows:

$$T = \begin{bmatrix} \sigma_{xx} & \sigma_{xy} & \sigma_{xz} \\ \sigma_{yx} & \sigma_{yy} & \sigma_{yz} \\ \sigma_{zx} & \sigma_{zy} & \sigma_{zz} \end{bmatrix} \tag{1.4}$$

The above array is called the "stress tensor." The first row includes the stresses in plane x, the second the stresses in plane y, and the third row includes the stresses in plane z. The first column includes stresses acting on the x axis, while the second and the third are stresses acting on the y and z axes, respectively.

By considering the equilibrium of shear stress-induced moments (Fig. 1.3) in point A

$$(\sigma_{zy} \cdot dA) \cdot da - (\sigma_{yz} \cdot dA) \cdot da = 0$$

we get $\sigma_{zy} = \sigma_{yz}$. Following the same concept, we get $\sigma_{zy} = \sigma_{yz}, \sigma_{zx} = \sigma_{xz}, \sigma_{xy} = \sigma_{yx}$, or in general:

$$\sigma_{ij} = \sigma_{ji} \quad i \neq j \tag{1.5}$$

Therefore only the six components $\sigma_{xx}, \sigma_{yy}, \sigma_{zz}, \sigma_{xy}, \sigma_{yz}, \sigma_{xz}$ are sufficient to describe the stress state of a point.

Then the stress tensor in Eq. (1.4) can be rewritten using only six components:

$$T = \begin{bmatrix} \sigma_{xx} & \sigma_{xy} & \sigma_{xz} \\ \sigma_{xy} & \sigma_{yy} & \sigma_{yz} \\ \sigma_{xz} & \sigma_{yz} & \sigma_{zz} \end{bmatrix} \tag{1.6}$$

It is also common to express the stress tensor as:

$$T = \begin{bmatrix} \sigma_{x} & \tau_{xy} & \tau_{xz} \\ \tau_{xy} & \sigma_{y} & \tau_{yz} \\ \tau_{xz} & \tau_{yz} & \sigma_{z} \end{bmatrix} \tag{1.7}$$

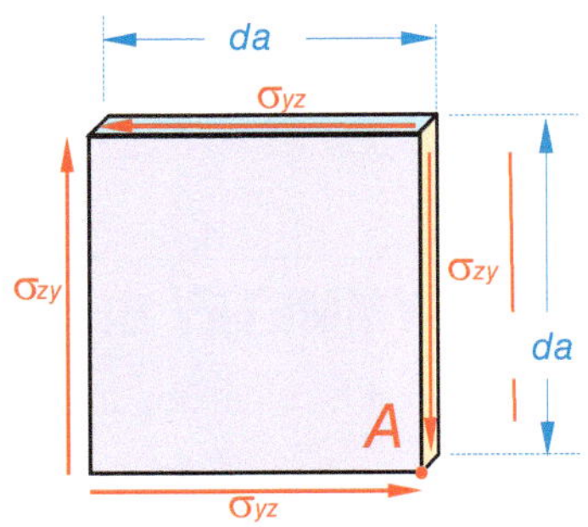

FIG. 1.3

Shear stress equilibrium.

1.4 STRESSES COMPONENTS ON AN ARBITRARY PLANE *ABC*

On a plane *ABC* of the material element in Fig. 1.4, a stress vector $\overrightarrow{\sigma_P}$ is acting. We aim now to correlate $\overrightarrow{\sigma_P}$ as a function of the other stress components of the stress tensor. The direction of the plane *ABC* is specified by a vector oriented perpendicular to the plane *ABC*.

To calculate $\overrightarrow{\sigma_P}$, we should specify the angles of the vector $\overrightarrow{ABC}$ with respect to the axes *x*, *y*, and *z*, as shown in Fig. 1.5. We can define the directional cosines *l*, *m*, and *n* as:

$$\begin{aligned} l &= \cos\ \theta_x \\ m &= \cos\ \theta_y \\ n &= \cos\ \theta_z \end{aligned} \tag{1.8}$$

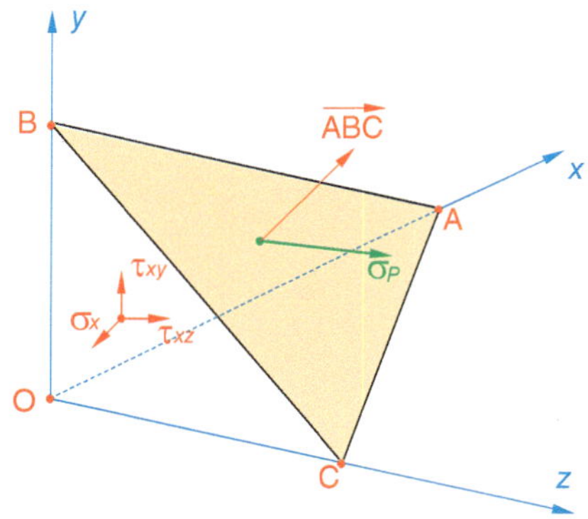

FIG. 1.4

Stress components on an arbitrary plane *ABC*.

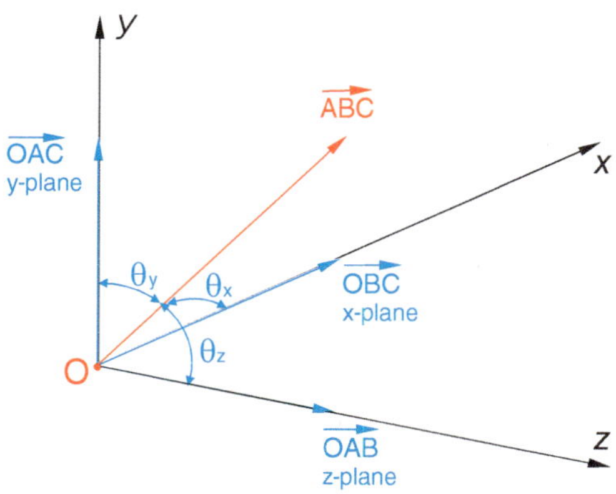

FIG. 1.5

Vectors of the planes $\overrightarrow{ABC}$, $\overrightarrow{OAB}$, $\overrightarrow{OBC}$, and $\overrightarrow{OAC}$.

Then, the magnitude of the planes $\overrightarrow{OAB}$, $\overrightarrow{OBC}$, and $\overrightarrow{OAC}$ can be calculated by the following equations:

$$
\begin{aligned}
(OBC) &= (ABC)\cos\ \theta_x \\
(OAC) &= (ABC)\cos\ \theta_y \\
(OAB) &= (ABC)\cos\ \theta_z
\end{aligned}
$$

or

$$
\begin{aligned}
(OBC) &= (ABC)\cdot l \\
(OAC) &= (ABC)\cdot m \\
(OAB) &= (ABC)\cdot n
\end{aligned}
\tag{1.9}
$$

With the aid of Eq. (1.9), the equilibrium of forces in the volume OABC (Fig. 1.4),

$$\vec{\sigma}_P\cdot(ABC) = \vec{\sigma}_x\cdot(OBC) + \vec{\sigma}_y\cdot(OAC) + \vec{\sigma}_z\cdot(OAB) \tag{1.10}$$

yields:

$$\vec{\sigma}_P\cdot(ABC) = \vec{\sigma}_x\cdot(ABC)\cdot l + \vec{\sigma}_y\cdot(ABC)\cdot m + \vec{\sigma}_z\cdot(ABC)\cdot n$$

or

$$\vec{\sigma}_P = \vec{\sigma}_x\cdot l + \vec{\sigma}_y\cdot m + \vec{\sigma}_z\cdot n \tag{1.11}$$

The component of $\vec{\sigma}_P$ on the axis x can be derived by the following equation:

$$\sigma_{Px} = \vec{\sigma}_P\cdot\vec{i} = \left(\vec{\sigma}_x\cdot l + \vec{\sigma}_y\cdot m + \vec{\sigma}_z\cdot n\right)\cdot\vec{i} \tag{1.12}$$

Taking into account the stress vector of the planes x, y, and z,

$$
\begin{aligned}
\vec{\sigma}_x &= \sigma_{xx}\vec{i} + \tau_{xy}\vec{j} + \tau_{xz}\vec{k} \\
\vec{\sigma}_y &= \tau_{xy}\vec{i} + \sigma_{yy}\vec{j} + \tau_{yz}\vec{k} \\
\vec{\sigma}_z &= \tau_{xz}\vec{i} + \tau_{yz}\vec{j} + \sigma_{zz}\vec{k}
\end{aligned}
\tag{1.13}
$$

Eq. (1.12) can be written as:

$$
\begin{aligned}
\sigma_{Px} = &\left(\sigma_{xx}\vec{i} + \tau_{xy}\vec{j} + \tau_{xz}\vec{k}\right)\cdot\vec{i}\cdot l + \\
&\left(\tau_{xy}\vec{i} + \sigma_{yy}\vec{j} + \tau_{yz}\vec{k}\right)\cdot\vec{i}\cdot m + \\
&\left(\tau_{xz}\vec{i} + \tau_{yz}\vec{j} + \sigma_{zz}\vec{k}\right)\cdot\vec{i}\cdot n
\end{aligned}
\tag{1.14}
$$

Considering the scalar product of unit vectors:

$$\begin{aligned} &\vec{i}\cdot\vec{i}=1 \quad \vec{j}\cdot\vec{i}=0 \quad \vec{k}\cdot\vec{i}=0 \\ &\vec{i}\cdot\vec{j}=0 \quad \vec{j}\cdot\vec{j}=1 \quad \vec{k}\cdot\vec{j}=0 \\ &\vec{i}\cdot\vec{k}=0 \quad \vec{j}\cdot\vec{k}=0 \quad \vec{k}\cdot\vec{k}=1 \end{aligned} \tag{1.15}$$

the expression for σ_{Px} in Eq. (1.14) can be simplified:

$$\sigma_{Px} = \sigma_{xx}\cdot l + \tau_{xy}\cdot m + \tau_{xz}\cdot n \tag{1.16}$$

With a similar procedure, the following components in the y and z directions can be obtained:

$$\sigma_{Py} = \tau_{xy}\cdot l + \sigma_{yy}\cdot m + \tau_{yz}\cdot n \tag{1.17}$$

$$\sigma_{Pz} = \tau_{xz}\cdot l + \tau_{yz}\cdot m + \sigma_{zz}\cdot n \tag{1.18}$$

Eqs. (1.16)–(1.18) can be written in a matrix format:

$$\begin{Bmatrix} \sigma_{Px} \\ \sigma_{Py} \\ \sigma_{Pz} \end{Bmatrix} = \begin{bmatrix} \sigma_x & \tau_{xy} & \tau_{xz} \\ \tau_{yx} & \sigma_y & \tau_{yz} \\ \tau_{zx} & \tau_{zy} & \sigma_z \end{bmatrix} \begin{Bmatrix} l \\ m \\ n \end{Bmatrix} \tag{1.19}$$

The above equation expresses the components of the stress vector on any arbitrary plane that has orientation l, m, and n.

1.5 NORMAL AND SHEAR STRESS ON AN ARBITRARY PLANE *ABC*

Let's now calculate the normal and shear stress components of a stress vector $\vec{\sigma_P}$ on an arbitrary plane *ABC* (Fig. 1.6). To this end, the following formula for the unit vector of $\overrightarrow{ABC}$ can be used:

$$\vec{N} = l\cdot\vec{i} + m\cdot\vec{j} + n\cdot\vec{k} \tag{1.20}$$

- Normal stress

The normal stress on the plane *ABC* is the projection of the vector $\vec{\sigma_P}$ on the vector $\vec{N}$. The magnitude of $\vec{\sigma_P}$ is the scalar product of $\vec{\sigma_P}$ and $\vec{N}$:

$$|\sigma_{PN}| = \vec{\sigma_P}\cdot\vec{N} = (\vec{\sigma_x}\cdot l + \vec{\sigma_y}\cdot m + \vec{\sigma_z}\cdot n)\left(l\vec{i} + m\vec{j} + n\vec{k}\right) \tag{1.21}$$

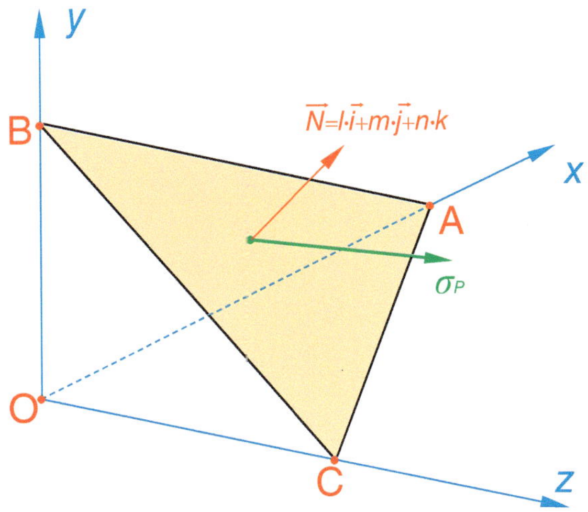

FIG. 1.6

Stress vector $\vec{\sigma}_P$ and plane unit vector $\vec{N}$ on an arbitrary plane *ABC*.

or

$$
\begin{aligned}
|\sigma_{PN}| = {} & l^2\vec{\sigma}_x \cdot \vec{i} + lm\vec{\sigma}_x \cdot \vec{j} + ln\vec{\sigma}_x \cdot \vec{k} + \\
& ml\vec{\sigma}_y \cdot \vec{i} + m^2\vec{\sigma}_y \cdot \vec{j} + mn\vec{\sigma}_y \cdot \vec{k} + \\
& nl\vec{\sigma}_z \cdot \vec{i} + nm\vec{\sigma}_z \cdot \vec{j} + n^2\vec{\sigma}_z \cdot \vec{k}
\end{aligned}
\tag{1.22}
$$

With the aid of Eq. (1.13), the following result can be obtained:

$$
\begin{aligned}
|\sigma_{PN}| = {} & l^2\sigma_x + lm\tau_{xy} + ln\tau_{xz} + ml\tau_{xy} + m^2\sigma_y + mn\tau_{yz} + nl\tau_{xz} + nm\tau_{yz} \\
& + n^2\sigma_{zz}
\end{aligned}
\tag{1.23}
$$

or

$$
|\sigma_{PN}| = l^2\sigma_x + m^2\sigma_y + n^2\sigma_{zz} + 2lm\tau_{xy} + 2ln\tau_{xz} + 2mn\tau_{yz} \tag{1.24}
$$

- Shear stress

From Fig. 1.7, we see that the magnitude of the shear stress $\vec{\sigma}_{PS}$ can be calculated with the aid of the Pythagorean theorem:

$$
|\sigma_{PS}| = \sqrt{|\sigma_P|^2 - |\sigma_{PN}|^2} \tag{1.25}
$$

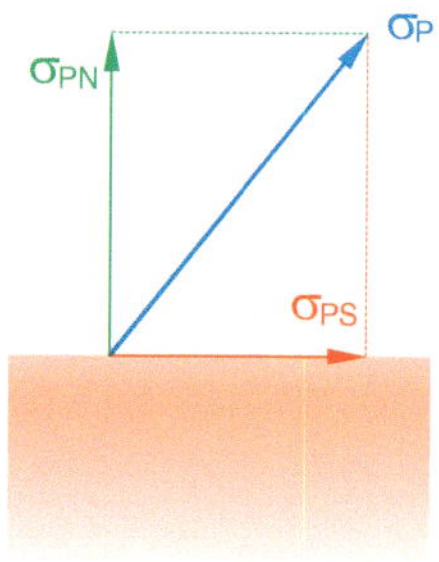

FIG. 1.7

Shear and normal stress components of $\vec{\sigma_P}$

The quantities $|\sigma_P|^2$ and $|\sigma_{PN}|^2$ can be calculated with the aid of Eqs. (1.19) and (1.24):

$$\begin{aligned} |\sigma_P|^2 &= |\sigma_{Px}|^2 + |\sigma_{Py}|^2 + |\sigma_{Pz}|^2 = \\ &= \left(\sigma_{xx} \cdot l + \tau_{xy} \cdot m + \tau_{xz} \cdot n\right)^2 + \\ &+ \left(\tau_{xy} \cdot l + \sigma_{yy} \cdot m + \tau_{yz} \cdot n\right)^2 + \\ &+ \left(\tau_{xz} \cdot l + \tau_{yz} \cdot m + \sigma_{zz} \cdot n\right)^2 \end{aligned} \tag{1.26}$$

and

$$|\sigma_{PN}|^2 = (l^2\sigma_x + m^2\sigma_y + n^2\sigma_{zz} + 2lm\tau_{xy} + 2ln\tau_{xz} + 2mn\tau_{yz})^2 \tag{1.27}$$

1.6 STRESS VECTOR TRANSFORMATION TO A ROTATED COORDINATE SYSTEM

Denoting by $\vec{i}, \vec{j}, \vec{k}$ the unit vectors of the reference coordinate system and $\vec{i}', \vec{j}', \vec{k}'$ the unit vectors of the rotated one (Fig. 1.8), the rotation matrix, including the directional cosines of the angles between the axes, is:

		$\vec{i}$ x	$\vec{j}$ y	$\vec{k}$ z
$\vec{i}'$	X	l_1	m_1	n_1
$\vec{j}'$	Y	l_2	m_2	n_2
$\vec{k}'$	Z	l_3	m_3	n_3

(1.28)

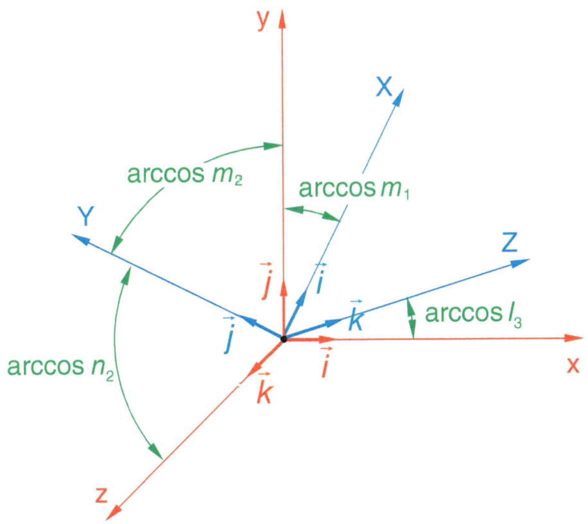

FIG. 1.8

Rotated coordinate system *X-Y-Z* versus reference coordinate system *x-y-z*.

The above matrix is called the transformation matrix and is written as:

$$[A] = \begin{bmatrix} l_1 & m_1 & n_1 \\ l_2 & m_2 & n_2 \\ l_3 & m_3 & n_3 \end{bmatrix} \tag{1.29}$$

With the aid of the matrix in Eq. (1.28), we can express the unit vectors of the reference system as a function of the unit vectors of the rotated system:

$$\begin{aligned} \vec{i} &= l_1 \vec{i'} + l_2 \vec{j'} + l_3 \vec{k'} \\ \vec{j} &= m_1 \vec{i'} + m_2 \vec{j'} + m_3 \vec{k'} \\ \vec{k} &= n_1 \vec{i'} + n_2 \vec{j'} + n_3 \vec{k'} \end{aligned} \tag{1.30}$$

The stress vector at a point *P* can be expressed versus the unit vectors of both coordinate systems:

Reference coordinate system:

$$\vec{\sigma_P} = \sigma_{P_x} \vec{i} + \sigma_{P_y} \vec{j} + \sigma_{P_z} \vec{k} \tag{1.31}$$

Rotated coordinate system:

$$\vec{\sigma}_p = \sigma_{PX}\vec{i'} + \sigma_{PY}\vec{j'} + \sigma_{PZ}\vec{k'} \tag{1.32}$$

Combining the above equations, the following formula can be obtained:

$$\sigma_{Px}\vec{i} + \sigma_{Py}\vec{j} + \sigma_{Pz}\vec{k} = \sigma_{PX}\vec{i'} + \sigma_{PY}\vec{j'} + \sigma_{PZ}\vec{k'} \tag{1.33}$$

With the aid of Eq. (1.30), the above equation yields:

$$\begin{aligned}\sigma_{PX}\vec{i'} + \sigma_{PY}\vec{j'} + \sigma_{PZ}\vec{k'} = {} & \sigma_{Px}\left(l_1\vec{i'} + l_2\vec{j'} + l_3\vec{k'}\right) + \\ & + \sigma_{Py}\left(m_1\vec{i'} + m_2\vec{j'} + m_3\vec{k'}\right) + \\ & + \sigma_{Pz}\left(n_1\vec{i'} + n_2\vec{j'} + n_3\vec{k'}\right)\end{aligned} \tag{1.34}$$

Rearranging the terms of the above equation, the following formula can be obtained:

$$\begin{aligned}\sigma_{PX}\vec{i'} + \sigma_{PY}\vec{j'} + \sigma_{PZ}\vec{k'} = {} & \left(\sigma_{Px}l_1 + \sigma_{Py}m_1 + \sigma_{Pz}n_1\right)\vec{i'} + \\ & + \left(\sigma_{Px}l_2 + \sigma_{Py}m_2 + \sigma_{Pz}n_2\right)\vec{j'} + \\ & + \left(\sigma_{Px}l_3 + \sigma_{Py}m_3 + \sigma_{Pz}n_3\right)\vec{k'}\end{aligned} \tag{1.35}$$

The coefficients of the unit vectors on both sides should be equal:

$$\left.\begin{aligned}\sigma_{PX} &= \sigma_{Px}l_1 + \sigma_{Py}m_1 + \sigma_{Pz}n_1 \\ \sigma_{PY} &= \sigma_{Px}l_2 + \sigma_{Py}m_2 + \sigma_{Pz}n_2 \\ \sigma_{PZ} &= \sigma_{Px}l_3 + \sigma_{Py}m_3 + \sigma_{Pz}n_3\end{aligned}\right\} \tag{1.36}$$

or in matrix form:

$$\begin{Bmatrix}\sigma_{PX} \\ \sigma_{PY} \\ \sigma_{PZ}\end{Bmatrix} = \begin{bmatrix} l_1 & m_1 & n_1 \\ l_2 & m_2 & n_2 \\ l_3 & m_3 & n_3\end{bmatrix}\begin{Bmatrix}\sigma_{Px} \\ \sigma_{Py} \\ \sigma_{Pz}\end{Bmatrix} \tag{1.37}$$

The above equation correlates the magnitude of the stress vector of the rotated system to the magnitude of the stress vector of the reference coordinate system.

1.7 STRESS TENSOR TRANSFORMATION TO A ROTATED COORDINATE SYSTEM

The transformation of the stress tensor of a rotated coordinate system *X-Y-Z* versus the stress tensor of the reference coordinate system *x-y-z* (Fig. 1.9) can be expressed by the following general function f:

$$\begin{bmatrix} \sigma_{XX} & \sigma_{XY} & \sigma_{XZ} \\ \sigma_{XY} & \sigma_{YY} & \sigma_{YZ} \\ \sigma_{XZ} & \sigma_{YZ} & \sigma_{ZZ} \end{bmatrix} = f\left(\begin{bmatrix} \sigma_{xx} & \sigma_{xy} & \sigma_{xz} \\ \sigma_{xy} & \sigma_{yy} & \sigma_{yz} \\ \sigma_{xz} & \sigma_{yz} & \sigma_{zz} \end{bmatrix}\right) \tag{1.38}$$

The direction cosines of the angles between the axes of both systems are included in Eq. (1.28). Eq. (1.29) provides the transformation matrix [A].

The arbitrary plane *ABC* can be defined by the vector $\overrightarrow{ABC} = (ABC) \cdot \overrightarrow{N}$, where $\overrightarrow{N}$ is the unit vector of $\overrightarrow{ABC}$ and is oriented perpendicular to the plane *ABC*.

The unit vector $\overrightarrow{N}$ can be expressed with respect to the axes of both the reference and the rotated coordinate system. The directional cosines of both systems are:

l, m, and n with respect to the axes x, y, and z.

l', m', and n' with respect to the axes X, Y, and Z.

According to Eq. (1.19), the stress components of a vector $\overrightarrow{\sigma_P}$ on a plane *ABC* can be expressed with respect to both coordinate systems:

$$\begin{Bmatrix} \sigma_{Px} \\ \sigma_{Py} \\ \sigma_{Pz} \end{Bmatrix} = \begin{bmatrix} \sigma_x & \tau_{xy} & \tau_{xz} \\ \tau_{yx} & \sigma_y & \tau_{yz} \\ \tau_{zx} & \tau_{zy} & \sigma_z \end{bmatrix} \begin{Bmatrix} l \\ m \\ n \end{Bmatrix} \quad \textit{for coordinate system } x - y - z \tag{1.39}$$

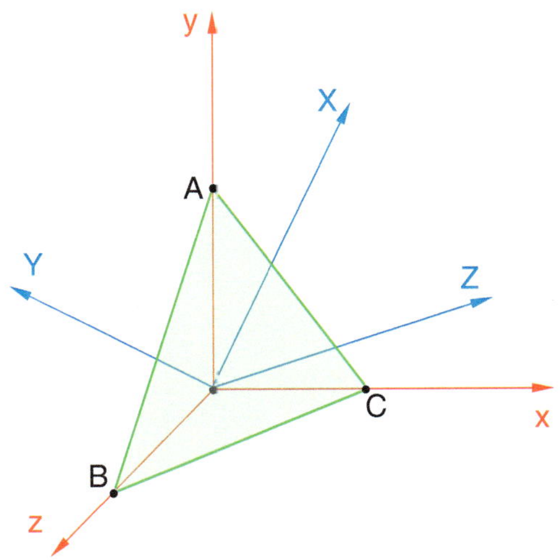

FIG. 1.9

Arbitrary plane *ABC* in the reference and the rotated coordinate systems.

$$\begin{Bmatrix} \sigma_{PX} \\ \sigma_{PY} \\ \sigma_{PZ} \end{Bmatrix} = \begin{bmatrix} \sigma_X & \tau_{XY} & \tau_{XZ} \\ \tau_{XY} & \sigma_Y & \tau_{YZ} \\ \tau_{XZ} & \tau_{YZ} & \sigma_Z \end{bmatrix} \begin{Bmatrix} l' \\ m' \\ n' \end{Bmatrix} \quad \textit{for coordinate system } X-Y-Z \tag{1.40}$$

With the aid of Eqs. (1.37) and (1.39), the above equations yield:

$$\begin{bmatrix} \sigma_X & \tau_{XY} & \tau_{XZ} \\ \tau_{XY} & \sigma_Y & \tau_{YZ} \\ \tau_{XZ} & \tau_{YZ} & \sigma_Z \end{bmatrix} \begin{Bmatrix} l' \\ m' \\ n' \end{Bmatrix} = \begin{bmatrix} l_1 & m_1 & n_1 \\ l_2 & m_2 & n_2 \\ l_3 & m_3 & n_3 \end{bmatrix} \begin{bmatrix} \sigma_x & \tau_{xy} & \tau_{xz} \\ \tau_{yx} & \sigma_y & \tau_{yz} \\ \tau_{zx} & \tau_{zy} & \sigma_z \end{bmatrix} \begin{Bmatrix} l \\ m \\ n \end{Bmatrix} \tag{1.41}$$

On the other hand, the vector $\overrightarrow{N}$ can be expressed with respect to both coordinate systems:

Reference coordinate system:

$$\overrightarrow{N} = l \cdot \overrightarrow{i} + m \cdot \overrightarrow{j} + n \cdot \overrightarrow{k} \tag{1.42}$$

Rotated coordinate system:

$$\overrightarrow{N} = l' \cdot \overrightarrow{i'} + m' \cdot \overrightarrow{j'} + n' \cdot \overrightarrow{k'} \tag{1.43}$$

Combining Eqs. (1.42) and (1.43), the following formula can be obtained:

$$l \cdot \overrightarrow{i} + m \cdot \overrightarrow{j} + n \cdot \overrightarrow{k} = l' \cdot \overrightarrow{i'} + m' \cdot \overrightarrow{j'} + n' \cdot \overrightarrow{k'} \tag{1.44}$$

The unit vectors $\overrightarrow{i}$, $\overrightarrow{j}$, and $\overrightarrow{k}$ can be expressed by the projections of the unit vectors $\overrightarrow{i'}$, $\overrightarrow{j'}$, and $\overrightarrow{k'}$ on the axes x, y, and z:

$$\left.\begin{aligned} \overrightarrow{i} &= l_1 \overrightarrow{i'} + l_2 \overrightarrow{j'} + l_3 \overrightarrow{k'} \\ \overrightarrow{j} &= m_1 \overrightarrow{i'} + m_2 \overrightarrow{j'} + m_3 \overrightarrow{k'} \\ \overrightarrow{k} &= n_1 \overrightarrow{i'} + n_2 \overrightarrow{j'} + n_3 \overrightarrow{k'} \end{aligned}\right\} \tag{1.45}$$

With the aid of the above equations, Eq. (1.44) yields:

$$l \cdot \left(l_1 \overrightarrow{i'} + l_2 \overrightarrow{j'} + l_3 \overrightarrow{k'}\right) + m \cdot \left(m_1 \overrightarrow{i'} + m_2 \overrightarrow{j'} + m_3 \overrightarrow{k'}\right) + n \cdot \left(n_1 \overrightarrow{i'} + n_2 \overrightarrow{j'} + n_3 \overrightarrow{k'}\right)$$
$$= l' \cdot \overrightarrow{i'} + m' \cdot \overrightarrow{j'} + n' \cdot \overrightarrow{k'}$$

or

$$\begin{aligned} &(ll_1 + mm_1 + nn_1) \cdot \overrightarrow{i'} + (ll_2 + mm_2 + nn_2) \cdot \overrightarrow{j'} + (ll_3 + mm_3 + nn_3) \cdot \overrightarrow{k'} \\ &= l' \cdot \overrightarrow{i'} + m' \cdot \overrightarrow{j'} + n' \cdot \overrightarrow{k'} \end{aligned} \tag{1.46}$$

The coefficients of the unit vectors in the left side of the above equation should be equal to the coefficients of the unit vectors on the right side:

$$\left.\begin{aligned} l' &= ll_1 + mm_1 + nn_1 \\ m' &= ll_2 + mm_2 + nn_2 \\ n' &= ll_3 + mm_3 + nn_3 \end{aligned}\right\} \tag{1.47}$$

Eq. (1.47) can be written in matrix format:

$$\begin{Bmatrix} l' \\ m' \\ n' \end{Bmatrix} = \begin{bmatrix} l_1 & m_1 & n_1 \\ l_2 & m_2 & n_2 \\ l_3 & m_3 & n_3 \end{bmatrix} \begin{Bmatrix} l \\ m \\ n \end{Bmatrix} = [A] \begin{Bmatrix} l \\ m \\ n \end{Bmatrix} \tag{1.48}$$

With the aid of the above equation, Eq. (1.41) yields:

$$\begin{bmatrix} \sigma_X & \tau_{XY} & \tau_{XZ} \\ \tau_{XY} & \sigma_Y & \tau_{YZ} \\ \tau_{XZ} & \tau_{YZ} & \sigma_Z \end{bmatrix} \begin{bmatrix} l_1 & m_1 & n_1 \\ l_2 & m_2 & n_2 \\ l_3 & m_3 & n_3 \end{bmatrix} \begin{Bmatrix} l \\ m \\ n \end{Bmatrix} = \begin{bmatrix} l_1 & m_1 & n_1 \\ l_2 & m_2 & n_2 \\ l_3 & m_3 & n_3 \end{bmatrix} \begin{bmatrix} \sigma_x & \tau_{xy} & \tau_{xz} \\ \tau_{yx} & \sigma_y & \tau_{yz} \\ \tau_{zx} & \tau_{zy} & \sigma_z \end{bmatrix} \begin{Bmatrix} l \\ m \\ n \end{Bmatrix} \tag{1.49}$$

We can now multiply both sides of Eq. (1.49) by the matrix $[l \quad m \quad n]$. Since

$$\begin{Bmatrix} l \\ m \\ n \end{Bmatrix} [l \quad m \quad n] = \begin{bmatrix} l^2 & 0 & 0 \\ 0 & m^2 & 0 \\ 0 & 0 & n^2 \end{bmatrix} = [M] \tag{1.50}$$

Eq. (1.49) can be transformed to the following equation:

$$\begin{bmatrix} \sigma_X & \tau_{XY} & \tau_{XZ} \\ \tau_{XY} & \sigma_Y & \tau_{YZ} \\ \tau_{XZ} & \tau_{YZ} & \sigma_Z \end{bmatrix} [A][M] = [A] \begin{bmatrix} \sigma_x & \tau_{xy} & \tau_{xz} \\ \tau_{yx} & \sigma_y & \tau_{yz} \\ \tau_{zx} & \tau_{zy} & \sigma_z \end{bmatrix} [M] \tag{1.51}$$

We can now multiply both sides by $[M]^{-1}$ and get the following result:

$$\begin{bmatrix} \sigma_X & \tau_{XY} & \tau_{XZ} \\ \tau_{XY} & \sigma_Y & \tau_{YZ} \\ \tau_{XZ} & \tau_{YZ} & \sigma_Z \end{bmatrix} [A] = [A] \begin{bmatrix} \sigma_x & \tau_{xy} & \tau_{xz} \\ \tau_{yx} & \sigma_y & \tau_{yz} \\ \tau_{zx} & \tau_{zy} & \sigma_z \end{bmatrix} \tag{1.52}$$

We know the matrix $[A]$ is an "orthogonal matrix." Orthogonal matrices are transformation matrices of vector components from a system x-y-z to a rotated system X-Y-Z that preserve the orthogonality of the transformed vectors. A property of orthogonal matrices is

$$[A][A]^T = [A]^T[A] = [I] \tag{1.53}$$

where $[I]$ is the unit matrix.

Taking into account the above property of orthogonal matrices, Eq. (1.52) yields:

$$\begin{bmatrix} \sigma_X & \tau_{XY} & \tau_{XZ} \\ \tau_{XY} & \sigma_Y & \tau_{YZ} \\ \tau_{XZ} & \tau_{YZ} & \sigma_Z \end{bmatrix} [A][A]^T = [A] \begin{bmatrix} \sigma_x & \tau_{xy} & \tau_{xz} \\ \tau_{yx} & \sigma_y & \tau_{yz} \\ \tau_{zx} & \tau_{zy} & \sigma_z \end{bmatrix} [A]^T \quad (1.54)$$

or

$$\begin{bmatrix} \sigma_X & \tau_{XY} & \tau_{XZ} \\ \tau_{XY} & \sigma_Y & \tau_{YZ} \\ \tau_{XZ} & \tau_{YZ} & \sigma_Z \end{bmatrix} = [A] \begin{bmatrix} \sigma_x & \tau_{xy} & \tau_{xz} \\ \tau_{yx} & \sigma_y & \tau_{yz} \\ \tau_{zx} & \tau_{zy} & \sigma_z \end{bmatrix} [A]^T \quad (1.55)$$

Eq. (1.55) can be written in an abbreviated format:

$$[\sigma_{XYZ}] = [A][\sigma_{xyz}][A]^T \quad (1.56)$$

EXERCISE 1

Prove the identity given in Eq. (1.53):

$$[A][A]^T = [A]^T[A] = [I]$$

The above identity means that the transpose matrix $[A]^T$ can be used as the inverse matrix $[A]^{-1}$, where $[A]$ is the transformation matrix

$$[A] = \begin{bmatrix} l_1 & m_1 & n_1 \\ l_2 & m_2 & n_2 \\ l_3 & m_3 & n_3 \end{bmatrix}$$

Solution

The transformation matrix $[A]$ is an orthogonal matrix. Orthogonal matrices are transformation matrices of vector components from a reference system x, y, z to a rotated one X, Y, Z that preserves the orthogonality of the transformed vectors. Indeed, the transformed unit vectors $\vec{i'}, \vec{j'}, \vec{k'}$ preserve the orthogonality.

We can derive the transposed matrix $[A]^T$ by exchanging the rows by the columns of the matrix $[A]$:

$$[A]^T = \begin{bmatrix} l_1 & l_2 & l_3 \\ m_1 & m_2 & m_3 \\ n_1 & n_2 & n_3 \end{bmatrix}$$

Then we calculate the product:

$$[A][A]^T = \begin{bmatrix} l_1 & m_1 & n_1 \\ l_2 & m_2 & n_2 \\ l_3 & m_3 & n_3 \end{bmatrix} \begin{bmatrix} l_1 & l_2 & l_3 \\ m_1 & m_2 & m_3 \\ n_1 & n_2 & n_3 \end{bmatrix} \quad (1.57)$$

(Continued)

EXERCISE 1—CONT'D

or

$$[A][A]^T = \begin{bmatrix} l_1^2 + m_1^2 + n_1^2 & l_1 l_2 + m_1 m_2 + n_1 n_2 & l_1 l_3 + m_1 m_3 + n_1 n_3 \\ l_1 l_2 + m_1 m_2 + n_1 n_2 & l_2^2 + m_2^2 + n_2^2 & l_2 l_3 + m_2 m_3 + n_2 n_3 \\ l_1 l_3 + m_1 m_3 + n_1 n_3 & l_2 l_3 + m_2 m_3 + n_2 n_3 & l_3^2 + m_3^2 + n_3^2 \end{bmatrix} \quad (1.58)$$

According to the Pythagorean theorem, the quantities $l^2 + m^2 + n^2 = 1$ of the diagonal of Eq. (1.58) are:

$$l^2 + m^2 + n^2 = 1$$

Let's a have a look at the other terms of the matrix in Eq. (1.58). With the aid of Eq. (1.30), the following transformation formulae can be obtained:

$$\begin{aligned} \vec{i'} &= l_1 \vec{i} - m_1 \vec{j} + n_1 \vec{k} \\ \vec{j'} &= l_2 \vec{i} + m_2 \vec{j} + n_2 \vec{k} \\ \vec{k'} &= l_3 \vec{i} + m_3 \vec{j} + n_3 \vec{k} \end{aligned} \quad (1.59)$$

Since the angle between orthogonal vectors is $\pi/2$, their scalar product is null:

$$\vec{i'} \cdot \vec{j'} = 0$$

or

$$\left(l_1 \vec{i} + m_1 \vec{j} + n_1 \vec{k}\right)\left(l_2 \vec{i} + m_2 \vec{j} + n_2 \vec{k}\right) = 0$$

or

$$l_1 l_2 + m_1 m_2 + n_1 n_2 = 0 \quad (1.60)$$

Following same procedure, the following results can be obtained:

$$\vec{i'} \cdot \vec{k'} = 0$$

or

$$\left(l_1 \vec{i} + m_1 \vec{j} + n_1 \vec{k}\right)\left(l_3 \vec{i} + m_3 \vec{j} + n_3 \vec{k}\right) = 0$$

or

$$l_1 l_3 + m_1 m_3 + n_1 n_3 = 0 \quad (1.61)$$

and

$$\vec{j'} \cdot \vec{k'} = 0$$

or

$$\left(l_2 \vec{i} + m_2 \vec{j} + n_2 \vec{k}\right)\left(l_3 \vec{i} + m_3 \vec{j} + n_3 \vec{k}\right) = 0$$

or

$$l_2 l_3 + m_2 m_3 + n_2 n_3 = 0 \quad (1.62)$$

EXERCISE 1—CONT'D

With the aid of Eq. (1.59) and Eqs. (1.60)–(1.62), Eq. (1.58) yields:

$$[A][A]^T = \begin{bmatrix} 1 & 0 & 0 \\ 0 & 1 & 0 \\ 0 & 0 & 1 \end{bmatrix} = [I] \tag{1.63}$$

If we inverse the sequence of the matrices in Eq. (1.57) and follow the same procedure, the following result can be obtained:

$$[A]^T [A] = [I]$$

EXERCISE 2

The stress vector $\vec{\sigma}_P$ of a plane ABC has magnitude 151 MPa and makes angles $\alpha_1 = 52$, $\beta_1 = 79$, and $\gamma_1 = 40.13$ with the axes x, y, and z. The unit vector $\overrightarrow{N}$ of the plane ABC makes angles $\alpha = 33$, $\beta = 65$, and $\gamma = 69.91$ with respect to the same axes x, y, and z, respectively. If the shear stress components are $\tau_{xy} = 49$ MN/m², $\tau_{yz} = -21$ MN/m², and $\tau_{zx} = 15$ MN/m², calculate the normal stresses σ_{xx}, σ_{yy}, and σ_{zz}

Solution

The magnitude of the stress vector is:

$$|\overrightarrow{\sigma_P}| = 151 MPa$$

We can project this vector on the x, y, and z axes (Fig. 10) in order to get the magnitude of the components σ_{Px}, σ_{Py}, σ_{Pz}:

$$\begin{aligned} \sigma_{Px} &= \left|\overrightarrow{\sigma_P}\right| \cdot \cos\ \alpha_1 = 151 \cdot \cos\ 52^\circ = 92.96\ MPa \\ \sigma_{Py} &= \left|\overrightarrow{\sigma_P}\right| \cdot \cos\ \beta_1 = 151 \cdot \cos\ 79^\circ = 28.81\ MPa \\ \sigma_{Pz} &= \left|\overrightarrow{\sigma_P}\right| \cdot \cos\ \chi_1 = 151 \cdot \cos\ 40.13^\circ = 115.45\ MPa \end{aligned} \tag{1.64}$$

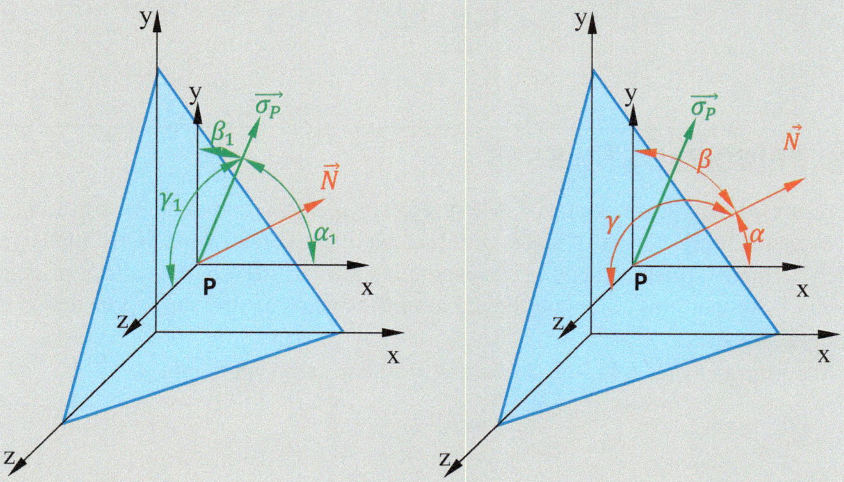

FIG. 1.10

Stress vector $\vec{\sigma}_P$ and cross-section vector $\overrightarrow{N}$.

(Continued)

EXERCISE 2—CONT'D

The directional cosines of the plane cross section are:

$$l = \cos\ \alpha$$
$$m = \cos\ \beta$$
$$n = \cos\ \gamma$$

Then the following equation (Eq. 1.19)

$$\begin{Bmatrix} \sigma_{Px} \\ \sigma_{Py} \\ \sigma_{Pz} \end{Bmatrix} = \begin{bmatrix} \sigma_x & \tau_{xy} & \tau_{xz} \\ \tau_{yx} & \sigma_y & \tau_{yz} \\ \tau_{zx} & \tau_{zy} & \sigma_z \end{bmatrix} \begin{Bmatrix} l \\ m \\ n \end{Bmatrix}$$

yields the following linear system:

$$\begin{aligned} \vec{\sigma}_{Px} &= \sigma_{xx} \cdot \cos\ \alpha + \tau_{xy} \cdot \cos\ \beta + \tau_{xz} \cdot \cos\ \gamma \\ \vec{\sigma}_{Py} &= \tau_{xy} \cdot \cos\ \alpha + \sigma_{yy} \cdot \cos\ \beta + \tau_{yz} \cdot \cos\ \gamma \\ \vec{\sigma}_{Pz} &= \tau_{xz} \cdot \cos\ \alpha + \tau_{yz} \cdot \cos\ \beta + \sigma_{zz} \cdot \cos\ \gamma \end{aligned} \tag{1.65}$$

Substituting the values we already know, we get:

$$\begin{aligned} 92.96 &= \sigma_{xx} \cdot \cos\ 33^\circ + 49 \cdot \cos\ 65^\circ + 15 \cdot \cos\ 69.91^\circ \\ 28.81 &= 49 \cdot \cos\ 33^\circ + \sigma_{yy} \cdot \cos\ 65^\circ - 21 \cdot \cos\ 69.91^\circ \\ 115.45 &= 15 \cdot \cos\ 33^\circ - 21 \cdot \cos\ 65^\circ + \sigma_{zz} \cdot \cos\ 69.91^\circ \end{aligned} \tag{1.66}$$

The above system of equations can be utilized to calculate σ_{xx}, σ_{yy}, and σ_{zz}:

$$\begin{aligned} \sigma_{xx} &= \frac{92.96 - 49 \cdot \cos\ 65^\circ - 15 \cdot \cos\ 69.91^\circ}{\cos\ 33^\circ} = 80\ MPa \\ \sigma_{yy} &= \frac{28.81 - 49 \cdot \cos\ 33^\circ + 21 \cdot \cos\ 69.91^\circ}{\cos\ 65^\circ} = -12\ MPa \\ \sigma_{zz} &= \frac{115.45 - 15 \cdot \cos\ 33^\circ + 21 \cdot \cos\ 65^\circ}{\cos\ 69.91^\circ} = 325.32\ MPa \end{aligned} \tag{1.67}$$

1.8 PRINCIPAL STRESSES

There are planes in the material where the shear stress components are null. On these planes, only normal stresses components exist (Fig. 1.11). These planes are called "principal planes," and the corresponding normal stresses are called "principal stresses." We can now calculate the value of the principal stress components on the principal planes.

According to Fig. 1.11, the following formulae can be derived:

$$\begin{aligned} \sigma_{Px} &= \sigma \cdot l \\ \sigma_{Py} &= \sigma \cdot m \\ \sigma_{Pz} &= \sigma \cdot n \end{aligned} \tag{1.68}$$

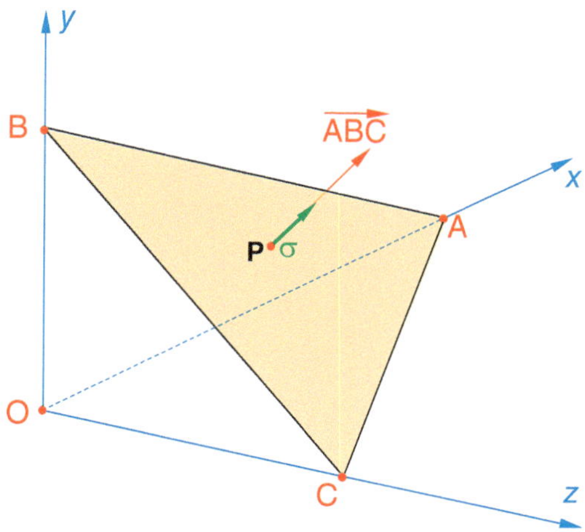

FIG. 1.11

Principal plane

Substitution of the above data into Eq, (1.19) yields:

$$\begin{Bmatrix} \sigma \cdot l \\ \sigma \cdot m \\ \sigma \cdot n \end{Bmatrix} = \begin{bmatrix} \sigma_x & \tau_{xy} & \tau_{xz} \\ \tau_{xy} & \sigma_y & \tau_{yz} \\ \tau_{xz} & \tau_{yz} & \sigma_z \end{bmatrix} \begin{Bmatrix} l \\ m \\ n \end{Bmatrix}$$

Alternatively, this can be written as:

$$\left.\begin{aligned} \sigma \cdot l &= \sigma_x l + \tau_{xy} m + \tau_{xz} n \\ \sigma \cdot m &= \tau_{xy} l + \sigma_y m + \tau_{yz} n \\ \sigma \cdot n &= \tau_{xz} l + \tau_{yz} m + \sigma_z n \end{aligned}\right\} \tag{1.69}$$

or

$$\begin{bmatrix} \sigma_x - \sigma & \tau_{xy} & \tau_{xz} \\ \tau_{xy} & \sigma_y - \sigma & \tau_{yz} \\ \tau_{yz} & \tau_{yz} & \sigma_z - \sigma \end{bmatrix} \begin{Bmatrix} l \\ m \\ n \end{Bmatrix} = \begin{Bmatrix} 0 \\ 0 \\ 0 \end{Bmatrix} \tag{1.70}$$

The condition for nontrivial solution of the above equation is:

$$\det \begin{bmatrix} \sigma_x - \sigma & \tau_{xy} & \tau_{xz} \\ \tau_{xy} & \sigma_y - \sigma & \tau_{yz} \\ \tau_{xz} & \tau_{yz} & \sigma_z - \sigma \end{bmatrix} = 0 \tag{1.71}$$

Eq. (1.71) yields the following cubic algebraic equation:

$$\begin{aligned}&(\sigma_x - \sigma)[(\sigma_y - \sigma)(\sigma_z - \sigma) - \tau_{yz}\tau_{yz}]\\&- \tau_{xy}[\tau_{xy}(\sigma_z - \sigma) - \tau_{yz}\tau_{xz}]\\&+ \tau_{xz}[\tau_{xy}\tau_{yz} - (\sigma_y - \sigma)\tau_{xz}] = 0\end{aligned} \tag{1.72}$$

or

$$\begin{aligned}&\sigma^3 - (\sigma_x + \sigma_y + \sigma_z)\sigma^2\\&+ (\sigma_x\sigma_y + \sigma_y\sigma_z + \sigma_x\sigma_z - \tau_{xy}^2 - \tau_{xz}^2 - \tau_{yz}^2)\sigma\\&- (\sigma_x\sigma_y\sigma_z + 2\tau_{xy}\tau_{yz}\tau_{xz} - \sigma_x\tau_{yz}^2 - \sigma_y\tau_{xz}^2 - \sigma_z\tau_{xy}^2) = 0\end{aligned} \tag{1.73}$$

We can see that the coefficients of the variable σ, which we will symbolize by I_1, I_2 and I_3, are stress invariants because they are independent of the selected coordinate system. Then we can express the cubic equation for the principal stress with the aid of these invariants:

$$\sigma^3 - I_1\sigma^2 + I_2\sigma - I_3 = 0 \tag{1.74}$$

where

$$\begin{aligned}&I_1 = \sigma_x + \sigma_y + \sigma_z\\&I_2 = \sigma_x\sigma_y + \sigma_y\sigma_z + \sigma_x\sigma_z - \tau_{xy}^2 - \tau_{yz}^2 - \tau_{xz}^2\\&I_3 = \sigma_x\sigma_y\sigma_z + 2\tau_{xy}\tau_{yz}\tau_{xz} - \sigma_x\tau_{yz}^2 - \sigma_y\tau_{xz}^2 - \sigma_z\tau_{xy}^2 = \begin{vmatrix}\sigma_x & \tau_{xy} & \tau_{xz}\\ \tau_{yx} & \sigma_y & \tau_{yz}\\ \tau_{zx} & \tau_{zy} & \sigma_z\end{vmatrix}\end{aligned} \tag{1.75}$$

The three real solutions of the cubic equation are the principal stresses: σ_1, σ_2 and σ_3. Each principal stress corresponds to a plane. If we set $\sigma_x = \sigma_1$, $\sigma_y = \sigma_2$, $\sigma_z = \sigma_3$, $\tau_{xy} = 0$, $\tau_{xz} = 0$, and $\tau_{yz} = 0$, the stress invariants can also be written as a function of σ_1, σ_2, and σ_3:

$$I_1 = \sigma_1 + \sigma_2 + \sigma_3 \tag{1.76}$$

$$I_2 = \sigma_1\sigma_2 + \sigma_2\sigma_3 + \sigma_1\sigma_3 \tag{1.77}$$

$$I_3 = \sigma_1\sigma_2\sigma_3 \tag{1.78}$$

For expressing the invariants I_1, I_2, and I_3 in terms of the principal σ_1, σ_2, σ_3 we take into account that the principal planes are subjected to normal stresses only.

1.9 ORIENTATION L, M, N OF THE PRINCIPAL PLANES

Since three principal stresses (σ_1, σ_2, and σ_3) exist, three principal planes corresponding to the above principal stresses can be found. Let's denote their directional cosines by the following notations:

$$\begin{aligned}&\text{Principal plane of } \sigma_1 \rightarrow l_1, m_1, n_1\\&\text{Principal plane of } \sigma_2 \rightarrow l_2, m_2, n_2\\&\text{Principal plane of } \sigma_3 \rightarrow l_3, m_3, n_3\end{aligned}$$

For $\sigma = \sigma_I$, Eq. (1.70) yields:

$$\left.\begin{aligned}l_1(\sigma_{xx} - \sigma_1) + m_1\tau_{xy} + n_1\tau_{xz} = 0\\l_1\tau_{xy} + m_1(\sigma_{yy} - \sigma_1) + n_1\tau_{xz} = 0\\l_1\tau_{xz} + m_1\tau_{xy} + n_1(\sigma_{zz} - \sigma_1) = 0\end{aligned}\right\} \tag{1.79}$$

The above three equations are not independent because the determinant in Eq. (1.71) is null. That means two of the above three equations are the same. In order to calculate l, m, and n, we need an additional equation. To this end, the Pythagorean theorem can be utilized:

$$l_1^2 + m_1^2 + n_1^2 = 1 \tag{1.80}$$

Using the three independent equations of the set of the four equations (Eq. 1.79) and (Eq. 1.80), we can obtain the directional cosines for the principal stress σ_I. We have to apply the same procedure for the principal stresses σ_2 and σ_3.

APPENDIX: SOLUTION OF THE CUBIC EQUATION

Usually, the solution of the cubic equation (Eq. 1.74) is obtained with the aid of numerical methods. However, an analytical solution exists. The analytical solution was first published by the Italian mathematician Girolamo Gardano in 1501 in his algebra book *Ars Magna*.

First, we assume a general cubic equation:

$$z^3 + a_2z^2 + a_1z + a_0 = 0$$

We start defining the parameters q and r, given by the formulae:

$$q = \frac{1}{3}a_1 - \frac{1}{9}a_2^2$$

$$r = \frac{1}{6}(a_1a_2 - 3a_0) - \frac{1}{27}a_2^3$$

Once we have the above parameters, we can get the following parameters s_1 and s_2:

$$s_1 = \left[r + (q^3 + r^2)^{1/2}\right]^{1/3}$$

$$s_2 = \left[r - (q^3 + r^2)^{1/2}\right]^{1/3}$$

With the aid of the parameters s_1 and s_2, the following solution of the cubic equation can be obtained:

$$z_1 = (s_1 + s_2) - \frac{a_2}{3}$$

$$z_2 = -\frac{1}{2}(s_1 + s_2) - \frac{a_2}{3} + \frac{i\sqrt{3}}{2}(s_1 - s_2)$$

$$z_3 = -\frac{1}{2}(s_1 + s_2) - \frac{a_2}{3} - \frac{i\sqrt{3}}{2}(s_1 - s_2)$$

where i is the imaginary part of a complex number. In the above equations, we can classify three cases because the roots z_1, z_2, and z_3 can be real or complex.

- $q^3 + r^2 > 0$: One real root and pair of complex conjugate roots.
- $q^3 + r^2 = 0$: All roots real and atleast two are equal.
- $q^3 + r^2 < 0$: All roots are real.

It should be noted that the stresses cannot be complex numbers.
This method for analytical solution can be found on the website:
http://www.hawaii.edu/suremath/jrootsCubic.html

1.10 OCTAHEDRAL STRESSES

An arbitrary plane ABC represented by the vector $\overrightarrow{N}$, and an arbitrary stress vector $\overrightarrow{\sigma_p}$ are shown in Fig. 1.12. If the three axes of the coordinate system are the principal axes, and the plane ABC has the same angles with respect to the three axes, then the plane ABC is called "octahedral plane." The normal stress of each octahedral plane is called "octahedral stress."

In summary, the conditions for defining the octahedral stresses are:

1. The axes x, y, and z are principal axes. Therefore:

$$\left.\begin{aligned} \sigma_x &= \sigma_X = \sigma_1 \\ \sigma_y &= \sigma_Y = \sigma_2 \\ \sigma_z &= \sigma_Z = \sigma_3 \\ \tau_{xy} &= \tau_{xz} = \tau_{yz} = 0 \end{aligned}\right\} \tag{1.81}$$

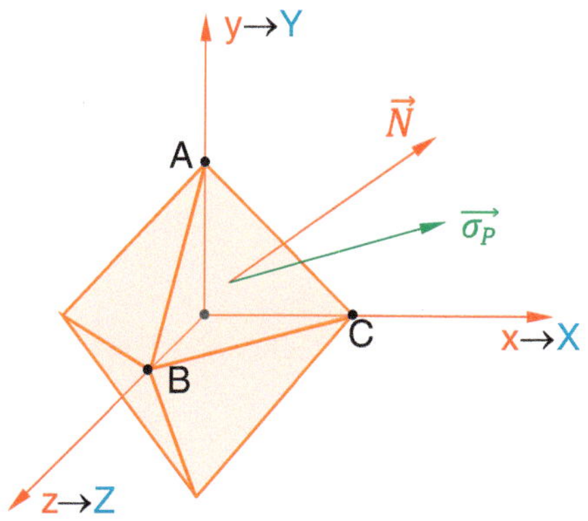

FIG. 1.12

Definition of octahedral planes.

2. The plane *ABC* has equal angles to the principal axes:

$$\left.\begin{array}{l} l = m = n \\ l^2 + m^2 + n^2 = 1 \end{array}\right\} \tag{1.82}$$

or

$$l^2 = m^2 = n^2 = \frac{1}{3} \tag{1.83}$$

The octahedral stress is the normal stress that fulfils the conditions 1 and 2. Therefore:

$$\sigma_{oct} = |\overrightarrow{\sigma}_{PN}| \tag{1.84}$$

Let's try to express the octahedral stresses versus the principal stresses. We recall Eq. (1.24):

$$|\sigma_{PN}| = l^2\sigma_x + m^2\sigma_y + n^2\sigma_{zz} + 2lm\tau_{xy} + 2ln\tau_{xz} + 2mn\tau_{yz}$$

Taking into account the above equation and the two conditions of the octahedral stress, the following expression can be obtained (note that shear stresses are equal to zero):

$$\sigma_{oct} = |\sigma_{PN}| = \frac{1}{3}(\sigma_1 + \sigma_2 + \sigma_3) \tag{1.85}$$

For the octahedral shear stress, Eq. (1.25) can be utilized:

$$\tau_{oct} = |\sigma_{PS}| = \sqrt{|\sigma_P|^2 - |\sigma_{PN}|^2} = \sqrt{|\sigma_P|^2 - \sigma_{oct}^2} \tag{1.86}$$

where (Eq. 1.26):

$$|\sigma_P|^2 = |\sigma_{Px}|^2 + |\sigma_{Py}|^2 + |\sigma_{Pz}|^2 = |\sigma_{PX}|^2 + |\sigma_{PY}|^2 + |\sigma_{PZ}|^2 \quad (1.87)$$

Since *X, Y*, and *Z* are the principal directions, the shear stresses are null. Then, the magnitude of the stress vector components can be obtained with the aid of Eq. (1.19):

$$\left.\begin{aligned} \sigma_{PX} &= \sigma_X l + 0 \cdot m + 0 \cdot n = \sigma_1 l \\ \sigma_{PY} &= 0 \cdot l + \sigma_Y m + 0 \cdot n = \sigma_2 m \\ \sigma_{PZ} &= 0 \cdot l + 0 \cdot m + \sigma_Z n = \sigma_3 n \end{aligned}\right\} \quad (1.88)$$

The above equations have been derived with the aid of the first condition (Eq. 1.82) for octahedral stresses. According to Eq. (1.87), the following magnitude of the stress vector $\vec{\sigma}_P$ can be obtained:

$$|\sigma_p|^2 = \sigma_1^2 l^2 + \sigma_2^2 m^2 + \sigma_3^2 n^2 \quad (1.89)$$

With the aid of the second condition (Eq. 1.83), the above equation yields:

$$|\sigma_P|^2 = \frac{1}{3}\left(\sigma_1^2 + \sigma_2^2 + \sigma_3^2\right) \quad (1.90)$$

Considering Eqs. (1.85) and (1.90), the octahedral shear stress can be obtained by Eq. (1.86):

$$\tau_{oct} = \sqrt{|\sigma_P|^2 - \sigma_{oct}^2} = \sqrt{\frac{1}{3}\left(\sigma_1^2 + \sigma_2^2 + \sigma_3^2\right) - \frac{1}{9}(\sigma_1 + \sigma_2 + \sigma_3)^2} \quad (1.91)$$

Reordering the terms of the above equation, the following expression can be obtained:

$$\tau_{oct} = \frac{1}{3}\sqrt{(\sigma_1 - \sigma_2)^2 + (\sigma_2 - \sigma_3)^2 + (\sigma_1 - \sigma_3)^2} \quad (1.92)$$

EXERCISE 3

The stress vector at a point *P* is given by the following matrix:

$$\sigma = \begin{bmatrix} \sigma_x & \tau_{xy} & \tau_{xz} \\ \tau_{xy} & \sigma_y & \tau_{yz} \\ \tau_{xz} & \tau_{yz} & \sigma_z \end{bmatrix} = \begin{bmatrix} 80 & 20 & 40 \\ 20 & 60 & 10 \\ 40 & 10 & 20 \end{bmatrix} MPa$$

(a) Determine the stress vector on a plane perpendicular to the vector $\vec{R} = \vec{i} + 2\vec{j} + \vec{k}$

(b) Determine the principal stresses σ_1, σ_2, and σ_3

(c) Determine the octahedral shear stress σ_{oct}

EXERCISE 3—CONT'D

Solution

(a) In order to calculate the stress vector on a plane perpendicular to $\vec{R}$, we should calculate first the direction cosines:

$$l = \frac{R_x}{|\vec{R}|} = \frac{1}{\sqrt{1^2+2^2+1^2}} = \frac{1}{\sqrt{6}}$$

(b) $$m = \frac{R_y}{|\vec{R}|} = \frac{2}{\sqrt{1^2+2^2+1^2}} = \frac{2}{\sqrt{6}}$$

$$n = \frac{R_z}{|\vec{R}|} = \frac{1}{\sqrt{1^2+2^2+1^2}} = \frac{1}{\sqrt{6}}$$

With the aid of Eq. (1.19), the following vector components can be obtained:

$$\begin{Bmatrix} \sigma_{Px} \\ \sigma_{Py} \\ \sigma_{Pz} \end{Bmatrix} = \begin{bmatrix} \sigma_x & \tau_{xy} & \tau_{xz} \\ \tau_{yx} & \sigma_y & \tau_{yz} \\ \tau_{zx} & \tau_{zy} & \sigma_z \end{bmatrix} \begin{Bmatrix} l \\ m \\ n \end{Bmatrix} = \begin{bmatrix} 80 & 20 & 40 \\ 20 & 60 & 10 \\ 40 & 10 & 20 \end{bmatrix} \begin{Bmatrix} 1/\sqrt{6} \\ 2/\sqrt{6} \\ 1/\sqrt{6} \end{Bmatrix} = \begin{Bmatrix} 65.3 \\ 61.2 \\ 32.6 \end{Bmatrix} \ MPa$$

(c) To calculate the principal stresses, the first step is to calculate the stress invariants

$$I_1 = \sigma_x + \sigma_y + \sigma_z = 80 + 60 + 20 = 160 \ MPa$$

$$I_2 = \sigma_x\sigma_y + \sigma_y\sigma_z + \sigma_x\sigma_z - \tau_{xy}^2 - \tau_{yz}^2 - \tau_{xz}^2 =$$
$$= 80 \cdot 60 + 60 \cdot 20 + 80 \cdot 20 - 20^2 - 10^2 - 40^2 =$$
$$= 5500 \ MPa$$

$$I_3 = \sigma_x\sigma_y\sigma_z + 2\tau_{xy}\tau_{yz}\tau_{xz} - \sigma_x\tau_{yz}^2 - \sigma_y\tau_{xz}^2 - \sigma_z\tau_{xy}^2 = 0 \ MPa$$

Then, the cubic equation (Eq. 1.74) can be used to derive the principal stresses:

$$\sigma^3 - I_1\sigma^2 + I_2\sigma - I_3 = 0$$

This can be easily solved because the third stress invariant is null:

$$\sigma^3 - 160\sigma^2 + 5500\sigma = 0$$

or

$$\sigma(\sigma^2 - 160\sigma + 5500) = 0$$

The above equation yields the following solutions:

$$\begin{cases} \sigma = 0 \\ \sigma^2 - 160\sigma + 5500 = 0 \end{cases}$$

or

$$\begin{cases} \sigma = 0 \ MPa \\ \sigma = \dfrac{160 + \sqrt{160^2 - 4 \cdot 5500}}{2} = 110 \ MPa \\ \sigma = \dfrac{160 - \sqrt{160^2 - 4 \cdot 5500}}{2} = 50 \ MPa \end{cases}$$

(d) We can calculate the octahedral shear stress with the aid of Eq. (1.92)

$$\tau_{oct} = \frac{1}{3}\sqrt{(\sigma_1 - \sigma_2)^2 + (\sigma_2 - \sigma_3)^2 + (\sigma_1 - \sigma_3)^2}$$

or

$$\tau_{oct} = \frac{1}{3}\sqrt{(110 - 50)^2 + (50 - 0)^2 + (110 - 0)^2} = 44.9 \ MPa$$

1.11 SPHERICAL AND DEVIATORIC STRESS TENSOR

The stress tensor can be analyzed into two components: the spherical and the deviatoric one. The spherical stress tensor (also called hydrostatic or mean stress tensor) is responsible for the volume change of a material element. The deviatoric stress tensor is responsible for the change of the shape of a material element. The change of the volume of a material element does not cause yielding. Only the deviatoric stress tensor is responsible for material yielding. If we define the mean stress σ_m by the formula:

$$\sigma_m = \frac{1}{3}(\sigma_1 + \sigma_2 + \sigma_3), \tag{1.93}$$

the stress tensor can be analyzed into the following two components:

$$\begin{bmatrix} \sigma_x & \tau_{xy} & \tau_{xz} \\ \tau_{yx} & \sigma_y & \tau_{yz} \\ \tau_{zx} & \tau_{zy} & \sigma_z \end{bmatrix} = \begin{bmatrix} \sigma_m & 0 & 0 \\ 0 & \sigma_m & 0 \\ 0 & 0 & \sigma_m \end{bmatrix} + [T_d] \tag{1.94}$$

where $[T_d]$ is the deviatoric stress tensor.

With the aid of the above equation, the following formula for the deviatoric tensor can be obtained:

$$[T_d] = \begin{bmatrix} \sigma_x - \sigma_m & \tau_{xy} & \tau_{xz} \\ \tau_{yx} & \sigma_y - \sigma_m & \tau_{yz} \\ \tau_{zx} & \tau_{zy} & \sigma_z - \sigma_m \end{bmatrix} \tag{1.95}$$

In terms of the principal stresses, the deviatoric stress tensor can be written as:

$$[T_d] = \begin{bmatrix} \sigma_1 - \sigma_m & 0 & 0 \\ 0 & \sigma_2 - \sigma_m & 0 \\ 0 & 0 & \sigma_3 - \sigma_m \end{bmatrix} \tag{1.96}$$

and has the following stress invariants:

$$J_1 = [(\sigma_1 - \sigma_m) + (\sigma_2 - \sigma_m) + (\sigma_3 - \sigma_m)] = 0 \tag{1.97}$$

$$J_2 = (\sigma_1 - \sigma_m)(\sigma_2 - \sigma_m) + (\sigma_2 - \sigma_m)(\sigma_3 - \sigma_m) + (\sigma_1 - \sigma_m)(\sigma_3 - \sigma_m) \tag{1.98}$$

$$J_3 = \begin{vmatrix} \sigma_1 - \sigma_m & 0 & 0 \\ 0 & \sigma_2 - \sigma_m & 0 \\ 0 & 0 & \sigma_3 - \sigma_m \end{vmatrix} = (\sigma_1 - \sigma_m)(\sigma_2 - \sigma_m)(\sigma_3 - \sigma_m) \tag{1.99}$$

If we replace σ_m from Eq. (1.93) to Eq. (1.98), we can express J_2 as a function of the invariants I_1 and of the stress tensor:

$$J_2 = I_2 - \frac{1}{3}I_1^2 \tag{1.100}$$

1.12 PLANE STRESS: STRESS TENSOR IN ROTATED COORDINATE SYSTEM

We are going now to calculate the stresses in a rotated coordinate system in plane stress conditions. The thin plate in Fig. 1.13 is subjected to in-plane loads that cause stresses σ_x, σ_y, and τ_{xy} in a material element. We are looking for the stress tensor in the rotated coordinate system X-Y-Z. The rotating angle θ between the reference and the rotated system is known.

With the aid of Eq. (1.28), the following transformation matrix can be obtained:

		$\vec{i}$ x	$\vec{j}$ y	$\vec{k}$ z
$\vec{i'}$	**X**	$l_1 = \cos(\theta)$	$m_1 = \cos\left(\frac{\pi}{2} - \theta\right)$	$n_1 = 0$
$\vec{j'}$	**Y**	$l_2 = \cos\left(\frac{\pi}{2} + \theta\right)$	$m_2 = \cos(\theta)$	$n_2 = 0$
$\vec{k'}$	**Z**	$l_3 = 0$	$m_3 = 0$	$n_3 = 1$

(1.101)

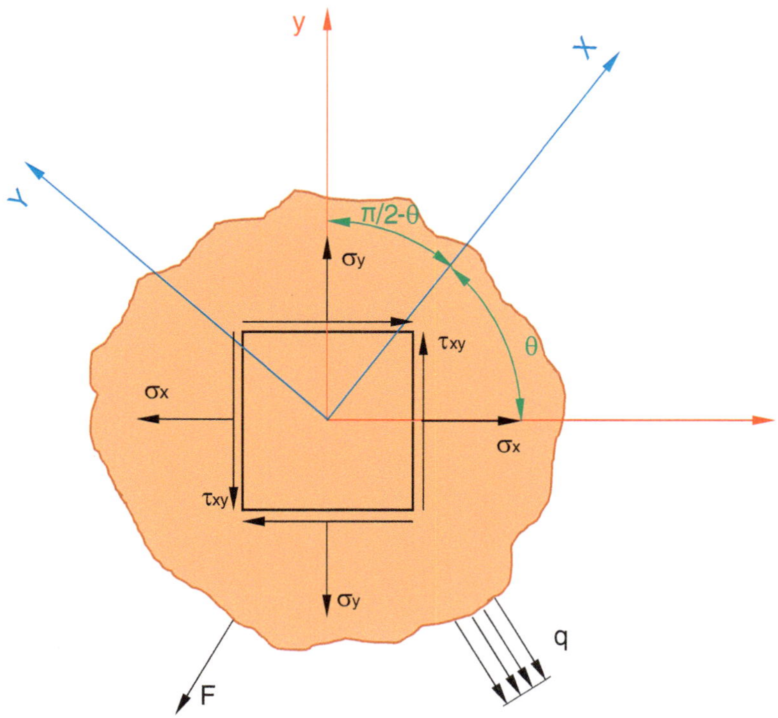

FIG. 1.13

Material element under plane stress conditions.

Eliminating the terms of the axes z and Z, the transformation matrix $[A]$ in the plane x-y is:

$$[A] = \begin{bmatrix} \cos(\theta) & \cos\left(\frac{\pi}{2} - \theta\right) \\ \cos\left(\frac{\pi}{2} + \theta\right) & \cos(\theta) \end{bmatrix} = \begin{bmatrix} \cos(\theta) & \sin(\theta) \\ -\sin(\theta) & \cos(\theta) \end{bmatrix} \tag{1.102}$$

With the aid of Eq. (1.55) for transformation of the stress tensor in a rotated coordinate system, the following equation can be obtained for the plane x-y:

$$\begin{bmatrix} \sigma_X & \tau_{XY} \\ \tau_{XY} & \sigma_Y \end{bmatrix} = [A] \begin{bmatrix} \sigma_x & \tau_{xy} \\ \tau_{xy} & \sigma_y \end{bmatrix} [A]^T$$

or

$$\begin{bmatrix} \sigma_X & \tau_{XY} \\ \tau_{XY} & \sigma_Y \end{bmatrix} = \begin{bmatrix} \cos(\theta) & \sin(\theta) \\ -\sin(\theta) & \cos(\theta) \end{bmatrix} \begin{bmatrix} \sigma_x & \tau_{xy} \\ \tau_{xy} & \sigma_y \end{bmatrix} \begin{bmatrix} \cos(\theta) & -\sin(\theta) \\ \sin(\theta) & \cos(\theta) \end{bmatrix}$$

The above equation yields the following stress components in the rotated plane X-Y:

$$\sigma_X = \frac{\sigma_x + \sigma_y}{2} + \frac{\sigma_x - \sigma_y}{2} \cos(2\theta) + \tau_{xy} \sin(2\theta) \tag{1.103}$$

$$\sigma_Y = \frac{\sigma_x + \sigma_y}{2} - \frac{\sigma_x - \sigma_y}{2} \cos(2\theta) - \tau_{xy} \sin(2\theta) \tag{1.104}$$

$$\tau_{XY} = -\frac{1}{2}(\sigma_x - \sigma_y)\sin(2\theta) + \tau_{xy} \cos(2\theta) \tag{1.105}$$

The stresses in the rotated plane are depicted in Fig. 1.14.

- **Condition of maximum normal stress σx**

  With the aid of Eq. (1.103), the condition for maximum σ_X

$$\frac{d\sigma_X}{d\theta} = 0 \tag{1.106}$$

yields:

$$-\frac{\sigma_x - \sigma_y}{2} 2 \sin 2\theta + \tau_{xy} 2 \cos 2\theta = 0$$

Solution of the above equation with respect to θ yields the corresponding angle of the plane of the maximum normal stress σ_X

$$\tan 2\theta = \frac{2\tau_{xy}}{\sigma_x - \sigma_y} \tag{1.107}$$

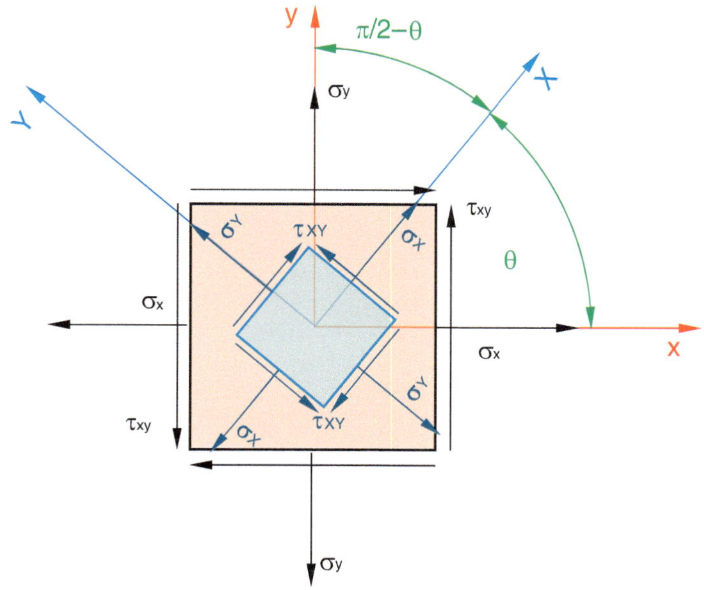

FIG. 1.14

Stress components on the rotated and the reference plane.

Using the trigonometric identity

$$\sin(2\theta) = \tan(2\theta)\cos(2\theta) \tag{1.108}$$

or

$$\sin\ 2\theta = \frac{2\tau_{xy}}{\sigma_x - \sigma_y}\ \cos\ 2\theta \tag{1.109}$$

the magnitude of the maximum normal stress can be obtained by Eq. (1.103):

$$\sigma_{X_{MAX}} = \frac{\sigma_x + \sigma_y}{2} + \frac{\sigma_x - \sigma_y}{2}\ \cos(2\theta) + \tau_{xy}\frac{2\tau_{xy}}{\sigma_x - \sigma_y}\ \cos(2\theta) \tag{1.110}$$

or

$$\sigma_{X_{MAX}} = \frac{\sigma_x + \sigma_y}{2} + \left(\frac{\sigma_x - \sigma_y}{2} + \frac{2\tau_{xy}^2}{\sigma_x - \sigma_y}\right)\cos(2\theta) \tag{1.111}$$

With the aid of the trigonometric identity $\cos^2(x) = 1 - \sin^2(x)$ and Eq. (1.109), the following equation can be resulted:

$$\cos^2(2\theta) = 1 - \sin^2(2\theta) = 1 - \frac{4\tau_{xy}^2}{(\sigma_x - \sigma_y)^2}\cos^2(2\theta)$$

or

$$\cos(2\theta) = \pm\frac{\sigma_x - \sigma_y}{\sqrt{(\sigma_x - \sigma_y)^2 + 4\tau_{xy}^2}} \tag{1.112}$$

Combining Eqs. (1.111) and (1.112), the following equation can be obtained:

$$\sigma_{X_{\max,\min}} = \sigma_{1,2} = \frac{\sigma_x + \sigma_y}{2} \pm \left(\frac{\sigma_x - \sigma_y}{2} + \frac{2\tau_{xy}^2}{\sigma_x - \sigma_y}\right)\frac{\sigma_x - \sigma_y}{\sqrt{(\sigma_x - \sigma_y)^2 + 4\tau_{xy}^2}} \tag{1.113}$$

or

$$\sigma_{1,2} = \frac{\sigma_x + \sigma_y}{2} \pm \sqrt{\frac{1}{4}(\sigma_x - \sigma_y)^2 + \tau_{xy}^2} \tag{1.114}$$

Therefore two values of principal stresses are identified:

$$\begin{cases} \sigma_1 = \dfrac{\sigma_x + \sigma_y}{2} + \sqrt{\dfrac{1}{4}(\sigma_x - \sigma_y)^2 + \tau_{xy}^2} \\ \sigma_2 = \dfrac{\sigma_x + \sigma_y}{2} - \sqrt{\dfrac{1}{4}(\sigma_x - \sigma_y)^2 + \tau_{xy}^2} \end{cases} \tag{1.115}$$

- **Condition for maximum shear stress τxy**

Now we follow the same steps for finding the plane and the magnitude of the maximum shear stress. The condition of maximization

$$\frac{d\tau_{XY}}{d\theta} = 0 \tag{1.116}$$

yields

$$-\frac{\sigma_x - \sigma_y}{2}\cos 2\theta - \tau_{xy}\sin 2\theta = 0 \tag{1.117}$$

or

$$\tan 2\theta = -\frac{\sigma_x - \sigma_y}{2\tau_{xy}} \tag{1.118}$$

The following trigonometric identity

$$\cos 2\theta = \frac{\sin 2\theta}{\tan 2\theta} = \frac{\sqrt{1 - \cos^2 2\theta}}{\tan 2\theta} \tag{1.119}$$

or

$$\cos^2 2\theta = \frac{1}{1 + \tan^2 2\theta} \tag{1.120}$$

can utilize Eq. (1.118). Then, a combination of Eqs. (1.118) and (1.120) yields:

$$\cos^2 2\theta = \frac{1}{1 + \left(\frac{\sigma_x - \sigma_y}{2\tau_{xy}}\right)^2} = \frac{4\tau_{xy}^2}{4\tau_{xy}^2 + (\sigma_x - \sigma_y)^2} \tag{1.121}$$

or

$$\cos 2\theta = \frac{2\tau_{xy}}{\sqrt{4\tau_{xy}^2 + (\sigma_x - \sigma_y)^2}} \tag{1.122}$$

Combining Eqs. (1.105), (1.108), (1.118), and (1.122), the maximum shear stress can be derived as follows:

$$\begin{aligned}
\tau_{XY} &= -\frac{1}{2}(\sigma_x - \sigma_y)\sin(2\theta) + \tau_{xy}\ \cos(2\theta) = \\
&= \cos(2\theta)\left[-\frac{1}{2}(\sigma_x - \sigma_y)\tan(2\theta) + \tau_{xy}\right] = \\
&= \frac{2\tau_{xy}}{\sqrt{4\tau_{xy}^2 + (\sigma_x - \sigma_y)^2}}\left[-\frac{1}{2}(\sigma_x - \sigma_y)\left(-\frac{\sigma_x - \sigma_y}{2\tau_{xy}}\right) + \tau_{xy}\right]
\end{aligned}$$

or

$$\tau_{XY_{MAX}} = \frac{1}{2}\sqrt{4\tau_{xy}^2 + (\sigma_x - \sigma_y)^2} \tag{1.123}$$

EXERCISE 4

A thin plate is subjected to plane stress conditions (Fig. 1.15). The normal and shear stresses are given: $\sigma_x = 70\ MPa$, $\sigma_y = 300\ MPa$, and $\tau_{xy} = 180\ MPa$. The material is ductile. It is assumed that the fracture takes place on the plane of $\tau_{\max}$ when the maximum shear stress reaches the value of the ultimate shear stress. The ultimate shear stress is $\tau_u = 0.5 \cdot S_u$ and the condition for fracture is $\tau_{\max} = \tau_u$.

a) Estimate whether the thin plate will fail.
b) Estimate the orientation of the crack in case of failure.

Solution

a) To fulfill the condition for fracture, we should calculate the maximum shear stress with the aid of Eq. (1.123):

$$\tau_{XY_{MAX}} = \frac{1}{2}\sqrt{4\tau_{xy}^2 + (\sigma_x - \sigma_y)^2}$$

(Continued)

EXERCISE 4—CONT'D

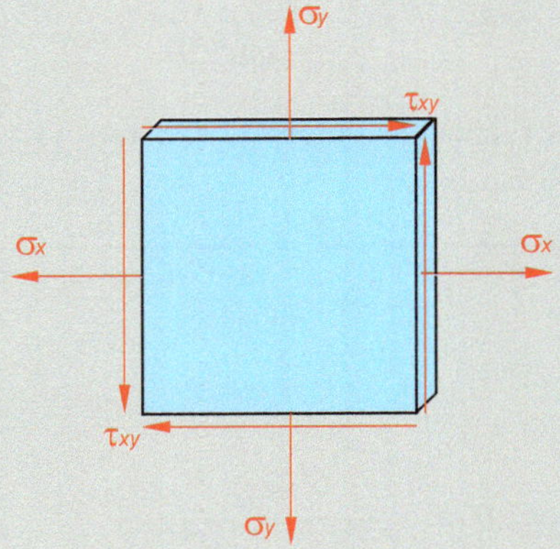

FIG. 1.15

Thin plate under plane stress conditions.

or

$$\tau_{XY_{MAX}} = \sqrt{180^2 + \frac{1}{4}(70 - 300)^2} = 213\ MPa$$

We see that the maximum shear stress is larger than the ultimate strength

$$(\tau_{\max} = 213) > (\tau_{\max} = 0.5 \cdot S_u = 185) \Rightarrow \text{Fracture}$$

Therefore the plate will fail.

b) Let's now calculate the orientation of the fracture plane. We use Eq. (1.118) in order to identify the plane of the maximum shear stress:

$$\tan 2\theta = -\frac{\sigma_x - \sigma_y}{2\tau_{xy}} = -\frac{70 - 300}{2 \cdot 180} = 0.6388$$

or

$$\theta = 16.3^{\circ}$$

The orientation of the crack is demonstrated in Fig. 1.16.

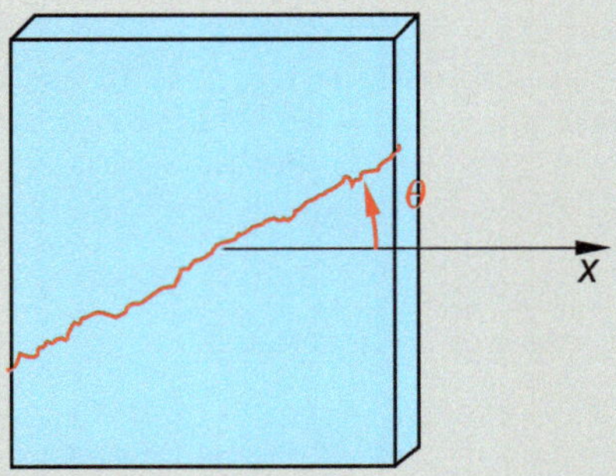

FIG. 1.16

Fracture orientation of a ductile plate under plane stress conditions.

EXERCISE 5

The stress components for the material element in Fig. 1.17 with respect to the coordinate system x, y, and z are given below:

$$_{xx} = 30\ MPa \qquad \sigma_{yy} = 12\ MPa \qquad \sigma_{zz} = -20\ MPa$$
$$_{xy} = -8\ MPa \qquad \tau_{xz} = 15\ MPa \qquad \tau_{yz} = 0\ MPa$$

a) Determine the components of the stress vector $\overrightarrow{\sigma_P}$ of the stresses acting on the plane OAB.
b) Determine the magnitude of the normal stress $|\sigma_{PN}|$ and shear stress $|\sigma_{PS}|$ on the plane OAB.
c) Determine the direction cosines of the stress vector $\overrightarrow{\sigma_P}$

Solution

a) In order to calculate the stress vector, we need to know the direction cosines of the plane OAB. This plane is a vector that can be derived as a sum of vectors of the planes OAC, ACB, and OCB (see Figs. 1.17 and 1.18):

$$\overrightarrow{OAB} = \overrightarrow{OAC} + \overrightarrow{OCB} + \overrightarrow{ACB}$$

The planes OAC, OCB, and ACB are oriented perpendicular to the x, y, and z axes. Each plane can be represented by its magnitude and the corresponding unit vector of the plane. The magnitude can be obtained by the formula providing the area of a triangle. Then:

$$\overrightarrow{OAB} = \left(\frac{1}{2}\cdot 7\cdot 10\right)\vec{j} + \left(\frac{1}{2}\cdot 10\cdot 10\right)\vec{i} - \left(\frac{1}{2}\cdot 10\cdot 7\right)\vec{k}$$

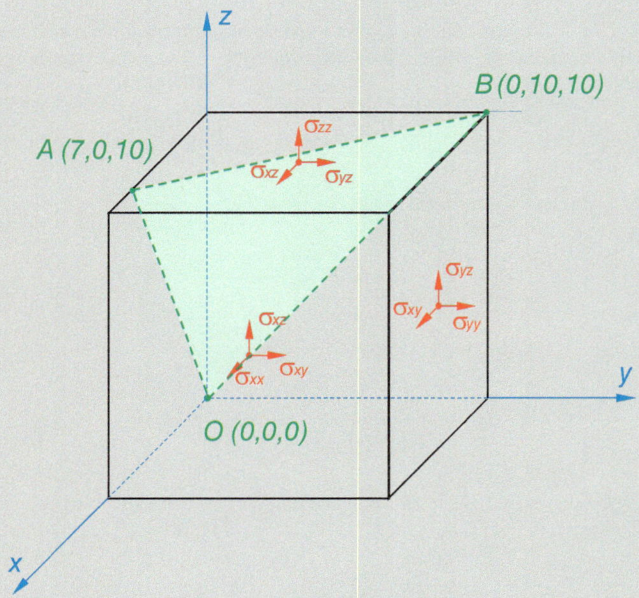

FIG. 1.17

Stress components on a material element.

(Continued)

EXERCISE 5—CONT'D

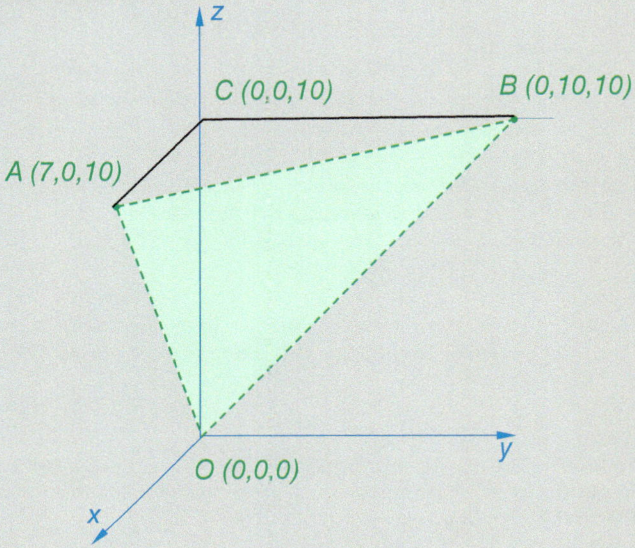

FIG. 1.18

Orientation of the plane *OAB*.

Note: The vector *ACB* has an opposite direction versus the direction of the *z* axis; that is why it is negative. The normal vectors should always go towards the inside of the prismatic volume. Above equation yields:

$$\overrightarrow{OAB} = 50\vec{i} + 35\vec{j} - 35\vec{k}$$

Then we can calculate the directional cosines by the following formulae:

$$l = \cos\,\theta_x = \frac{50}{\sqrt{50^2 + 35^2 + (-35)^2}} = 0.710$$

$$m = \cos\,\theta_y = \frac{35}{\sqrt{50^2 + 35^2 + (-35)^2}} = 0.497$$

$$n = \cos\,\theta_z = \frac{-35}{\sqrt{50^2 + 35^2 + (-35)^2}} = -0.497$$

The above direction cosines can be utilized in the following equation for calculating the stress vector components:

$$\begin{Bmatrix} \sigma_{Px} \\ \sigma_{Py} \\ \sigma_{Pz} \end{Bmatrix} = \begin{bmatrix} \sigma_x & \tau_{xy} & \tau_{xz} \\ \tau_{xy} & \sigma_y & \tau_{yz} \\ \tau_{xz} & \tau_{yz} & \sigma_z \end{bmatrix} \begin{Bmatrix} l \\ m \\ n \end{Bmatrix}$$

EXERCISE 5—CONT'D

or

$$\begin{Bmatrix} \sigma_{Px} \\ \sigma_{Py} \\ \sigma_{Pz} \end{Bmatrix} = \begin{bmatrix} 30 & -8 & 15 \\ -8 & 12 & 0 \\ 15 & 0 & -20 \end{bmatrix} \begin{Bmatrix} 0.710 \\ 0.497 \\ -0.497 \end{Bmatrix} = \begin{Bmatrix} 9.869 \\ 0.284 \\ 20.59 \end{Bmatrix} MPa$$

b) For calculating the normal component, we recall Eq. (1.24)

$$\begin{aligned} |\sigma_{PN}| &= l^2\sigma_x + m^2\sigma_y + n^2\sigma_{zz} + 2lm\tau_{xy} + 2ln\tau_{xz} + 2mn\tau_{yz} = \\ &= 0.710^2 \cdot 30 + 0.497^2 \cdot 12 + 0.497^2 \cdot (-20) + \\ &+ 2 \cdot 0.710 \cdot 0.497 \cdot (-8) + 2 \cdot (-0.497) \cdot 0.710 \cdot 15 + 2 \cdot (-0.497) \cdot 0.497 \cdot 10 \end{aligned}$$

or

$$|\sigma_{PN}| = -3.085 \ MPa$$

For the shear component, we recall Eqs. (1.25) and (1.26):

$$\begin{aligned} |\sigma_{PS}| &= \sqrt{|\sigma_P|^2 - |\sigma_{PN}|^2} = \sqrt{\sigma_{Px}^2 + \sigma_{Py}^2 + \sigma_{Pz}^2 - |\sigma_{PN}|^2} = \\ &= \sqrt{9.869^2 + 0.284^2 + 20.59^2 - (-3.085)^2} \end{aligned}$$

or

$$|\sigma_{PS}| = 22.62 \ MPa$$

c) The directional cosines of $\overrightarrow{\sigma_P}$ vector can be calculated by the following formulae:

$$\cos \varphi_x = \frac{\sigma_{Px}}{\sqrt{\sigma_{Px}^2 + \sigma_{Py}^2 + \sigma_{Pz}^2}} = 0.432$$

$$\cos \varphi_y = \frac{\sigma_{Py}}{\sqrt{\sigma_{Px}^2 + \sigma_{Py}^2 + \sigma_{Pz}^2}} = 0.012$$

$$\cos \varphi_z = \frac{\sigma_{Pz}}{\sqrt{\sigma_{Px}^2 + \sigma_{Py}^2 + \sigma_{Pz}^2}} = 0.901$$

1.13 DIFFERENTIAL EQUATIONS OF MOTION IN CARTESIAN COORDINATE SYSTEM

The differential equations of motion are the equilibrium equations of stresses in a material element. Let's assume the material element with dimensions *dx*, *dy*, and *dz* in Fig. 1.19. The stresses represented in Fig. 1.19 are the ones on the planes behind the material element, corresponding to point *O*. In Fig. 1.20, we can see the stresses on the visible planes, corresponding to the point *P*.

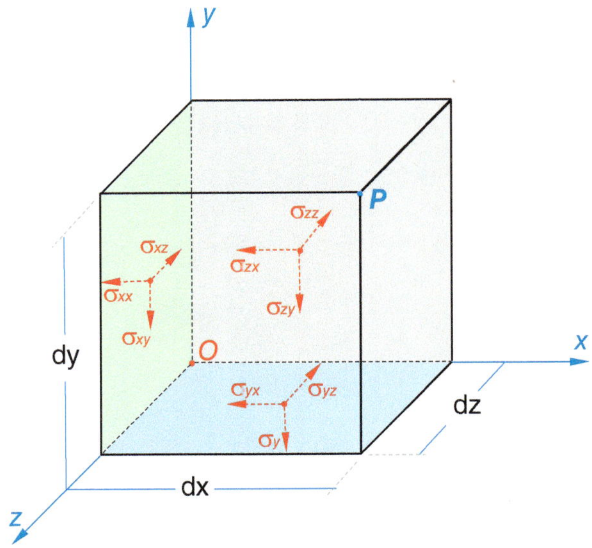

FIG. 1.19

Stress components corresponding to the planes x, y, and z of point O.

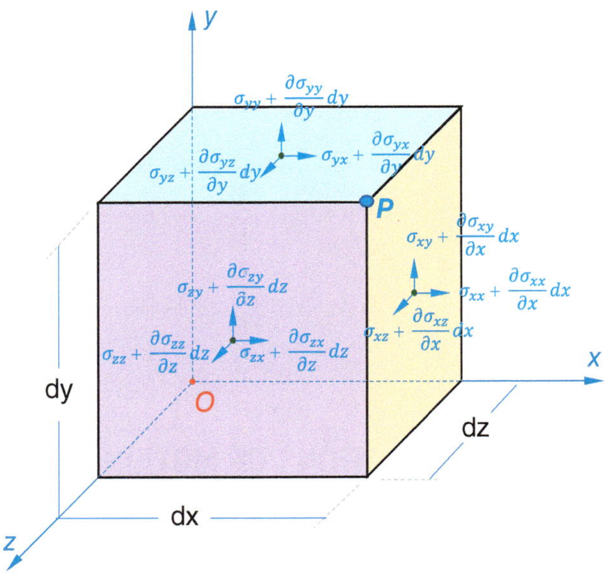

FIG. 1.20

Stress components corresponding to the planes x, y, and z of point P.

The stress tensors at points O and P are given by Eqs. (1.124) and (1.125), respectively:

$$[T_\sigma]^O = \begin{bmatrix} -\sigma_{xx} & -\sigma_{xy} & -\sigma_{xz} \\ -\sigma_{xy} & -\sigma_{yy} & -\sigma_{yz} \\ -\sigma_{xz} & -\sigma_{yz} & -\sigma_{zz} \end{bmatrix} \tag{1.124}$$

$$[T_\sigma]^P = \begin{bmatrix} \sigma_{xx} + \frac{\partial \sigma_{xx}}{\partial x} dx & \sigma_{xy} + \frac{\partial \sigma_{xy}}{\partial y} dy & \sigma_{xz} + \frac{\partial \sigma_{xz}}{\partial z} dz \\ \sigma_{xy} + \frac{\partial \sigma_{xy}}{\partial x} dx & \sigma_{yy} + \frac{\partial \sigma_{yy}}{\partial y} dy & \sigma_{yz} + \frac{\partial \sigma_{yz}}{\partial z} dz \\ \sigma_{xz} + \frac{\partial \sigma_{xz}}{\partial x} dx & \sigma_{yz} + \frac{\partial \sigma_{yz}}{\partial y} dy & \sigma_{zz} + \frac{\partial \sigma_{zz}}{\partial z} dz \end{bmatrix} \tag{1.125}$$

In addition to these stresses, the material element is also subjected to body forces B_x, B_y, and B_z. Body forces can be gravitational forces or magnetic ones. The equilibrium equation of forces in x direction:

$$\sum F_x = 0 \tag{1.126}$$

yields:

$$\left(\sigma_{xx} + \frac{\partial \sigma_{xx}}{\partial x} dx\right) dy\ dz + \left(\sigma_{yx} + \frac{\partial \sigma_{yx}}{\partial y} dy\right) dx\ dz + \left(\sigma_{zx} + \frac{\partial \sigma_{zx}}{\partial z} dz\right) dy\ dz + \\ + B_x dV - \sigma_{xx} dy\ dz - \sigma_{yx} dx\ dz - \sigma_{zx} dx\ dy = 0 \tag{1.127}$$

If we divide both sides of the above equation by $dV = dx\ \ dy\ \ dz$, the following motion equation can be obtained:

$$\frac{\partial \sigma_{xx}}{\partial x} + \frac{\partial \sigma_{yx}}{\partial y} + \frac{\partial \sigma_{zx}}{\partial z} + B_x = 0 \tag{1.128}$$

Taking the equilibrium equations of forces in y and z directions and following the same procedure, the following motion equations can be derived:

$$\frac{\partial \sigma_{xy}}{\partial x} + \frac{\partial \sigma_{yy}}{\partial y} + \frac{\partial \sigma_{zy}}{\partial z} + B_y = 0 \tag{1.129}$$

$$\frac{\partial \sigma_{xz}}{\partial x} + \frac{\partial \sigma_{yz}}{\partial y} + \frac{\partial \sigma_{zz}}{\partial z} + B_z = 0 \tag{1.130}$$

1.14 DIFFERENTIAL EQUATIONS OF MOTION IN POLAR COORDINATE SYSTEM

The formulation of the differential equations of motion in a polar coordinate system is suitable for axisymmetric solids (e.g., shafts, axes, pipes) under axisymmetric loads. We assume a material element in polar coordinates as shown in Fig. 1.21. This figure demonstrates the stresses on the r-planes corresponding to points O and P. In Figs. 1.22 and 1.23 we can see the stresses on the z-plane and θ-plane, respectively.

The equilibrium equation in r direction:

$$\sum F_r = 0 \tag{1.131}$$

yields:

$$\begin{aligned} &-\sigma_{rr}(r\,d\theta)dz + \left(\sigma_{rr} + \frac{\partial \sigma_{rr}}{\partial r}dr\right)(r+dr)d\theta\ dz- \\ &-\sigma_{zr}(r\,d\theta)dr + \left(\sigma_{zr} + \frac{\partial \sigma_{zr}}{\partial z}dz\right)r\,d\theta\ dr+ \\ &+B_r\,dV + F_r^{\theta} = 0 \end{aligned} \tag{1.132}$$

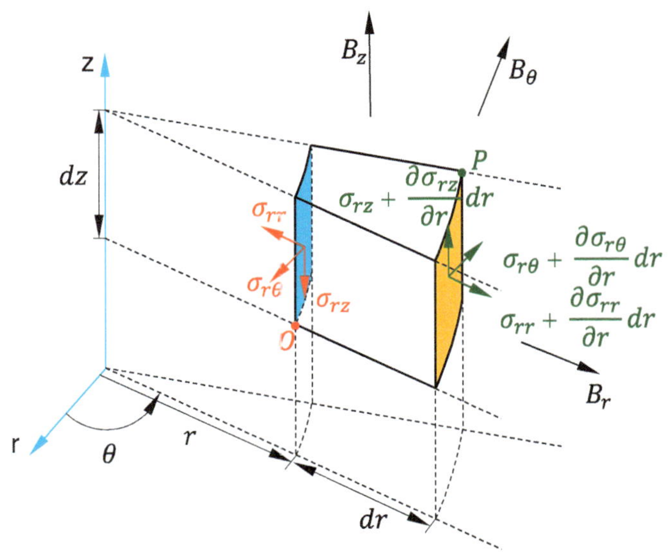

FIG. 1.21

Stress components on r planes.

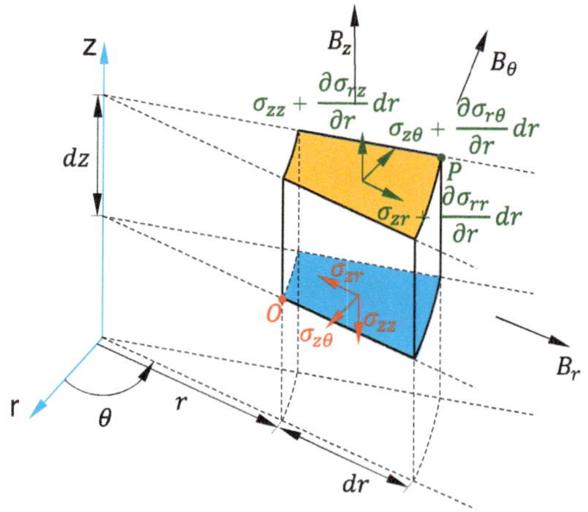

FIG. 1.22

Stress components on z planes.

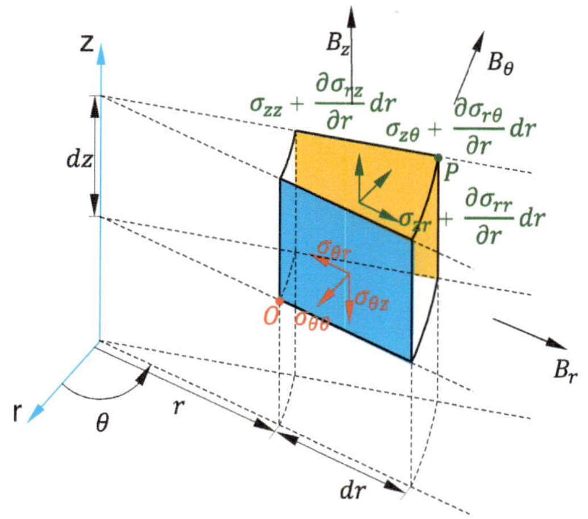

FIG. 1.23

Stress components on θ planes.

In the above equation, F_r^θ is the contribution of the forces of the θ planes in the r direction. This term includes the contribution in the r direction of the shear forces $F_r^\theta(I)$ and the normal forces $F_r^\theta(II)$ of the θ plane (Fig. 1.24)

$$F_r^\theta = F_r^\theta(I) + F_r^\theta(II) \tag{1.133}$$

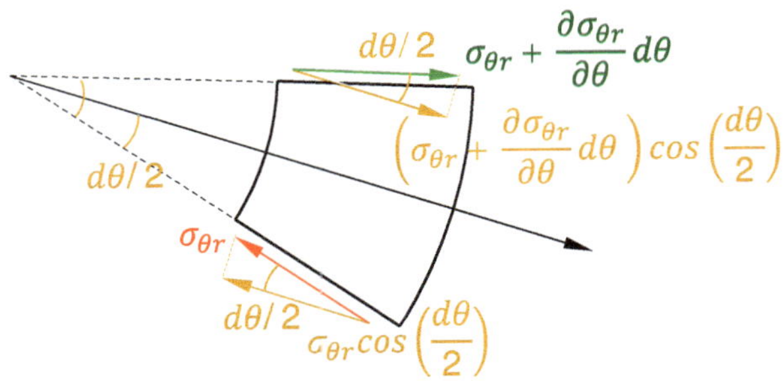

FIG. 1.24

Contribution of shear forces of θ-planes in the r-direction.

a) Contribution in r direction of shear forces in θ planes:
According to Fig. 1.24, the resultant on the r direction of the components of shear forces of θ planes are given by the following equation:

$$F^{\theta}_{r(I)} = -\sigma_{\theta r} \cos \frac{d\theta}{2} dr\, dz + \left(\sigma_{\theta r} + \frac{\partial \sigma_{\theta r}}{\partial \theta} d\theta\right) \cos \frac{d\theta}{2} dr\, dz \tag{1.134}$$

Since $d\theta$ is a small angle, $\cos \frac{d\theta}{2} \simeq 1$

Therefore Eq. (1.134) yields:

$$F^{\theta}_{r(I)} = -\sigma_{\theta r} dr\, dz + \left(\sigma_{\theta r} + \frac{\partial \sigma_{\theta r}}{\partial \theta} d\theta\right) dr\, dz \tag{1.135}$$

or

$$F^{\theta}_{r(I)} = \frac{\partial \sigma_{\theta r}}{\partial \theta} d\theta\, dr\, dz \tag{1.136}$$

b) Contribution in the r direction of normal forces in θ planes:

According to Fig. 1.25, the resultant on the r-direction of the components of normal forces of the θ plane can be obtained as:

$$F^{\theta}_{r(II)} = -\sigma_{\theta\theta} \sin \frac{d\theta}{2} dr\, dz - \left(\sigma_{\theta\theta} + \frac{\partial \sigma_{\theta\theta}}{\partial \theta} d\theta\right) \sin \frac{d\theta}{2} dr\, dz \tag{1.137}$$

since $d\theta$ is a small angle $\sin \frac{d\theta}{2} \simeq \frac{d\theta}{2}$

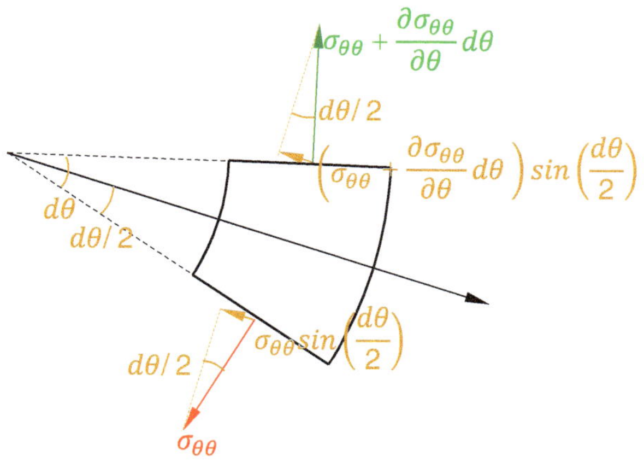

FIG. 1.25

Contribution of normal forces of the θ planes in the r direction.

Therefore Eq. (1.137) can be written as:

$$\begin{aligned} F^{\theta}_{r(II)} &= -\sigma_{\theta\theta}\frac{d\theta}{2}\,dr\,dz - \left(\sigma_{\theta\theta} + \frac{\partial\sigma_{\theta\theta}}{\partial\theta}d\theta\right)\frac{d\theta}{2}\,dr\,dz = \\ &= -\sigma_{\theta\theta}\frac{d\theta}{2}\,dr\,dz - \sigma_{\theta\theta}\frac{d\theta}{2}\,dr\,dz + \frac{\partial\sigma_{\theta\theta}}{\partial\theta}\frac{(d\theta)^2}{2}\,dr\,dz \end{aligned} \tag{1.138}$$

Since $(d\theta)^2$ is a higher-order differential, it can be neglected, that is, $d\theta \approx 0$. Then, Eq. (1.138) can be simplified:

$$F^{\theta}_{r(II)} = -\sigma_{\theta\theta}d\theta\;dr\;dz \tag{1.139}$$

With the aid of Eqs. (1.133), (1.136), and (1.139), Eq. (1.132) yields:

$$\begin{aligned} &-\sigma_{rr}(r\,d\theta)dz + \left(\sigma_{rr} + \frac{\partial\sigma_{rr}}{\partial r}dr\right)(r + dr)d\theta\,dz- \\ &-\sigma_{zr}(r\,d\theta)dr + \left(\sigma_{zr} + \frac{\partial\sigma_{zr}}{\partial z}dz\right)r\,d\theta\,dr+ \\ &+B_r\,dV + \left(\frac{\partial\sigma_{\theta r}}{\partial\theta} - \sigma_{\theta\theta}\right)d\theta\,dr\,dz = 0 \end{aligned} \tag{1.140}$$

Making some simple algebraic operations and dividing both sides by $r{\cdot}dV$, the following equation can be obtained:

$$\frac{\partial\sigma_{rr}}{\partial r} + \frac{1}{r}\frac{\partial\sigma_{r\theta}}{\partial\theta} + \frac{\partial\sigma_{rz}}{\partial z} + \frac{(\sigma_{rr} - \sigma_{\theta\theta})}{r} + B_r = 0 \tag{1.141}$$

Following the same procedure for the axes θ and z, the remaining motion equations can be derived:

$$\frac{\partial \sigma_{\theta r}}{\partial r} + \frac{1}{r}\frac{\partial \sigma_{\theta\theta}}{\partial \theta} + \frac{\partial \sigma_{\theta z}}{\partial z} + \frac{2\sigma_{r\theta}}{r} + B_\theta = 0 \tag{1.142}$$

$$\frac{\partial \sigma_{zr}}{\partial r} + \frac{1}{r}\frac{\partial \sigma_{z\theta}}{\partial \theta} + \frac{\partial \sigma_{zz}}{\partial z} + \frac{\sigma_{zz}}{r} + B_z = 0 \tag{1.143}$$

REFERENCES

[1] Boresi Arthur P, Richard J. Schmidt, Advanced Mechanics of Materials. Wiley; 2002.

[2] Armenakas Anthony. Advanced mechanics of materials and applied elasticity. CRC press; 2006.

[3] Fenner RT. Engineering elasticity. Ellis Horwood ltd; 1986.

[4] Bittencourt Marco. Computational solid mechanics. CRC press; 2015.

[5] Ugural Ansel, Fenster Saul. Advanced Mechanics of Materials and Applied Elasticity. Pearson; 2019.

[6] Solecki Roman, Conant RJay. Acvanced Mechanics of Materials. Oxford University Press; 2003.

[7] Cook Robert. Advanced Mechanics of Materials. Pearson; 1998.

[8] Bickford William. Advanced Mechanics of Materials. Longman Pub Group; 1998.

[9] Timoshenko SP, Goodier JN. Theory of elasticity. McGraw-Hill; 1970.

[10] Anand Lallit, Govindjee Sanjay. Continuum Mechanics of Solids. OUP Oxford; 2020.

[11] Pavlou GG. Essentials of the Finite Element Method. Academic Press; 2015.

CHAPTER

Strain Theory

2

Chapter Outline

2.1 DEFINITION

The points P and Q (Fig. 2.1) in the solid before deformation (*green domain*) are moved to the locations P* and Q* in the deformed solid (*light blue domain*). Therefore the infinitesimal line segment PQ is deformed. P*Q* is the deformed line segment with length ds*.

The ratio

$$\varepsilon_E = \frac{ds^* - ds}{ds} = \frac{ds^*}{ds} - 1 \tag{2.1}$$

is the definition of the *Engineering Normal Strain* ε_E. This definition is of common use in engineering practice and is valid for small strains. However, the above definition is not unique. An alternative definition for strain is given by the following equation:

$$M = \frac{1}{2}\left[\left(\frac{ds^*}{ds}\right)^2 - 1\right] \tag{2.2}$$

Advanced Mechanics of Solids and Structures. DOI: http://doi.org/10.1016/B978-0-443-23778-2.00002-0

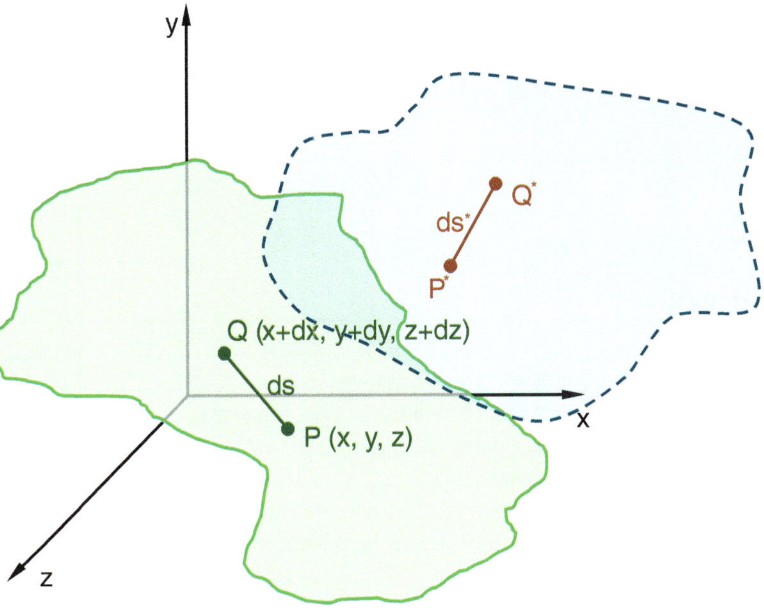

FIG. 2.1

Deformation of a solid.

The strain M is known as *Total Green Normal Strain*. This definition is suitable for large strains.

Eq. (2.2) can be written in the following format:

$$M = \frac{1}{2}\frac{(ds^* + ds)(ds^* - ds)}{ds^2}$$

or

$$M = \frac{1}{2}\frac{(ds^* + ds)}{ds}\varepsilon_E = \frac{1}{2}\frac{(ds^* + 2ds - ds)}{ds}\varepsilon_E = \frac{1}{2}\left(\varepsilon_E + \frac{2ds}{ds}\right)\varepsilon_E$$

The above equation yields:

$$M = \varepsilon_E + \frac{1}{2}\varepsilon_E^2 \tag{2.3}$$

Obviously, for small strains $\varepsilon_E^2 \to 0$, the Green and the Engineering strain are equal

$$M = \varepsilon_E \tag{2.4}$$

2.2 CORRELATION OF NORMAL STRAINS TO DISPLACEMENTS

Let's now correlate the normal strains to displacements. Considering the material element of Fig. 2.2, we assume that the displacement components of the points A and C are:

$$\text{Displacement components of A:} \begin{cases} u(x) \\ v(x) \\ w(x) \end{cases}$$

$$\text{Displacement components of C:} \begin{cases} u(x+\delta x) = u + \dfrac{\partial u}{\partial x}\delta x \\ v(x+\delta x) = v + \dfrac{\partial v}{\partial x}\delta x \\ w(x+\delta x) = w + \dfrac{\partial w}{\partial x}\delta x \end{cases}$$

Therefore the following coordinates of the points A, A′, C, and C′ can be obtained:

$$A[x, y, z]$$

$$A'[x+u, y+v, z+w]$$

$$C[(x+\delta x), y, z]$$

$$C'\left[(x+\delta x) + \left(u + \frac{\partial u}{\partial x}\delta x\right), y + \left(v + \frac{\partial v}{\partial x}\delta x\right), z + \left(w + \frac{\partial w}{\partial x}\delta x\right)\right]$$

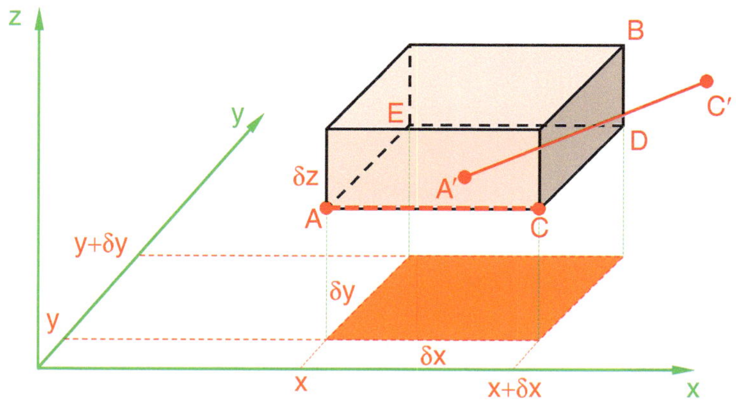

FIG. 2.2

Correlation of strains to displacements.

With the aid of the above coordinates, the components of the segment A′C′ on the *x*, *y*, and *z* axes can easily be obtained:

$$(A'C')_x = \left(x + \delta x + u + \frac{\partial u}{\partial x}\delta x\right) - (x + u) = \delta x + \frac{\partial u}{\partial x}\delta x \tag{2.5}$$

$$(A'C')_y = \left(y + v + \frac{\partial v}{\partial x}\delta x\right) - (y + v) = \frac{\partial v}{\partial x}\delta x \tag{2.6}$$

$$(A'C')_z = \left(z + w + \frac{\partial w}{\partial x}\delta x\right) - (z + w) = \frac{\partial w}{\partial x}\delta x \tag{2.7}$$

Geometric representation of the quantities in the *x*-*y* plane are shown in Fig. 2.3. With the aid of the above components of the segment A′C′, the strain component in the *x*-direction can be obtained by the following equation

$$\varepsilon_{xx} = \frac{1}{2}\left[\left(\frac{A'C'}{AC}\right)^2 - 1\right] \tag{2.8}$$

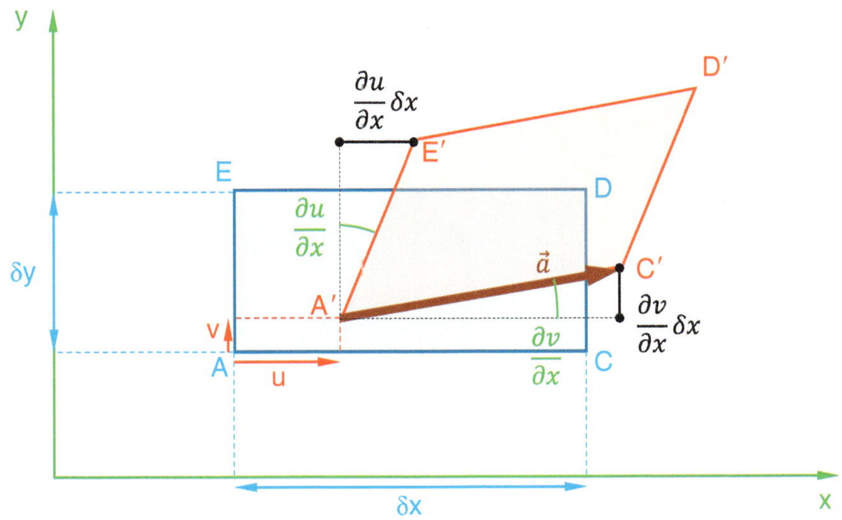

FIG. 2.3

Projection of the segment A′C′ on the axes x and y.

where

$$
\begin{aligned}
(A'C')^2 &= (A'C')_x^2 + (A'C')_y^2 + (A'C')_z^2 = \\
&= \left(\delta x + \frac{\partial u}{\partial x}\delta x\right)^2 + \left(\frac{\partial v}{\partial x}\delta x\right)^2 + \left(\frac{\partial w}{\partial x}\delta x\right)^2 = \\
&= (\delta x)^2\left[1 + 2\frac{\partial u}{\partial x} + \left(\frac{\partial u}{\partial x}\right)^2 + \left(\frac{\partial v}{\partial x}\right)^2 + \left(\frac{\partial w}{\partial x}\right)^2\right]
\end{aligned}
\tag{2.9}
$$

Substituting this into Eq. (2.8), and setting $AC = \delta x$, the following formula for the strain component ε_{xx} can be obtained:

$$
\varepsilon_{xx} = \frac{1}{2}\left[\left(\frac{(\delta x)^2\left[1 + 2\frac{\partial u}{\partial x} + \left(\frac{\partial u}{\partial x}\right)^2 + \left(\frac{\partial v}{\partial x}\right)^2 + \left(\frac{\partial w}{\partial x}\right)^2\right]}{(\delta x)^2}\right) - 1\right] \tag{2.10}
$$

Making some algebraic operations, the following results can be obtained:

$$
\varepsilon_{xx} = \frac{\partial u}{\partial x} + \frac{1}{2}\left[\left(\frac{\partial u}{\partial x}\right)^2 + \left(\frac{\partial v}{\partial x}\right)^2 + \left(\frac{\partial w}{\partial x}\right)^2\right] \tag{2.11}
$$

Following the same procedure for the strain components ε_{yy} and ε_{zz}, the following equations can be obtained:

$$
\varepsilon_{yy} = \frac{\partial v}{\partial y} + \frac{1}{2}\left[\left(\frac{\partial u}{\partial y}\right)^2 + \left(\frac{\partial v}{\partial y}\right)^2 + \left(\frac{\partial w}{\partial y}\right)^2\right] \tag{2.12}
$$

$$
\varepsilon_{xx} = \frac{\partial w}{\partial z} + \frac{1}{2}\left[\left(\frac{\partial u}{\partial z}\right)^2 + \left(\frac{\partial v}{\partial z}\right)^2 + \left(\frac{\partial w}{\partial z}\right)^2\right] \tag{2.13}
$$

We can see that for small strains, the second-order terms (products of derivatives) are negligible compared to the first-order ones and can be neglected:

$$
\varepsilon_{xx} = \frac{\partial u}{\partial x} \tag{2.14}
$$

$$
\varepsilon_{yy} = \frac{\partial v}{\partial y} \tag{2.15}
$$

$$
\varepsilon_{xx} = \frac{\partial w}{\partial z} \tag{2.16}
$$

2.3 DEFINITION OF SHEAR STRAINS AND CORRELATION TO DISPLACEMENTS

Let's now define the shear strain components for large strains. Referring to Fig. 2.4, the definition of the shear strain in the *x-y* plane is given by the following equation:

$$\gamma_{xy} = \tan\gamma \frac{\text{Area of}(A'C'D'E')}{\text{Area of}(ACDE)}, \tag{2.17}$$

where

$$\gamma = \varphi_1 + \varphi_2 = \frac{\pi}{2} - \Psi \tag{2.18}$$

Geometric representation of angle γ is shown in Fig. 2.5. In Eq. (2.18), Ψ is the angle between the vectors A′C′ and A′E′ that have been noted as $\vec{a}$ and $\vec{b}$, respectively.

The angle Ψ can be calculated with the aid of the direction cosines of the vectors $\vec{a}$ and $\vec{b}$. To this end, we recall the following formulae:

$$\frac{\vec{a}}{|\vec{a}|} = l_1\vec{i} + m_1\vec{j} + n_1\vec{k} \tag{2.19}$$

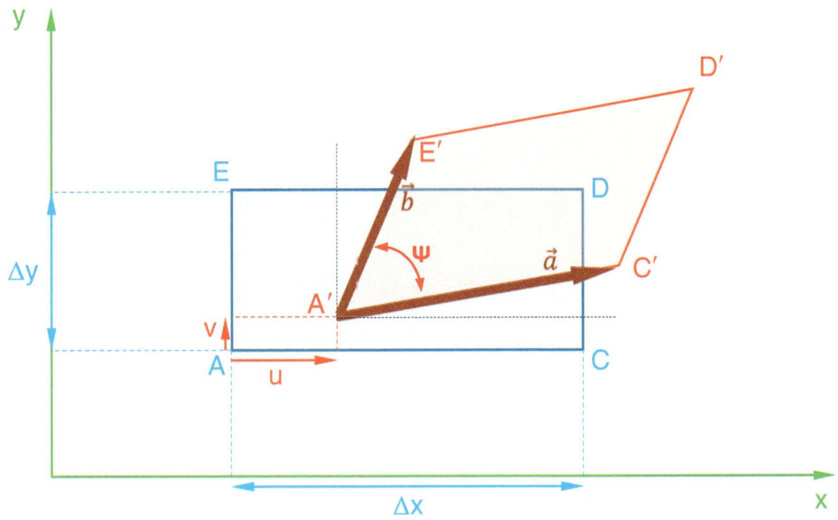

FIG. 2.4

Definition of the shear strain ε_{xy}.

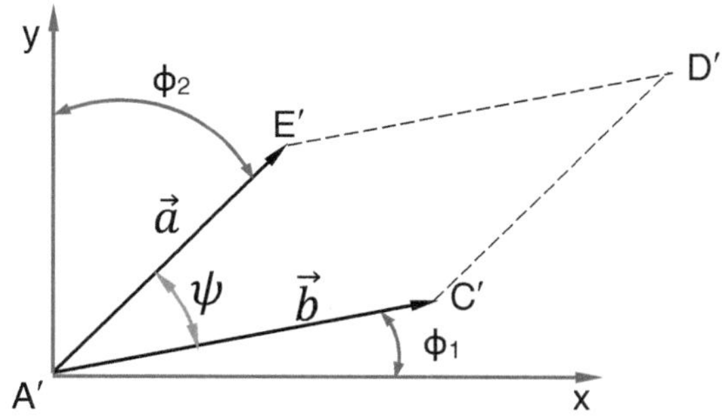

FIG. 2.5

Calculation of the angle Ψ with the aid of the direction cosines of the vectors $\vec{a}$ and $\vec{b}$.

$$\frac{\vec{b}}{|\vec{b}|} = l_2\vec{i} + m_2\vec{j} + n_2\vec{k} \tag{2.20}$$

With the aid of the above equations, the scalar product of the vectors $\vec{a}$ and $\vec{b}$

$$\vec{a}\cdot\vec{b} = |\vec{a}| \cdot |\vec{b}| \cdot \cos \Psi$$

yields

$$|\vec{a}|\left(l_1\vec{i} + m_1\vec{j} + n_1\vec{k}\right) \cdot |\vec{b}|\left(l_2\vec{i} + m_2\vec{j} + n_2\vec{k}\right) = |\vec{a}| \cdot |\vec{b}| \cdot \cos \Psi$$

or

$$\cos \Psi = l_1 l_2 + m_1 m_2 + n_1 n_2 \tag{2.21}$$

Taking into account Eqs. (2.5)–(2.7), the following formulae for the direction cosines of the vector $\vec{a}$ can be obtained:

$$l_1 = \frac{(A'C')_x}{A'C'} = \frac{\delta x + \dfrac{\partial u}{\partial x}\delta x}{A'C'} \tag{2.22}$$

$$m_1 = \frac{(A'C')_y}{A'C'} = \frac{\dfrac{\partial v}{\partial x}\delta x}{A'C'} \tag{2.23}$$

$$n_1 = \frac{(A'C')_z}{A'C'} = \frac{\dfrac{\partial w}{\partial x}\delta x}{A'C'} \tag{2.24}$$

Similarly, the direction cosines of the vector $\vec{b}$ are:

$$l_2 = \frac{(A'E')_x}{A'E'} = \frac{\frac{\partial u}{\partial y}\delta y}{A'E'} \tag{2.25}$$

$$m_2 = \frac{(A'E')_y}{A'E'} = \frac{\delta y + \frac{\partial v}{\partial y}\delta y}{A'E'} \tag{2.26}$$

$$n_2 = \frac{(A'E')_z}{A'E'} = \frac{\frac{\partial w}{\partial y}\delta y}{A'E'} \tag{2.27}$$

Therefore Eq. (2.21) yields:

$$\cos\Psi = \frac{\delta x\ \delta y\left[\frac{\partial u}{\partial y} + \frac{\partial v}{\partial x} + \left(\frac{\partial u}{\partial x}\right)\left(\frac{\partial u}{\partial y}\right) + \left(\frac{\partial v}{\partial x}\right)\left(\frac{\partial v}{\partial y}\right) + \left(\frac{\partial w}{\partial x}\right)\left(\frac{\partial w}{\partial y}\right)\right]}{(A'C')(A'E')} \tag{2.28}$$

According to Fig. 2.5, the area of A′C′D′E′ is:

$$Area\ \ A'C'D'E' = (A'C')(A'E')\cos\ \gamma$$

Then, Eq. (2.17) yields:

$$\gamma_{xy} = \frac{\sin\gamma}{\cos\gamma}\frac{(A'C')(A'E')\cos\gamma}{(AC)(AE)}$$

or

$$\gamma_{xy} = \sin\gamma\ \frac{A'C'}{AC}\frac{A'E'}{AE} \tag{2.29}$$

Because $\gamma = \frac{\pi}{2} - \Psi$, the trigonometric identity $\sin\gamma = \cos\Psi$ can be used in the above formula. Then, the following results can be obtained:

$$\gamma_{xy} = \cos\Psi\frac{A'C'}{\delta x}\frac{A'E'}{\delta y}$$

With the aid of Eq. (2.28), the above equation yields:

$$\gamma_{xy} = \frac{\partial u}{\partial y} + \frac{\partial v}{\partial x} + \left(\frac{\partial u}{\partial x}\right)\left(\frac{\partial u}{\partial y}\right) + \left(\frac{\partial v}{\partial x}\right)\left(\frac{\partial v}{\partial y}\right) + \left(\frac{\partial w}{\partial x}\right)\left(\frac{\partial w}{\partial y}\right) \tag{2.30}$$

With the same procedure, the following results can be obtained for the shear strains γ_{yz} and γ_{xz}.

$$\gamma_{yz} = \frac{\partial v}{\partial z} + \frac{\partial w}{\partial y} + \left(\frac{\partial u}{\partial y}\right)\left(\frac{\partial u}{\partial z}\right) + \left(\frac{\partial v}{\partial y}\right)\left(\frac{\partial v}{\partial z}\right) + \left(\frac{\partial w}{\partial y}\right)\left(\frac{\partial w}{\partial z}\right) \tag{2.31}$$

$$\gamma_{xz} = \frac{\partial u}{\partial z} + \frac{\partial w}{\partial x} + \left(\frac{\partial u}{\partial x}\right)\left(\frac{\partial u}{\partial z}\right) + \left(\frac{\partial v}{\partial x}\right)\left(\frac{\partial v}{\partial z}\right) + \left(\frac{\partial w}{\partial x}\right)\left(\frac{\partial w}{\partial z}\right) \tag{2.32}$$

In the above equations, $u = u(x, y, z)$, $v = v(x, y, z)$, and $w = w(x, y, z)$ are the displacements in the directions x, y, and z, respectively. For small displacements, the second-order terms tend to zero and can be neglected. Therefore, for small displacements, the following formulae for shear strains can be obtained:

$$\gamma_{xy} = \frac{\partial u}{\partial y} + \frac{\partial v}{\partial x} \tag{2.33}$$

$$\gamma_{yz} = \frac{\partial v}{\partial z} + \frac{\partial w}{\partial y} \tag{2.34}$$

$$\gamma_{xz} = \frac{\partial u}{\partial z} + \frac{\partial w}{\partial x} \tag{2.35}$$

The shear strain γ_{xy} expresses the relative rotation of the vector $\overrightarrow{AE}$ with respect to $\overrightarrow{AC}$ (Fig. 2.3). An alternative definition of the shear strain is based on the average relative rotation of $\overrightarrow{AE}$ versus $\overrightarrow{AC}$:

$$\varepsilon_{xy} = \frac{1}{2}\gamma_{xy} = \frac{1}{2}\left(\frac{\partial u}{\partial y} + \frac{\partial v}{\partial x}\right) \tag{2.36}$$

Similarly,

$$\varepsilon_{yz} = \frac{1}{2}\gamma_{yz} = \frac{1}{2}\left(\frac{\partial v}{\partial z} + \frac{\partial w}{\partial y}\right) \tag{2.37}$$

$$\varepsilon_{xz} = \frac{1}{2}\gamma_{xz} = \frac{1}{2}\left(\frac{\partial u}{\partial z} + \frac{\partial w}{\partial x}\right) \tag{2.38}$$

Like stresses, the strain components compose the strain tensor:

$$\varepsilon = \begin{bmatrix} \varepsilon_{xx} & \varepsilon_{xy} & \varepsilon_{xz} \\ \varepsilon_{xy} & \varepsilon_{yy} & \varepsilon_{yz} \\ \varepsilon_{xz} & \varepsilon_{yz} & \varepsilon_{zz} \end{bmatrix} \tag{2.39}$$

The corresponding strain vectors in the x, y, and z planes are:

$$\vec{\varepsilon}_x = \varepsilon_{xx}\vec{i} + \varepsilon_{xy}\vec{j} + \varepsilon_{xz}\vec{k} \tag{2.40}$$

$$\vec{\varepsilon}_y = \varepsilon_{xy}\vec{i} + \varepsilon_{yy}\vec{j} + \varepsilon_{yz}\vec{k} \tag{2.41}$$

$$\vec{\varepsilon}_z = \varepsilon_{xz}\vec{i} + \varepsilon_{yz}\vec{j} + \varepsilon_{zz}\vec{k} \tag{2.42}$$

2.4 SMALL STRAINS IN POLAR COORDINATES

For axi-symmetric problems, it is convenient to use the strain-displacement equations in polar coordinates.

$$\varepsilon_{rr} = \frac{\partial u_r}{\partial r} \tag{2.43}$$

$$\varepsilon_{\theta\theta} = \frac{u_r}{r} + \frac{1}{r}\frac{\partial u_\theta}{\partial \theta} \tag{2.44}$$

$$\varepsilon_{zz} = \frac{\partial u_z}{\partial z} \tag{2.45}$$

$$\varepsilon_{r\theta} = \varepsilon_{\theta r} = \frac{1}{2r}\frac{\partial u_r}{\partial \theta} + \frac{1}{2}\frac{\partial u_\theta}{\partial r} - \frac{u_\theta}{2r} \tag{2.46}$$

$$\varepsilon_{\theta z} = \varepsilon_{z\theta} = \frac{1}{2}\frac{\partial u_\theta}{\partial z} + \frac{1}{2r}\frac{\partial u_z}{\partial \theta} \tag{2.47}$$

$$\varepsilon_{rz} = \varepsilon_{zr} = \frac{1}{2}\frac{\partial u_r}{\partial z} + \frac{1}{2}\frac{\partial u_z}{\partial r} \tag{2.48}$$

In the above equations, the variables u_r, u_θ are the displacements in the radial and tangential direction r and θ, respectively.

2.5 PROPERTIES OF THE STRAIN TENSOR

2.5.1 STRAIN TENSOR ON A PLANE OF ARBITRARY ORIENTATION

Following the same vector transformation like in stress theory, the components of the strain tensor $\vec{\varepsilon}_P$ for any plane ABC with known direction cosines l, m, and n (Fig. 2.6) are:

$$\begin{Bmatrix} \varepsilon_{Px} \\ \varepsilon_{Py} \\ \varepsilon_{Pz} \end{Bmatrix} = \begin{bmatrix} \varepsilon_{xx} & \varepsilon_{xy} & \varepsilon_{xz} \\ \varepsilon_{xy} & \varepsilon_{yy} & \varepsilon_{yz} \\ \varepsilon_{xz} & \varepsilon_{yz} & \varepsilon_{zz} \end{bmatrix} \begin{Bmatrix} l \\ m \\ n \end{Bmatrix} \tag{2.49}$$

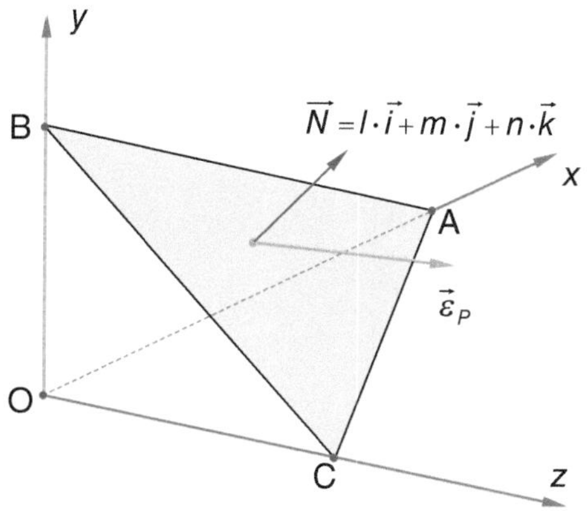

FIG. 2.6

Strain vector on an arbitrary plane ABC with unit vector $\vec{N} = l \cdot \vec{i} + m \cdot \vec{j} + n \cdot \vec{k}$.

Therefore the magnitude of the normal strain can be calculated by the following scalar product:

$$|\varepsilon_{PN}| = \vec{\varepsilon}_P \cdot \vec{N} = \left(\vec{\varepsilon}_x l + \vec{\varepsilon}_y m + \vec{\varepsilon}_z n\right)\left(l\vec{i} + m\vec{j} + n\vec{k}\right)$$

With the aid of Eqs. (2.40)–(2.42), the above equation yields:

$$|\varepsilon_{PN}| = \varepsilon_{xx} l^2 + \varepsilon_{yy} m^2 + \varepsilon_{zz} n^2 + 2\varepsilon_{xy} lm + 2\varepsilon_{xz} lm + 2\varepsilon_{yz} mn \tag{2.50}$$

2.5.2 TRANSFORMATION OF A STRAIN TENSOR IN A ROTATED COORDINATE SYSTEM

Fig. 2.7 demonstrates a rotated coordinate system *XYZ* versus the reference coordinate system *xyz*.

The angles between the axes of both systems are known, and their direction cosines are given in the following array:

		$\vec{i}$ x	$\vec{j}$ y	$\vec{k}$ z
$\vec{i'}$	X	l_1	m_1	n_1
$\vec{j'}$	Y	l_2	m_2	n_2
$\vec{k'}$	Z	l_3	m_3	n_3

(2.51)

The corresponding transformation matrix is:

$$A = \begin{bmatrix} l_1 & m_1 & n_1 \\ l_2 & m_2 & n_2 \\ l_3 & m_3 & n_3 \end{bmatrix} \tag{2.52}$$

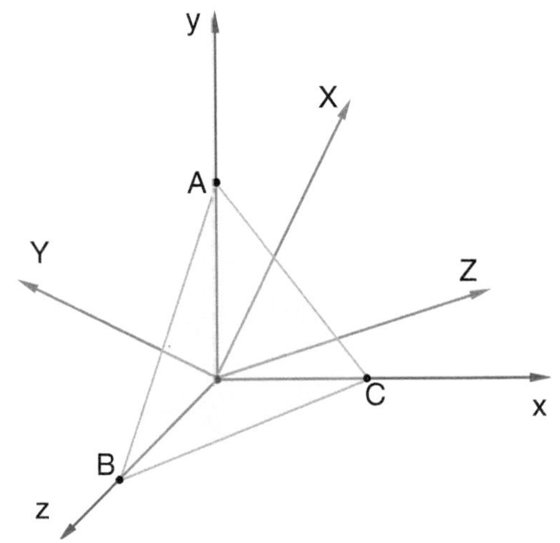

FIG. 2.7

Rotated coordinate system XYZ versus the reference coordinate system xyz.

Following the same operations as the transformation of the stress tensor, the strain tensor in the rotated coordinate system can be obtained by the following formula:

$$\begin{bmatrix} \varepsilon_{XX} & \varepsilon_{XY} & \varepsilon_{XZ} \\ \varepsilon_{XY} & \varepsilon_{YY} & \varepsilon_{YZ} \\ \varepsilon_{XZ} & \varepsilon_{YZ} & \varepsilon_{ZZ} \end{bmatrix} = [A] \begin{bmatrix} \varepsilon_{xx} & \varepsilon_{xy} & \varepsilon_{xz} \\ \varepsilon_{xy} & \varepsilon_{yy} & \varepsilon_{yz} \\ \varepsilon_{xz} & \varepsilon_{yz} & \varepsilon_{zz} \end{bmatrix} [A]^T \tag{2.53}$$

The elements of the strain tensor in the rotated coordinate system are given below:

$$\varepsilon_{XX} = \varepsilon_{xx} l_1^2 + \varepsilon_{yy} m_1^2 + \varepsilon_{zz} n_1^2 + 2\varepsilon_{xy} l_1 m_1 + 2\varepsilon_{xz} l_1 n_1 + 2\varepsilon_{yz} m_1 n_1 \tag{2.54}$$

$$\varepsilon_{YY} = \varepsilon_{xx} l_2^2 + \varepsilon_{yy} m_2^2 + \varepsilon_{zz} n_2^2 + 2\varepsilon_{xy} l_2 m_2 + 2\varepsilon_{xz} l_2 n_2 + 2\varepsilon_{yz} m_2 n_2 \tag{2.55}$$

$$\varepsilon_{ZZ} = \varepsilon_{xx} l_3^2 + \varepsilon_{yy} m_3^2 + \varepsilon_{zz} n_3^2 + 2\varepsilon_{xy} l_3 m_3 + 2\varepsilon_{xz} l_3 n_3 + 2\varepsilon_{yz} m_3 n_3 \tag{2.56}$$

$$\begin{aligned} \varepsilon_{XY} &= \varepsilon_{xx} l_1 l_2 + \varepsilon_{yy} m_1 m_2 + \varepsilon_{zz} n_1 n_2 + \\ &+ \varepsilon_{xy}(l_1 m_2 + l_2 m_1) + \varepsilon_{xz}(l_1 n_2 + l_2 n_1) + \varepsilon_{yz}(m_1 n_2 + m_2 n_1) \end{aligned} \tag{2.57}$$

$$\begin{aligned} \varepsilon_{XZ} &= \varepsilon_{xx} l_1 l_3 + \varepsilon_{yy} m_1 m_3 + \varepsilon_{zz} n_1 n_3 + \\ &+ \varepsilon_{xy}(l_1 m_3 + l_3 m_1) + \varepsilon_{xz}(l_1 n_3 + l_3 n_1) + \varepsilon_{yz}(m_1 n_3 + m_3 n_1) \end{aligned} \tag{2.58}$$

$$\begin{aligned}\varepsilon_{YZ} = \varepsilon_{xx} l_2 l_3 + \varepsilon_{yy} m_2 m_3 + \varepsilon_{zz} n_2 n_3 + \\ + \varepsilon_{xy}\left(l_2 m_3 + l_3 m_2\right) + \varepsilon_{xz}\left(l_2 n_3 + l_3 n_2\right) + \varepsilon_{yz}\left(m_2 n_3 + m_3 n_2\right)\end{aligned} \tag{2.59}$$

The above strain tensor transformation has important engineering applications in strain measurements. Strain measurement on three prescribed directions on a surface can be used to calculate the strains (and stresses) on the directions x-y of the reference coordinate system with the aid of Eqs. (2.54) and (2.55). The strain measurements take place with strain gauges.

2.6 STRAIN GAUGES

Strain gauges are thin tapes containing a long wire. The strain gauges are perfectly bonded on surfaces in order to measure the normal strain. The deformation on a small surface area causes elongation of the wire that changes its electric resistance. With the aid of an electronic circuit, the change of the electric resistance provides the values of the local strain in the direction of the strain gauge. There are single-strain gauges (Fig. 2.8) or rosettes of three-strain gauges in prescribed directions (e.g., Fig. 2.9).

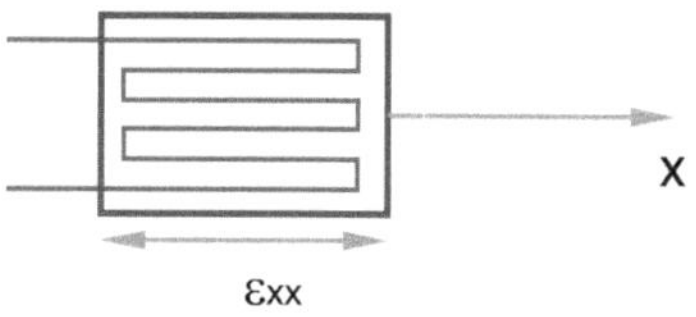

FIG. 2.8

Schematic demonstration of a strain gauge.

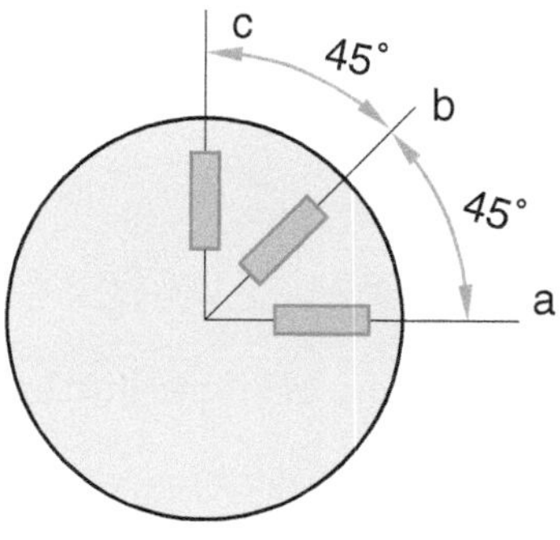

FIG. 2.9

Rosette of three-strain gauges with a 45-degree angle (45-degree rosette).

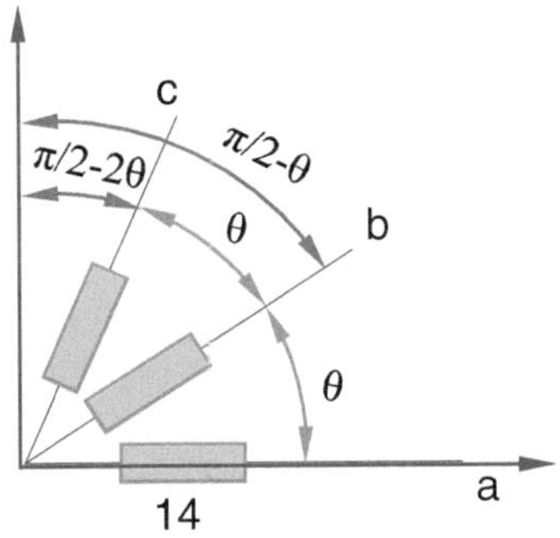

FIG. 2.10

Rosette of three-strain gauges in an arbitrary orientation θ.

Let's assume three strain gauges a, b, and c with orientation angles 0, θ, 2θ, respectively (Fig. 2.10). This rosette can be measuring the normal strains ε_a, ε_b, and ε_c. The aim now is the calculation of the strains ε_{xx}, ε_{yy}, and ε_{zz}.

Taking into account the orientation angles (Fig. 2.10) of the strain gauges a, b, and c, the following direction cosines of the transformation matrix [A] (see Eq. 2.51) can be obtained:

$l_a = \cos(0) = 1$	$l_b = \cos(\theta)$	$l_c = \cos(2\theta)$
$m_a = \cos\left(\frac{\pi}{2}\right) = 0$	$m_b = \cos\left(\frac{\pi}{2} - \theta\right)$	$m_c = \cos\left(\frac{\pi}{2} - 2\theta\right) = \sin(2\theta)$
$n_a = \cos\left(\frac{\pi}{2}\right) = 0$	$n_b = \cos\left(\frac{\pi}{2}\right) = 0$	$n_c = \cos\left(\frac{\pi}{2}\right) = 0$

(2.60)

Using Eq. (2.54), the strains ε_a, ε_b, and ε_c can be correlated to the strains ε_{xx}, ε_{yy}, and ε_{zz}:

$$\varepsilon_a = \varepsilon_{xx} l_a^2 + \varepsilon_{yy} m_a^2 + \varepsilon_{zz} n_a^2 + 2\varepsilon_{xy} l_a m_a + 2\varepsilon_{xz} l_a n_a + 2\varepsilon_{yz} m_a n_a \tag{2.61}$$

$$\varepsilon_b = \varepsilon_{xx} l_b^2 + \varepsilon_{yy} m_b^2 + \varepsilon_{zz} n_b^2 - 2\varepsilon_{xy} l_b m_b + 2\varepsilon_{xz} l_b n_b + 2\varepsilon_{yz} m_b n_b \tag{2.62}$$

$$\varepsilon_c = \varepsilon_{xx} l_c^2 + \varepsilon_{yy} m_c^2 + \varepsilon_{zz} n_c^2 + 2\varepsilon_{xy} l_c m_c + 2\varepsilon_{xz} l_c n_c + 2\varepsilon_{yz} m_c n_c \tag{2.63}$$

With the aid of the direction cosines of the transformation matrix, the above equations yield:

$$\varepsilon_a = \varepsilon_{xx} \tag{2.64}$$

$$\varepsilon_b = \varepsilon_{xx} \cos^2\theta + \varepsilon_{yy} \sin^2\theta + 2\varepsilon_{xy} \cos\theta \sin\theta \tag{2.65}$$

$$\varepsilon_c = \varepsilon_{xx}\cos^2 2\theta + \varepsilon_{yy}\sin^2 2\theta + 2\varepsilon_{xy}\cos 2\theta \sin 2\theta \tag{2.66}$$

Solution of the above system with respect to the strains ε_{xx}, ε_{yy}, and ε_{zz} provides the following results:

$$\varepsilon_{xx} = \varepsilon_a \tag{2.67}$$

$$\varepsilon_{yy} = \frac{(\varepsilon_a - 2\varepsilon_b)\sin 4\theta + 2\varepsilon_c \sin 2\theta}{4\sin^2\theta \cdot \sin 2\theta} \tag{2.68}$$

$$\varepsilon_{xy} = \frac{2\varepsilon_a(\sin^2\theta \cos^2 2\theta - \sin^2 2\theta \cos^2\theta) + 2\left(\varepsilon_b \sin^2\theta - \varepsilon_c \sin^2\theta\right)}{4\sin^2\theta \cdot \sin 2\theta} \tag{2.69}$$

For the rosette of strain gauges with $\theta = 45°$ (Fig. 2.9), the above equations yield:

$$\varepsilon_{xx} = \varepsilon_a \tag{2.70}$$

$$\varepsilon_{yy} = \varepsilon_c \tag{2.71}$$

$$\varepsilon_{xy} = \varepsilon_b - 1/2(\varepsilon_a + \varepsilon_c) \tag{2.72}$$

For the rosette of strain gauges with $\theta = 60°$ (Fig. 2.11), the following results can be obtained:

$$\varepsilon_{xx} = \varepsilon_a \tag{2.73}$$

$$\varepsilon_{yy} = 1/3\left(2(\varepsilon_b + \varepsilon_c) - \varepsilon_a\right) \tag{2.74}$$

$$\varepsilon_{zz} = 1/\sqrt{3}\,(\varepsilon_b - \varepsilon_c) \tag{2.75}$$

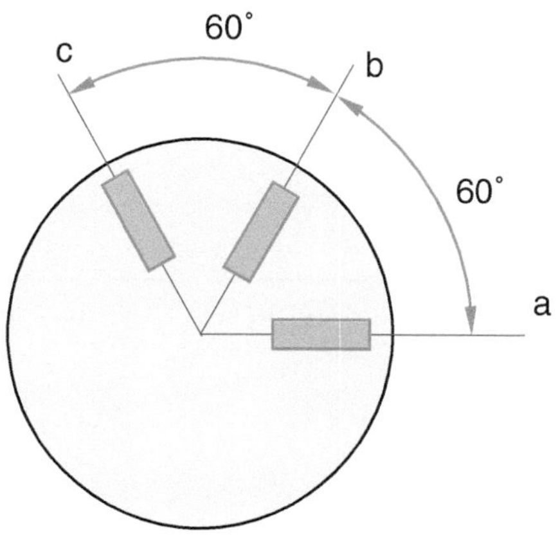

FIG. 2.11

Rosette of strain gauges, 60-degree angle.

EXERCISE 7

A square thin plate with dimensions 1 m × 1 m is loaded in a state of plane stress and is deformed elastically as shown in Fig. 2.12.

A. Derive the expressions for the displacements u and v for any point (x,y) of the plate.
B. Taking into account the small-displacement theory, determine the corresponding strain components $\varepsilon_{xx}, \varepsilon_{yy}$, and ε_{xy} at any point (x,y) of the plate.
C. Determine the strain at point C in a direction $\xi' - \xi$ with angle of orientation $\theta = 30°$.

Solution

(a) Because all sides of the plate remain linear after deformation, the displacements u and v can be expressed by linear functions:

$$u = a_1 + b_1 x + c_1 y + d_1 xy$$
$$v = a_2 + b_2 x + c_2 y + d_2 xy$$

The coefficients a_i, b_i, c_i, and d_i (i = 1,2) are unknown parameters. To obtain these parameters, we apply the above equations on the points A, B, C, and D where the displacements are known:

Point A: $x = 0$, y = 0, u = 0, v = 0

$$0 = a_1 + b_1 0 + c_1 0 + d_1 0 \Rightarrow \boxed{a_1 = 0}$$
$$0 = a_2 + b_2 0 + c_2 0 + d_2 0 \Rightarrow \boxed{a_2 = 0}$$

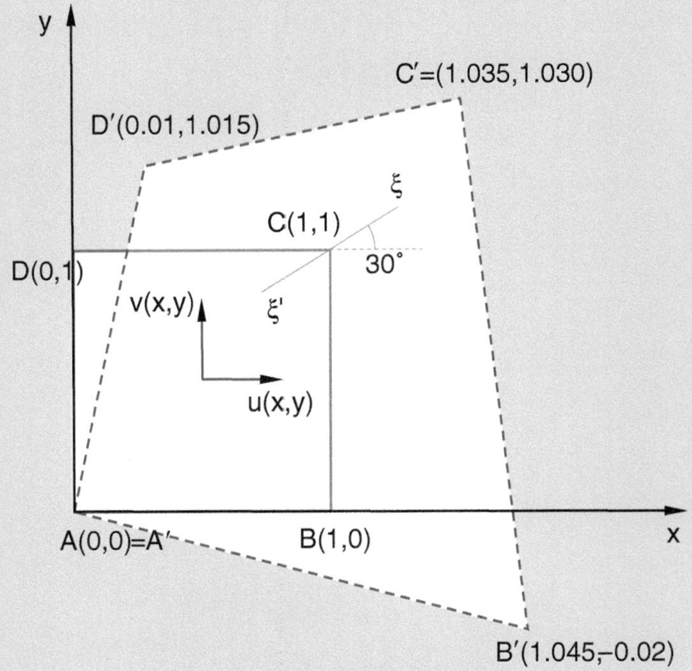

FIG. 2.12

Deformation of a thin plate under plain stress conditions.

EXERCISE 7—CONT'D

Point B: $x = 1$, $y = 0$, $u = 1.045 - 1.000 = 0.045$, $v = -0.02 - 0.00 = -0.02$

$$0.045 = b_1 \cdot 1 + c_1 0 + d_1 0 \Rightarrow \boxed{b_1 = 0.045}$$
$$-0.02 = b_2 \cdot 1 + c_2 0 + d_2 0 \Rightarrow \boxed{b_2 = -0.02}$$

Point C: $x = 1$, $y = 1$, $u = 1.035 - 1.000 = 0.035$, $v = 1.03 - 1.00 = 0.03$

$$0.035 = 0.045 \cdot 1 + 0.01 \cdot 1 + d_1 \cdot 1 \Rightarrow \boxed{d_1 = -0.02}$$
$$0.03 = -0.02 \cdot 1 + 0.015 \cdot 1 + d_2 \cdot 1 \Rightarrow \boxed{d_2 = 0.035}$$

Point D: $x = 0$, $y = 1$, $u = 0.01 - 0.00 = 0.01$, $v = 1.015 - 1.000 = 0.015$

$$0.01 = 0.045 \cdot 0 + c_1 \cdot 1 + d_1 0 \Rightarrow \boxed{c_1 = 0.01}$$
$$0.015 = -0.02 \cdot 0 + c_2 \cdot 1 + d_2 0 \Rightarrow \boxed{c_2 = 0.015}$$

With the aid of the calculated parameters, the displacement distribution can be obtained by:

$$u(x, y) = 0.045x + 0.01y - 0.02xy$$
$$v(x, y) = -0.02x + 0.015y + 0.035xy$$

(a) With the aid of the derived displacement functions, the following formulae can be used for calculating the strains:

$$\begin{aligned}
\varepsilon_{xx} &= \frac{\partial u}{\partial x} = \frac{\partial}{\partial x}(0.045x + 0.01y - 0.02xy) \\
\varepsilon_{yy} &= \frac{\partial v}{\partial y} = \frac{\partial}{\partial y}(-0.02x + 0.015y + 0.035xy) \\
\varepsilon_{xy} &= \frac{1}{2}\left(\frac{\partial v}{\partial x} + \frac{\partial u}{\partial y}\right) \\
&= \frac{1}{2}\left(\frac{\partial}{\partial x}(-0.02x + 0.015y + 0.035xy) + \frac{\partial u}{\partial y}(0.045x + 0.01y - 0.02xy)\right)
\end{aligned}$$

or

$$\begin{aligned}
\varepsilon_{xx} &= 0.045 - 0.02y \\
\varepsilon_{yy} &= 0.015 + 0.035x \\
\varepsilon_{xy} &= -0.005 - 0.01x + 0.0175y
\end{aligned}$$

(b) Taking into account the coordinates of the point C, the strain components at the reference coordinate system x-y-z at the point C can be calculated with the aid of the above formulae:

$$\begin{aligned}
\varepsilon_{xx}(1,1) &= 0.045 - 0.02 \cdot 1 = 0.025 \\
\varepsilon_{yy}(1,1) &= 0.015 + 0.035 \cdot 1 = 0.05 \\
\varepsilon_{xy}(1,1) &= -0.005 - 0.01 \cdot 1 + 0.0175 \cdot 1 = 0.0025
\end{aligned}$$

The direction cosines of the orientation $\xi' - \xi$ are:

$$l = \cos 30^\circ = \frac{\sqrt{3}}{2}$$

$$m = \cos(-60^\circ) = \frac{1}{2}$$

$$n = \cos 90^\circ = 0$$

Then the strain at point C in the direction $\xi' - \xi$ can be obtained by the following Eq. (2.54):

$$\varepsilon_{\xi'-\xi} = \varepsilon_{xx}l^2 + \varepsilon_{yy}m^2 + \varepsilon_{zz}n^2 + 2\varepsilon_{xy}lm + 2\varepsilon_{xz}ln + 2\varepsilon_{yz}mn$$

For the conditions $\varepsilon_{xz} = 0$, $\varepsilon_{yz} = 0$, and $\varepsilon_{zz} = 0$, the above equation yields:

$$\varepsilon_{\xi'-\xi} = \varepsilon_{xx}l^2 + \varepsilon_{yy}m^2 + 2\varepsilon_{xy}lm$$

or

$$\varepsilon_{\xi'-\xi} = \left(\frac{\sqrt{3}}{2}\right)^2 \cdot 0.025 + \left(\frac{1}{2}\right)^2 \cdot 0.05 + 2\frac{\sqrt{3}}{2}\frac{1}{2}0.0025 = 0.0334$$

2.7 STRAIN COMPATIBILITY EQUATIONS

In the chapter on stress theory (Chapter 1), we derived the set of the stress equilibrium equations. However, the equilibrium equations are not always sufficient to analyze the stress field of a solid or structure. To this end, the strain compatibility conditions should be considered in addition to the stress equilibrium equations. We can see that the strain components of a solid are not independent but fulfill certain conditions, the strain compatibility conditions. Let's simplify the problem by considering the plain-strain case (the displacement w in the z-direction is constant):

$$\text{Plane-strain:} \begin{cases} u(x, y) \\ v(x, y) \\ w = \text{constant} \end{cases}$$

Let's recall Eqs. (2.14), (2.15), and (2.36) for the plane-strain conditions:

$$\varepsilon_{xx} = \frac{\partial u}{\partial x}$$

$$\varepsilon_{yy} = \frac{\partial v}{\partial y}$$

$$\varepsilon_{xy} = \frac{1}{2}\left(\frac{\partial u}{\partial y} + \frac{\partial v}{\partial x}\right)$$

$$\varepsilon_{zz} = \varepsilon_{xz} = \varepsilon_{yz} = 0 \text{ (because w = constant)}$$

On the above equations, we can take the following derivatives:

$$\frac{\partial^2 \varepsilon_{xx}}{\partial y^2} = \frac{\partial^3 u}{\partial x\, \partial y^2} \tag{2.76}$$

$$\frac{\partial^2 \varepsilon_{yy}}{\partial x^2} = \frac{\partial^3 v}{\partial x^2 \partial y} \tag{2.77}$$

$$2\frac{\partial^2 \varepsilon_{xy}}{\partial x\, \partial y} = \frac{\partial^3 u}{dx\, \partial y^2} + \frac{\partial^3 v}{\partial x^2 \partial y} \tag{2.78}$$

Combining the above equations, the following equations can be obtained.

$$\frac{\partial^2 \varepsilon_{xx}}{\partial y^2} + \frac{\partial^2 \varepsilon_{yy}}{\partial x^2} = 2\frac{\partial^2 \varepsilon_{xy}}{\partial x\, \partial y} \tag{2.79}$$

Following the same procedure, we get:

$$\frac{\partial^2 \varepsilon_{xx}}{\partial z^2} + \frac{\partial^2 \varepsilon_{zz}}{\partial x^2} = 2\frac{\partial^2 \varepsilon_{xz}}{\partial x\, \partial z} \tag{2.80}$$

$$\frac{\partial^2 \varepsilon_{yy}}{\partial z^2} + \frac{\partial^2 \varepsilon_{zz}}{\partial y^2} = 2\frac{\partial^2 \varepsilon_{yz}}{\partial y\, \partial z} \tag{2.81}$$

$$\frac{\partial^2 \varepsilon_{zz}}{\partial x\, \partial y} + \frac{\partial^2 \varepsilon_{xy}}{\partial z^2} = \frac{\partial^2 \varepsilon_{yz}}{\partial x\, \partial z} + \frac{\partial^2 \varepsilon_{xz}}{\partial y\, \partial z} \tag{2.82}$$

$$\frac{\partial^2 \varepsilon_{yy}}{\partial x\, \partial z} + \frac{\partial^2 \varepsilon_{xz}}{\partial y^2} = \frac{\partial^2 \varepsilon_{xy}}{\partial y\, \partial z} + \frac{\partial^2 \varepsilon_{yz}}{\partial x\, \partial y} \tag{2.83}$$

$$\frac{\partial^2 \varepsilon_{xx}}{\partial y\, \partial z} + \frac{\partial^2 \varepsilon_{xy}}{\partial x^2} = \frac{\partial^2 \varepsilon_{xy}}{\partial x\, \partial y} + \frac{\partial^2 \varepsilon_{xz}}{\partial x\, \partial z} \tag{2.84}$$

2.8 VOLUMETRIC STRAIN

Generally, the stresses are changing the volume and the shape of a material element. The shear stresses are responsible for changing the shape, and the hydrostatic stresses (that involve normal stresses) are responsible for changing the volume. A suitable parameter for characterizing the volumetric change is the volumetric strain Δ:

$$\Delta = \frac{\text{change in volume}}{\text{initial volume}} \tag{2.85}$$

Considering Fig. 2.13, the initial and final volume of a material element are:

Initial volume is: $dV_0 = dx \cdot dy \cdot dz$
Final volume is: $dV = (dx + \varepsilon_x dx)(dy + \varepsilon_y dy)(dz + \varepsilon_z dz)$

Therefore the volumetric strain can be obtained by the following equation:

$$\Delta = \frac{dV - dV_0}{dV_0} = \frac{dV}{dV_0} - 1 \tag{2.86}$$

Substituting values dV_o and dV, the following result can be obtained:

$$\Delta = \frac{(dx + \varepsilon_x dx)(dy + \varepsilon_y dy)(dz + \varepsilon_z dz)}{dx \cdot dy \cdot dz} - 1$$

or

$$\Delta = \frac{dx(1 + \varepsilon_x)dy(1 + \varepsilon_y)dz(1 + \varepsilon_z)}{dx \cdot dy \cdot dz} - 1$$

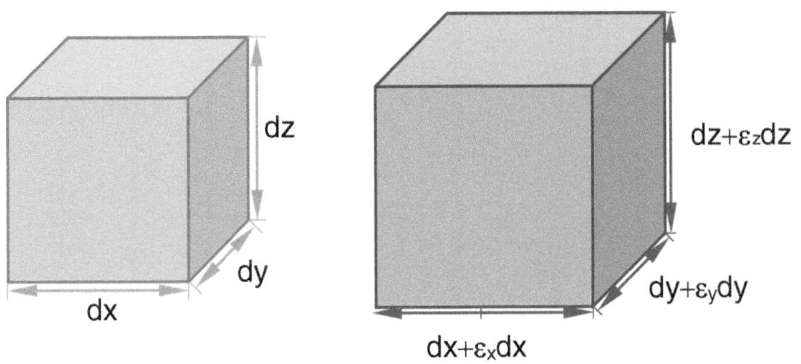

FIG. 2.13

Volumetric change.

Finally:

$$\Delta = (1 + \varepsilon_x)(1 + \varepsilon_y)(1 + \varepsilon_z) - 1$$

that yields:

$$\Delta = \varepsilon_x + \varepsilon_y + \varepsilon_z + \varepsilon_x\varepsilon_y + \varepsilon_x\varepsilon_z + \varepsilon_y\varepsilon_z \tag{2.87}$$

The second-order terms (products of strains) are negligible because they are small strains. Then we can write:

$$\Delta = \varepsilon_x + \varepsilon_y + \varepsilon_z \tag{2.88}$$

REFERENCES

[1] Boresi Arthur P, Schmidt Richard J. Advanced Mechanics of Materials. Wiley; 2002.
[2] Armenakas Anthony. Advanced mechanics of materials and applied elasticity. CRC press; 2006.
[3] Fenner RT. Engineering elasticity. Ellis Horwood ltd; 1986.
[4] Bittencourt Marco. Computational solid mechanics. CRC press; 2015.
[5] Ugural Ansel, Fenster Saul. Advanced Mechanics of Materials and Applied Elasticity. Pearson; 2019.
[6] Solecki Roman, Conant RJay. Advanced Mechanics of Materials. Oxford University Press; 2003.
[7] Cook Robert. Advanced Mechanics of Materials. Pearson; 1998.
[8] Bickford William. Advanced Mechanics of Materials. Longman Pub Group; 1998.
[9] Timoshenko SP, Goodier JN. Theory of elasticity. McGraw-Hill; 1970.
[10] Anand Lallit, Govindjee Sanjay. Continuum Mechanics of Solids. OUP Oxford; 2020.
[11] Pavlou GG. Essentials of the Finite Element Method. Academic Press; 2015.

CHAPTER

3 Constitutive Equations for Isotropic and Anisotropic Solids

Chapter Outline

In the previous chapters the stress and strain theory were examined separately. The aim of this chapter is to examine the relation between the stresses and strains (generalized Hooke's law) and the role of the material behavior. The applied stresses are the reason of the strains. The strains are correlated with stresses via the constitutive equations. In uniaxial stress state (Fig. 3.1), the normal strain ε is correlated to the normal stress σ with Hooke's law:

$$\sigma = E\varepsilon \tag{3.1}$$

where E is a material property called elastic modulus (or Young's modulus).

If a material with different properties in the directions X_1, X_2, and X_3 (anisotropic material) is subjected to multiaxial stress state (Fig. 3.2) under stresses $\{\sigma\} = [\sigma_{11} \ \ \sigma_{22} \ \ \sigma_{33} \ \ \sigma_{23} \ \ \sigma_{13} \ \ \sigma_{12}]^T$ and strains $\{\varepsilon\} = [\varepsilon_{11} \ \ \varepsilon_{22} \ \ \varepsilon_{33} \ \ \varepsilon_{23} \ \ \varepsilon_{13} \ \ \varepsilon_{12}]^T$, the constitutive equation is a matrix equation that can be abbreviated as follows:

$$\{\sigma\} = [C]\{\varepsilon\} \tag{3.2}$$

where

$$\begin{cases} \{\sigma\} = [\sigma_{11} \ \ \sigma_{22} \ \ \sigma_{33} \ \ \sigma_{23} \ \ \sigma_{13} \ \ \sigma_{12}]^T \\ \{\varepsilon\} = [\varepsilon_{11} \ \ \varepsilon_{22} \ \ \varepsilon_{33} \ \ \varepsilon_{23} \ \ \varepsilon_{13} \ \ \varepsilon_{12}]^T \end{cases} \tag{3.3}$$

Advanced Mechanics of Solids and Structures. DOI: http://doi.org/10.1016/B978-0-443-23778-2.00003-2

FIG. 3.1

Uniaxial stress state.

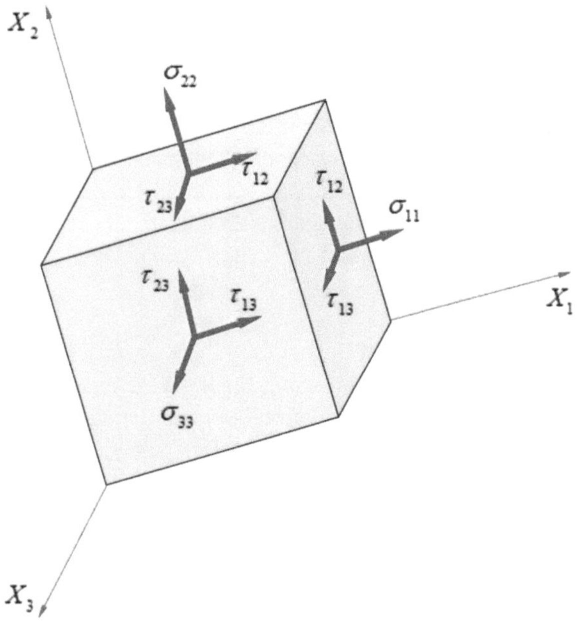

FIG. 3.2

Multiaxial stress state.

3.1 CONSTITUTIVE EQUATIONS

The abbreviated Eq. (3.2) can be expressed analytically by the following equation:

$$\begin{Bmatrix} \sigma_{11} \\ \sigma_{22} \\ \sigma_{33} \\ \tau_{23} \\ \tau_{13} \\ \tau_{12} \end{Bmatrix} = \begin{bmatrix} C_{11} & C_{12} & C_{13} & 0 & 0 & 0 \\ C_{21} & C_{22} & C_{23} & 0 & 0 & 0 \\ C_{31} & C_{32} & C_{33} & 0 & 0 & 0 \\ 0 & 0 & 0 & C_{44} & 0 & 0 \\ 0 & 0 & 0 & 0 & C_{55} & 0 \\ 0 & 0 & 0 & 0 & 0 & C_{66} \end{bmatrix} \begin{Bmatrix} \varepsilon_{11} \\ \varepsilon_{22} \\ \varepsilon_{23} \\ \varepsilon_{23} \\ \varepsilon_{13} \\ \varepsilon_{12} \end{Bmatrix} \tag{3.4}$$

In the above equation, the matrix $[C]$ is the stiffness matrix. In terms of strains versus stresses, Eq. (3.2) yields

$$\{\varepsilon\} = [C]^{-1}\{\sigma\} = [S]\{\sigma\} \tag{3.5}$$

or

$$\begin{Bmatrix} \varepsilon_{11} \\ \varepsilon_{22} \\ \varepsilon_{33} \\ \varepsilon_{23} \\ \varepsilon_{13} \\ \varepsilon_{12} \end{Bmatrix} = \begin{bmatrix} S_{11} & S_{12} & S_{13} & 0 & 0 & 0 \\ S_{21} & S_{22} & S_{23} & 0 & 0 & 0 \\ S_{31} & S_{32} & S_{33} & 0 & 0 & 0 \\ 0 & 0 & 0 & S_{44} & 0 & 0 \\ 0 & 0 & 0 & 0 & S_{55} & 0 \\ 0 & 0 & 0 & 0 & 0 & S_{66} \end{bmatrix} \begin{Bmatrix} \sigma_{11} \\ \sigma_{22} \\ \sigma_{33} \\ \tau_{23} \\ \tau_{13} \\ \tau_{12} \end{Bmatrix} \tag{3.6}$$

In the above equation, the matrix $[S]$ is the compliance matrix. The stiffness matrix is the inverse compliance matrix:

$$[C] = [S]^{-1} \tag{3.7}$$

The elements of the compliance matrix (compliance coefficients) can be calculated by the material properties in the X_1, X_2, and X_3 directions. Table 3.1 provides the compliance coefficients versus the basic material properties. The symbols E_i, G_{ij}, and ν_{ij} ($i = 1, 2, 3$ and $j = 1, 2, 3$) denote the elastic modulus, shear modulus, and Poisson's ratio, respectively.

Usually anisotropic materials in engineering are the composite materials. They consist of a matrix material that is usually a polymer, reinforced by fibers. The fibers are materials with very high strength (usually carbon or glass or synthetic fibers). In the nomenclature for the symbols for composite materials, the orientation of the fibers is aligned with the direction of the axis X_1. According to the Betti-Maxwell reciprocal theorem, the material properties in anisotropic elasticity are correlated:

$$\left.\begin{matrix} \dfrac{\nu_{12}}{E_1} = \dfrac{\nu_{21}}{E_2} \\ \dfrac{\nu_{13}}{E_1} = \dfrac{\nu_{31}}{E_3} \\ \dfrac{\nu_{23}}{E_2} = \dfrac{\nu_{32}}{E_3} \end{matrix}\right\} \Rightarrow \begin{Bmatrix} S_{21} = S_{12} \\ S_{31} = S_{13} \\ S_{32} = S_{23} \end{Bmatrix} \tag{3.8}$$

Table 3.1 Compliance coefficients.

$S_{11} = \dfrac{1}{E_1}$	$S_{21} = -\dfrac{\nu_{12}}{E_1}$	$S_{31} = -\dfrac{\nu_{13}}{E_1}$	$S_{44} = \dfrac{1}{G_{23}}$
$S_{12} = -\dfrac{\nu_{21}}{E_2}$	$S_{22} = \dfrac{1}{E_2}$	$S_{32} = -\dfrac{\nu_{23}}{E_2}$	$S_{55} = \dfrac{1}{G_{13}}$
$S_{13} = -\dfrac{\nu_{31}}{E_3}$	$S_{23} = -\dfrac{\nu_{32}}{E_3}$	$S_{33} = \dfrac{1}{E_3}$	$S_{66} = \dfrac{1}{G_{12}}$

Eq. (3.7), with the aid of Eq. (3.6), yields the following results for the stiffness coefficients:

$$C_{11} = \frac{S_{22}S_{33} - S_{23}^2}{S} \tag{3.9}$$

$$C_{12} = \frac{S_{13}S_{23} - S_{12}S_{33}}{S} \tag{3.10}$$

$$C_{13} = \frac{S_{12}S_{23} - S_{13}S_{22}}{S} \tag{3.11}$$

$$C_{21} = C_{12} \tag{3.12}$$

$$C_{22} = \frac{S_{33}S_{11} - S_{13}^2}{S} \tag{3.13}$$

$$C_{23} = \frac{S_{12}S_{13} - S_{23}S_{11}}{S} \tag{3.14}$$

$$C_{31} = C_{13} \tag{3.15}$$

$$C_{32} = C_{23} \tag{3.16}$$

$$C_{33} = \frac{S_{11}S_{22} - S_{12}^2}{S} \tag{3.17}$$

$$C_{44} = \frac{1}{S_{44}} \tag{3.18}$$

$$C_{55} = \frac{1}{S_{55}} \tag{3.19}$$

$$C_{66} = \frac{1}{S_{66}} \tag{3.20}$$

$$S = S_{11}S_{22}S_{33} - S_{11}S_{23}^2 - S_{22}S_{13}^2 - S_{33}S_{12}^2 + 2S_{12}S_{23}S_{13} \tag{3.21}$$

3.1.1 THERMAL STRESSES

Temperature changes in an anisotropic material element can cause significant stresses. The Hooke's law that includes the temperature change effects is provided by the constitutive Eq. 3.6:

$$\begin{Bmatrix} \varepsilon_{11} - \alpha_1 \Delta T \\ \varepsilon_{22} - \alpha_2 \Delta T \\ \varepsilon_{33} - \alpha_3 \Delta T \\ \varepsilon_{23} \\ \varepsilon_{13} \\ \varepsilon_{12} \end{Bmatrix} = \begin{bmatrix} S_{11} & S_{12} & S_{13} & 0 & 0 & 0 \\ S_{21} & S_{22} & S_{23} & 0 & 0 & 0 \\ S_{31} & S_{32} & S_{33} & 0 & 0 & 0 \\ 0 & 0 & 0 & S_{44} & 0 & 0 \\ 0 & 0 & 0 & 0 & S_{55} & 0 \\ 0 & 0 & 0 & 0 & 0 & S_{66} \end{bmatrix} \begin{Bmatrix} \sigma_{11} \\ \sigma_{22} \\ \sigma_{33} \\ \tau_{23} \\ \tau_{13} \\ \tau_{12} \end{Bmatrix} \tag{3.22}$$

where a_i (i = 1, 2, 3) is the thermal expansion coefficient in the direction i and ΔT is the temperature change. The above equation can be formulated in terms of stresses:

$$\begin{Bmatrix} \sigma_{11} \\ \sigma_{22} \\ \sigma_{33} \\ \tau_{23} \\ \tau_{13} \\ \tau_{12} \end{Bmatrix} = \begin{bmatrix} C_{11} & C_{12} & C_{13} & 0 & 0 & 0 \\ C_{21} & C_{22} & C_{23} & 0 & 0 & 0 \\ C_{31} & C_{32} & C_{33} & 0 & 0 & 0 \\ 0 & 0 & 0 & C_{44} & 0 & 0 \\ 0 & 0 & 0 & 0 & C_{55} & 0 \\ 0 & 0 & 0 & 0 & 0 & C_{66} \end{bmatrix} \begin{Bmatrix} \varepsilon_{11} - \alpha_1 \Delta T \\ \varepsilon_{22} - \alpha_2 \Delta T \\ \varepsilon_{33} - \alpha_3 \Delta T \\ \varepsilon_{23} \\ \varepsilon_{13} \\ \varepsilon_{12} \end{Bmatrix} \tag{3.23}$$

3.1.2 PLANE STRESS CONSTITUTIVE EQUATION

The composite materials are laminated. A typical layer (lamina) is demonstrated in Fig. 3.3. Each anisotropic lamina is very thin; therefore plain stress conditions can be assumed.

Assuming plane stress condition, $\sigma_{33} = \tau_{13} = \tau_{23} = 0$ Eq. (3.6) for the anisotropic layer of Fig. 3.3 can be simplified:

$$\begin{Bmatrix} \varepsilon_{11} \\ \varepsilon_{22} \\ \varepsilon_{33} \\ \varepsilon_{23} \\ \varepsilon_{13} \\ \varepsilon_{12} \end{Bmatrix} = \begin{bmatrix} S_{11} & S_{12} & S_{13} & 0 & 0 & 0 \\ S_{21} & S_{22} & S_{23} & 0 & 0 & 0 \\ S_{31} & S_{32} & S_{33} & 0 & 0 & 0 \\ 0 & 0 & 0 & S_{44} & 0 & 0 \\ 0 & 0 & 0 & 0 & S_{55} & 0 \\ 0 & 0 & 0 & 0 & 0 & S_{66} \end{bmatrix} \begin{Bmatrix} \sigma_{11} \\ \sigma_{22} \\ 0 \\ 0 \\ 0 \\ \tau_{12} \end{Bmatrix} \tag{3.24}$$

The above equation yields $\varepsilon_{13} = \varepsilon_{23} = 0$. Then, Eq. (3.24) takes the form:

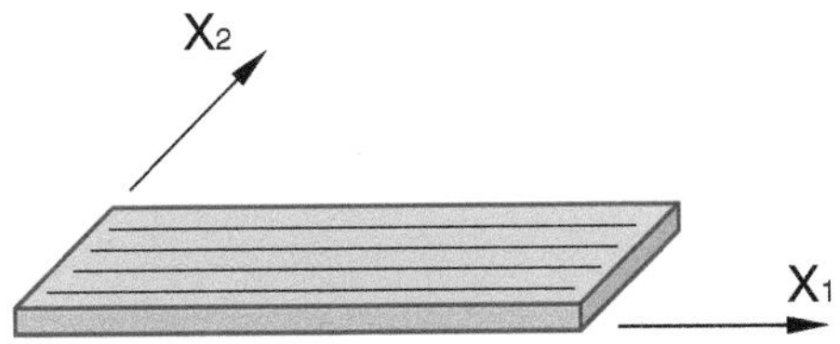

FIG. 3.3

Lamina of a composite material.

$$\begin{Bmatrix} \varepsilon_{11} \\ \varepsilon_{22} \\ \varepsilon_{12} \end{Bmatrix} = \begin{bmatrix} S_{11} & S_{12} & 0 \\ S_{12} & S_{22} & 0 \\ 0 & 0 & S_{66} \end{bmatrix} \begin{Bmatrix} \sigma_{11} \\ \sigma_{22} \\ \tau_{12} \end{Bmatrix} \tag{3.25}$$

The above equation expresses Hooke's law for plane stress. The compliance parameters are given below:

$$S_{11} = \frac{1}{E_1} \tag{3.26}$$

$$S_{12} = -\frac{\nu_{12}}{E_1} = -\frac{\nu_{21}}{E_2} \tag{3.27}$$

$$S_{22} = \frac{1}{E_2} \tag{3.28}$$

$$S_{66} = \frac{1}{G_{12}} \tag{3.29}$$

Eq. (3.25) can be written in terms of stresses:

$$\begin{Bmatrix} \sigma_{11} \\ \sigma_{22} \\ \tau_{12} \end{Bmatrix} = \begin{bmatrix} Q_{11} & Q_{12} & 0 \\ Q_{12} & Q_{22} & 0 \\ 0 & 0 & Q_{66} \end{bmatrix} \begin{Bmatrix} \varepsilon_{11} \\ \varepsilon_{22} \\ \varepsilon_{12} \end{Bmatrix} \tag{3.30}$$

The parameters Q_{ij} can be calculated by the inversion of the compliance matrix S:

$$Q_{11} = \frac{E_1}{1-\nu_{12}\nu_{21}} \tag{3.31}$$

$$Q_{12} = \frac{\nu_{21}E_1}{1-\nu_{12}\nu_{21}} = \frac{\nu_{12}E_2}{1-\nu_{12}\nu_{21}} \tag{3.32}$$

$$Q_{22} = \frac{E_2}{1 - \nu_{12}\nu_{21}} \tag{3.33}$$

$$Q_{66} = G_{12} \tag{3.34}$$

Isotropic materials have the same mechanical properties in any direction. If we set the conditions

$$E_1 = E_2 = E \tag{3.35}$$

$$\nu_{12} = \nu_{21} = \nu \tag{3.36}$$

$$G_{12} = G \tag{3.37}$$

in Eqs. (3.31)–(3.34), Eq. (3.30) provides the constitutive equation for stress:

$$\begin{Bmatrix} \sigma_{11} \\ \sigma_{22} \\ \tau_{12} \end{Bmatrix} = \begin{bmatrix} \dfrac{E}{1-\nu^2} & \dfrac{\nu E}{1-\nu^2} & 0 \\ \dfrac{\nu E}{1-\nu^2} & \dfrac{E}{1-\nu^2} & 0 \\ 0 & 0 & G \end{bmatrix} \begin{Bmatrix} \varepsilon_{11} \\ \varepsilon_{22} \\ \varepsilon_{12} \end{Bmatrix} \tag{3.38}$$

3.2 TRANSFORMATION OF STRESS AND STRAIN COMPONENTS OF A LAMINA TO A ROTATED COORDINATE SYSTEM

In composite materials we usually know the material properties in the local coordinate system X_1, X_2 (Fig. 3.4). Axis X_1 in this coordinate system, is aligned with the orientation of the fibers that makes angle θ with the x axis. However, the loads are usually given in the global coordinate system x-y.

Therefore the stress components should be transformed from the global to the local coordinate system according to the following formula:

$$\begin{Bmatrix} \sigma_{11} \\ \sigma_{22} \\ \tau_{12} \end{Bmatrix} = \begin{bmatrix} \cos^2\theta & \sin^2\theta & 2\sin\theta\cos\theta \\ \sin^2\theta & \cos^2\theta & -2\sin\theta\cos\theta \\ -\sin\theta\cos\theta & \sin\theta\cos\theta & \cos^2\theta - \sin^2\theta \end{bmatrix} \begin{Bmatrix} \sigma_{xx} \\ \sigma_{yy} \\ \tau_{xy} \end{Bmatrix} \tag{3.39}$$

Eq. (3.39) can be written as:

$$\begin{Bmatrix} \sigma_{11} \\ \sigma_{22} \\ \tau_{12} \end{Bmatrix} = [T] \begin{Bmatrix} \sigma_{xx} \\ \sigma_{yy} \\ \tau_{xy} \end{Bmatrix} \tag{3.40}$$

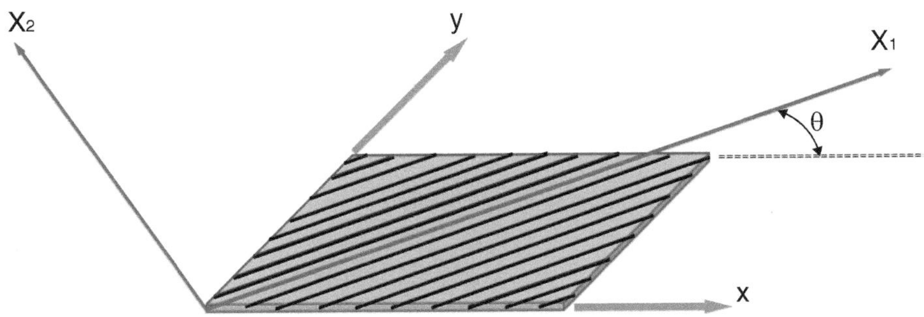

FIG. 3.4

Local and global coordinate system in an anisotropic lamina.

In the above equation, $[T]$ is a transformation matrix

$$[T] = \begin{bmatrix} m^2 & n^2 & 2mn \\ n^2 & m^2 & -2mn \\ -mn & mn & m^2 - n^2 \end{bmatrix} \tag{3.41}$$

where $m = \cos\theta$ and $n = \sin\theta$.

If we need to express Eq. (3.39) in terms of stresses, we should inverse the transformation matrix:

$$[T]^{-1} = \begin{bmatrix} m^2 & n^2 & -2mn \\ n^2 & m^2 & 2mn \\ mn & -mn & m^2 - n^2 \end{bmatrix} \tag{3.42}$$

Then, the following equation can be obtained:

$$\begin{Bmatrix} \sigma_{xx} \\ \sigma_{yy} \\ \tau_{xy} \end{Bmatrix} = \begin{bmatrix} m^2 & n^2 & -2mn \\ n^2 & m^2 & 2mn \\ mn & -mn & m^2 - n^2 \end{bmatrix} \begin{Bmatrix} \sigma_{11} \\ \sigma_{22} \\ \tau_{12} \end{Bmatrix} \tag{3.43}$$

The strains are vectors like stresses, and the transformation from global to the local coordinate system is the same:

$$\begin{Bmatrix} \varepsilon_{11} \\ \varepsilon_{22} \\ \varepsilon_{12} \end{Bmatrix} = \begin{bmatrix} \cos^2\theta & \sin^2\theta & 2\sin\theta\cos\theta \\ \sin^2\theta & \cos^2\theta & -2\sin\theta\cos\theta \\ -\sin\theta\cos\theta & \sin\theta\cos\theta & \cos^2\theta - \sin^2\theta \end{bmatrix} \begin{Bmatrix} \varepsilon_{xx} \\ \varepsilon_{yy} \\ \varepsilon_{xy} \end{Bmatrix} \tag{3.44}$$

With the aid of Eqs. (3.39) and (3.44), Eq. (3.25) can correlate the strain to the stresses in the x-y coordinate system:

$$[T]\begin{Bmatrix}\varepsilon_{xx}\\ \varepsilon_{yy}\\ \varepsilon_{xy}\end{Bmatrix} = \begin{bmatrix}S_{11} & S_{12} & 0\\ S_{12} & S_{22} & 0\\ 0 & 0 & S_{66}\end{bmatrix}[T]\begin{Bmatrix}\sigma_{xx}\\ \sigma_{yy}\\ \tau_{xy}\end{Bmatrix}$$

or

$$\begin{Bmatrix}\varepsilon_{xx}\\ \varepsilon_{yy}\\ \varepsilon_{xy}\end{Bmatrix} = [T]^{-1}\begin{bmatrix}S_{11} & S_{12} & 0\\ S_{12} & S_{22} & 0\\ 0 & 0 & S_{66}\end{bmatrix}[T]\begin{Bmatrix}\sigma_{xx}\\ \sigma_{yy}\\ \tau_{xy}\end{Bmatrix} \tag{3.45}$$

The following notations can be used for the product $[T]^{-1}[S][T]$:

$$[T]^{-1}[S][T] = [\overline{S}] = \begin{bmatrix}\overline{S}_{11} & \overline{S}_{12} & \overline{S}_{16}\\ \overline{S}_{12} & \overline{S}_{22} & \overline{S}_{26}\\ \overline{S}_{16} & \overline{S}_{26} & \overline{S}_{66}\end{bmatrix}$$

The matrix $[\overline{S}]$ is called transformed compliance matrix. Then, Eq. (3.45) takes the form:

$$\begin{Bmatrix}\varepsilon_{xx}\\ \varepsilon_{yy}\\ \varepsilon_{xy}\end{Bmatrix} = \begin{bmatrix}\overline{S}_{11} & \overline{S}_{12} & \overline{S}_{16}\\ \overline{S}_{12} & \overline{S}_{22} & \overline{S}_{26}\\ \overline{S}_{16} & \overline{S}_{26} & \overline{S}_{66}\end{bmatrix}\begin{Bmatrix}\sigma_{xx}\\ \sigma_{yy}\\ \tau_{xy}\end{Bmatrix} \tag{3.46}$$

where

$$\overline{S}_{11} = S_{11}m^4 + (2S_{12} + S_{66})n^2m^2 + S_{22}n^4 \tag{3.47}$$

$$\overline{S}_{12} = (S_{11} + S_{22} - 2S_{66})n^2m^2 + S_{12}(n^4 + m^4) \tag{3.48}$$

$$\overline{S}_{16} = (2S_{11} - 2S_{12} - S_{66})nm^3 - (2S_{22} - 2S_{12} - S_{66})n^3m \tag{3.49}$$

$$\overline{S}_{22} = S_{11}n^4 + 2(S_{12} + S_{66})n^2m^2 + S_{22}m^4 \tag{3.50}$$

$$\overline{S}_{26} = 2(S_{11} - S_{12} - S_{66})mn^3 - 2(S_{22} - S_{12} - S_{66})nm^3 \tag{3.51}$$

$$\overline{S}_{66} = 2(S_{11} + S_{22} - 2S_{12} - S_{66})n^2m^2 + S_{66}(n^4 + m^4) \tag{3.52}$$

Eq. (3.46) can be written in terms of stresses:

$$\begin{Bmatrix} \sigma_x \\ \sigma_y \\ \tau_{xy} \end{Bmatrix} = \begin{bmatrix} \overline{Q}_{11} & \overline{Q}_{12} & \overline{Q}_{16} \\ \overline{Q}_{12} & \overline{Q}_{22} & \overline{Q}_{26} \\ \overline{Q}_{16} & \overline{Q}_{26} & \overline{Q}_{66} \end{bmatrix} \begin{Bmatrix} \varepsilon_x \\ \varepsilon_y \\ \varepsilon_{xy} \end{Bmatrix} \tag{3.53}$$

where

$$\overline{Q}_{11} = Q_{11}m^4 + 2(Q_{12} + 2Q_{66})n^2m^2 + Q_{22}n^4 \tag{3.54}$$

$$\overline{Q}_{12} = (Q_{11} + Q_{22} - 4Q_{66})n^2m^2 + Q_{12}(n^4 + m^4) \tag{3.55}$$

$$\overline{Q}_{16} = (Q_{11} - Q_{12} - 2Q_{66})nm^3 + (Q_{12} - Q_{ww} + 2Q_{66})n^3m \tag{3.56}$$

$$\overline{Q}_{22} = Q_{11}n^4 + 2(Q_{12} + 2Q_{66})n^2m^2 + Q_{22}m^4 \tag{3.57}$$

$$\overline{Q}_{26} = (Q_{11} - Q_{12} - 2Q_{66})mn^3 + (Q_{12} - Q_{22} + Q_{66})nm^3 \tag{3.58}$$

$$\overline{Q}_{66} = (Q_{11} + Q_{22} - 2Q_{12} - 2Q_{66})n^2m^2 + Q_{66}(n^4 + m^4) \tag{3.59}$$

The matrix $[\overline{Q}]$ is called transformed stiffness matrix.

EXERCISE 1

Determine the material properties E_x, E_y, ν_{xy}, ν_{yx}, and G_{xy} in the global coordinate system versus the properties E_1, E_2, ν_{12}, and G_{12} of the local coordinate system. How does the fiber orientation angle θ affect the results?

Solution

We start calculating the equivalent elasticity modulus E_x.

Assuming uniaxial stress state ($\sigma_{xx} \neq 0$, $\sigma_{yy} = 0$, $\tau_{xy} = 0$) on the material element in Fig. 3.5, Hooke's law in the x-direction is:

$$\sigma_{xx} = E_x \varepsilon_{xx} \tag{3.60}$$

Using the conditions $\sigma_{yy} = 0$ and $\tau_{xy} = 0$ in Eq. (3.46)

$$\begin{Bmatrix} \varepsilon_{xx} \\ \varepsilon_{yy} \\ \varepsilon_{xy} \end{Bmatrix} = \begin{bmatrix} \overline{S}_{11} & \overline{S}_{12} & \overline{S}_{16} \\ \overline{S}_{12} & \overline{S}_{22} & \overline{S}_{26} \\ \overline{S}_{16} & \overline{S}_{26} & \overline{S}_{66} \end{bmatrix} \begin{Bmatrix} \sigma_{xx} \\ 0 \\ 0 \end{Bmatrix} \tag{3.61}$$

the following finding can be obtained:

$$\varepsilon_{xx} = \overline{S}_{11}\sigma_{xx} \tag{3.62}$$

EXERCISE 1—CONT'D

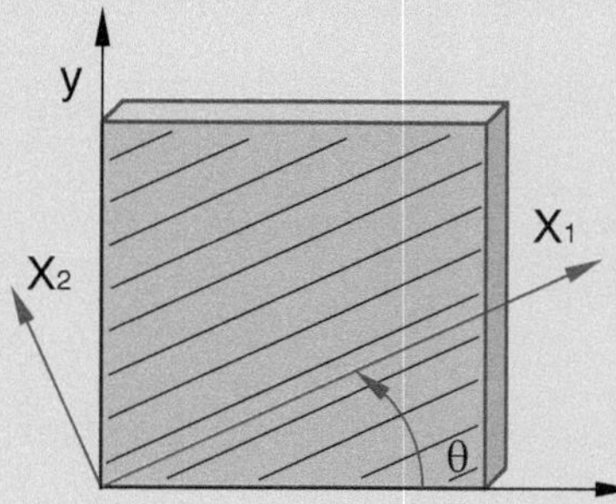

FIG 3.5

Transformation of the mechanical properties from the local to the global coordinate system for a lamina.

The combination of Eqs. (3.60) and (3.62) yields:

$$E_x = \frac{1}{\overline{S}_{11}} \tag{3.63}$$

With the aid of Eqs. (3.47) and (3.26)–(3.29), the following result can be obtained:

$$E_x = \frac{1}{\overline{S}_{11}} = \frac{E_1}{m^4 + \left(\frac{E_1}{G_{12}} - 2\nu_{12}\right)n^2m^2 + \frac{E_1}{E_2}n^4} \tag{3.64}$$

Let's now calculate Poisson's ratio in the x-y direction.

Poisson's ratio ν_{xy} is defined by the following formula:

$$\nu_{xy} = -\frac{\varepsilon_{yy}}{\varepsilon_{xx}} \tag{3.65}$$

The strain ε_{yy} can be obtained by Eq. (3.61):

$$\varepsilon_{yy} = \overline{S}_{12}\sigma_{xx} \tag{3.66}$$

With the aid of Eqs. (3.62) and (3.66), Eq. (3.65) yields:

$$\nu_{xy} = -\frac{\overline{S}_{12}}{\overline{S}_{11}} \tag{3.67}$$

Taking into account Eqs. (3.47), (3.48), and (3.26)–(3.29), the following result can be obtained:

$$\nu_{xy} = \frac{\nu_{12}(n^4 + m^4) - \left(1 + \frac{E_1}{E_2} - \frac{E_1}{G_{12}}\right)n^2m^2 +}{m^4 + \left(\frac{E_1}{G_{12}} - 2\nu_{12}\right)n^2m^2 + \frac{E_1}{E_2}n^4} \tag{3.68}$$

For the derivation of the formula for E_y, the conditions $\sigma_{xx} = 0$, $\sigma_{yy} \neq 0$, and $\tau_{xy} = 0$ should be applied in Eq. (3.46):

$$\begin{Bmatrix} \varepsilon_{xx} \\ \varepsilon_{yy} \\ \varepsilon_{xy} \end{Bmatrix} = \begin{bmatrix} \overline{S}_{11} & \overline{S}_{12} & \overline{S}_{16} \\ \overline{S}_{12} & \overline{S}_{22} & \overline{S}_{26} \\ \overline{S}_{16} & \overline{S}_{26} & \overline{S}_{66} \end{bmatrix} \begin{Bmatrix} 0 \\ \sigma_{yy} \\ 0 \end{Bmatrix} \tag{3.69}$$

(Continued)

EXERCISE 1—CONT'D

The above equation yields

$$\varepsilon_{yy} = \overline{S}_{22}\sigma_{yy} \tag{3.70}$$

With the aid of the above equation, the elastic modulus in y-direction

$$E_y = \frac{\sigma_{yy}}{\varepsilon_{yy}} \tag{3.71}$$

can be obtained by the following formula:

$$E_y = \frac{1}{\overline{S}_{22}} \tag{3.72}$$

With the aid of Eqs. (3.50) and (3.26)–(3.29), the above equation yields:

$$E_y = \frac{E_2}{m^4 + \left(\frac{E_1}{G_{12}} - 2\nu_{12}\right)n^2m^2 + \frac{E_2}{E_1}n^4} \tag{3.73}$$

Poisson's ratio ν_{yx} can be derived with the same procedure as the calculation of ν_{xy}. Then, the following result can be obtained:

$$\nu_{yx} = -\frac{\varepsilon_x}{\varepsilon_y} = -\frac{\overline{S}_{12}}{\overline{S}_{22}} = \frac{\nu_{21}(n^4 + m^4) - \left(1 - \frac{E_2}{E_1} - \frac{E_2}{G_{12}}\right)n^2m^2 +}{m^4 + \left(\frac{E_1}{G_{12}} - 2\nu_{12}\right)n^2m^2 + \frac{E_2}{E_1}n^4} \tag{3.74}$$

For the derivation of the shear modulus G_{xy}, the conditions $\sigma_{xx} = 0$, $\sigma_{yy} = 0$, and $\tau_{xy} \neq 0$ should be applied in Eq. (3.46):

$$\begin{Bmatrix} \varepsilon_{xx} \\ \varepsilon_{yy} \\ \varepsilon_{xy} \end{Bmatrix} = \begin{bmatrix} \overline{S}_{11} & \overline{S}_{12} & \overline{S}_{16} \\ \overline{S}_{12} & \overline{S}_{22} & \overline{S}_{26} \\ \overline{S}_{16} & \overline{S}_{26} & \overline{S}_{66} \end{bmatrix} \begin{Bmatrix} 0 \\ 0 \\ \tau_{xy} \end{Bmatrix} \tag{3.75}$$

The above equation yields:

$$\varepsilon_{xy} = \overline{S}_{66}\,\tau_{yy} \tag{3.76}$$

With the aid of the above equation, the following definition of the shear modulus

$$G_{xy} = \frac{\tau_{xy}}{\varepsilon_{xy}} \tag{3.77}$$

yields:

$$G_{xy} = \frac{1}{\overline{S}_{66}} \tag{3.78}$$

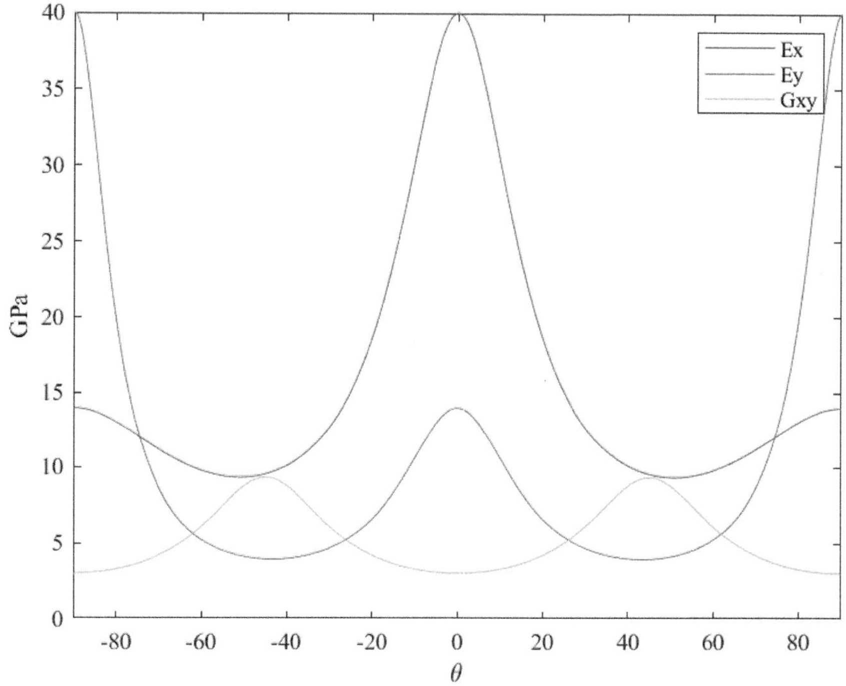

FIG. 3.6

Graphical representation of the distribution of E_x, E_y and G_{xy} versus θ.

The parameter $\bar{S}_{66}$ can be obtained by Eq. (3.52) with the aid of Eqs. (3.26)–(3.29). Then, the following result can be obtained:

$$G_{xy} = \frac{1}{\bar{S}_{66}} = \frac{G_{12}}{n^4 + m^4 + 2\left[\frac{G_{12}}{E_1}(1 + 2\nu_{12}) - 2\frac{G_{12}}{E_1} - 1\right]n^2m^2} \tag{3.79}$$

Eqs. (3.64), (3.68), (3.73), (3.74), and (3.79) provide the mechanical properties E_x, ν_{xy}, E_y, ν_{yx} and G_{xy} versus the trigonometric functions $m = \cos\theta$ and $n = \sin\theta$ of the fiber orientation angle θ. For $-90^o \le \theta \le 90^o$, the distribution of the material properties E_x, E_y, and G_{xy} have been demonstrated graphically in Fig. 3.6.

EXERCISE 9

The 55 × 55 mm anisotropic plate of Fig. 3.7 is subjected to uniaxial stress $\sigma_x = 160MPa$. The orientation of the L-axis is $\theta = 30°$. The material properties with respect to L-T coordinate system are $E_L = 155GPa$, $E_T = 12.10GPa$, $G_{LT} = 4.4GPa$, and $\nu_{LT} = 0.248$. Determine the deformed shape of the plate.

(Continued)

EXERCISE 9—CONT'D

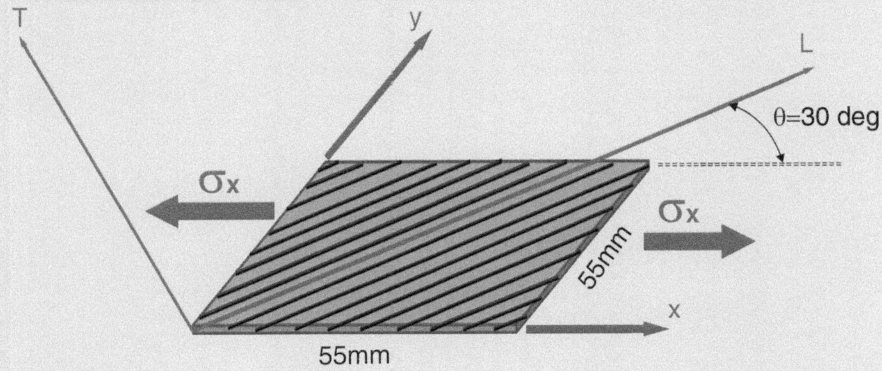

FIG. 3.7

Anisotropic plate under uniaxial stress.

Solution

$$m = \cos\theta = \frac{\sqrt{3}}{2}$$
$$n = \sin\theta = \frac{1}{2}$$

$$S_{11} = \frac{1}{E_L} = 6.45\ TPa^{-1}$$
$$S_{22} = \frac{1}{E_T} = 82.6\ TPa^{-1}$$
$$S_{12} = -\frac{\nu_{LT}}{E_L} = -1.6\ TPa^{-1}$$
$$S_{66} = \frac{1}{G_{LT}} = 227\ TPa^{-1}$$

$$\overline{S}_{11}(30^\circ) = S_{11}m^4 + (2S_{12} + S_{66})n^2m^2 + S_{22}n^4 = 50.8\ TPa^{-1}$$
$$\overline{S}_{12}(30^\circ) = (S_{11} + S_{22} - S_{66})n^2m^2 + S_{12}(n^4 + m^4) = -26.9\ TPa^{-1}$$
$$\overline{S}_{16}(30^\circ) = (2S_{11} - 2S_{12} - S_{66})nm^3 + (2S_{22} - 2S_{12} - S_{66})n^3m = -62.2\ TPa^{-1}$$
$$\overline{S}_{22}(30^\circ) = S_{11}n^4 + (2S_{12} + S_{66})n^2m^2 + S_{22}m^4 = 88.9\ TPa^{-1}$$
$$\overline{S}_{26}(30^\circ) = (2S_{11} - 2S_{12} - S_{66})mn^3 + (2S_{22} - 2S_{12} - S_{66})nm^3 = -3.77\ TPa^{-1}$$
$$\overline{S}_{66}(30^\circ) = 2(2S_{11} + 2S_{22} - 4S_{12} - S_{66})n^2m^2 + S_{66}(n^4 + m^4) = 123.6\ TPa^{-1}$$

Because $\sigma_y = \tau_{xy} = 0$, we can easily calculate the strains.

$$\varepsilon_x = \overline{S}_{11}(30^\circ) \cdot \sigma_x = 8128\ \mu mm/mm$$
$$\varepsilon_y = \overline{S}_{12}(30^\circ) \cdot \sigma_x = -4304\ \mu mm/mm$$
$$\varepsilon_{xy} = \overline{S}_{16}(30^\circ) \cdot \sigma_x = -9952\ \mu rad = -0.57^\circ$$

EXERCISE 9—CONT'D

$$\varepsilon_{xy} = \tan\gamma \frac{\text{Area of }(A'C'D'E')}{\text{Area of }(ACDE)}$$

Because the strains are small, the ratio between the areas $(A'C'D'E')$ and $(ACDE)$ is approximately 1; therefore $\varepsilon_{xy} = \tan\gamma$. Because the angle of deformation γ is small, $\varepsilon_{xy} \simeq \tan\gamma \simeq \gamma$. We can now calculate the elongation in the x and y directions:

$$\delta L_x = \varepsilon_x L_x = (0.008128)(55) = 0.447\ mm$$
$$\delta L_y = \varepsilon_y L_y = (-0.004304)(55) = -0.237\ mm$$

And the new deformed shape of our orthotropic plate:

$$x\text{: } L_x + \delta L_x = 55 + 0.447 = 55.447\ mm$$
$$y\text{: } L_y + \delta L_y = 55 - 0.237 = 54.763\ mm$$
$$x - y\text{: } 90 - \gamma = 90^\circ - 0.57^\circ = 89.43^\circ$$

3.3 CONSTITUTIVE EQUATIONS FOR ISOTROPIC MATERIALS

The constitutive equation for isotropic materials under plane stress conditions can be derived by the elasticity equation—Eq. (3.25)—for anisotropic materials. To this end, the conditions $E_1 = E_2 = E$, $\nu_{12} = \nu_{21} = \nu$, and $G_{12} = G$ and the notations $\varepsilon_1 = \varepsilon_x$, $\varepsilon_2 = \varepsilon_y$, and $\varepsilon_{12} = \varepsilon_{xy}$ can be used in Eqs. (3.26)–(3.29) in order to obtain the parameters S_{11}, S_{12}, and S_{22}:

$$S_{11} = S_{22} = \frac{1}{E} \tag{3.80}$$

$$S_{12} = -\nu\frac{1}{E} \tag{3.81}$$

$$S_{66} = \frac{1}{G} \tag{3.82}$$

With the aid of Eqs. (3.80)–(3.82), Eq. (3.25) can provide the constitutive equation for isotropic materials under plane stress conditions:

$$\begin{Bmatrix} \varepsilon_x \\ \varepsilon_y \\ \gamma_{xy} \end{Bmatrix} = \begin{bmatrix} \frac{1}{E} & -\frac{\nu}{E} & 0 \\ -\frac{\nu}{E} & \frac{1}{E} & 0 \\ 0 & 0 & \frac{1}{G} \end{bmatrix} \begin{Bmatrix} \sigma_x \\ \sigma_y \\ \tau_{xy} \end{Bmatrix} \tag{3.83}$$

Or alternatively, in terms of stresses:

$$\begin{Bmatrix} \sigma_x \\ \sigma_y \\ \tau_{xy} \end{Bmatrix} = \begin{bmatrix} \frac{E}{1-\nu^2} & \frac{\nu E}{1-\nu^2} & 0 \\ \frac{\nu E}{1-\nu^2} & \frac{E}{1-\nu^2} & 0 \\ 0 & 0 & G \end{bmatrix} \begin{Bmatrix} \varepsilon_x \\ \varepsilon_y \\ \gamma_{xy} \end{Bmatrix} \tag{3.84}$$

The elastic modulus E, Poisson's ratio ν, and shear modulus G can be correlated by the following formula:

$$G = \frac{E}{2(1+\nu)} \tag{3.85}$$

With the aid of Eq. (3.85), Eq. (3.84) can be written alternatively in the following form:

$$\begin{Bmatrix} \sigma_x \\ \sigma_y \\ \tau_{xy} \end{Bmatrix} = \begin{bmatrix} \frac{E}{1-\nu^2} & \frac{\nu E}{1-\nu^2} & 0 \\ \frac{\nu E}{1-\nu^2} & \frac{E}{1-\nu^2} & 0 \\ 0 & 0 & \frac{E}{2(1+\nu)} \end{bmatrix} \begin{Bmatrix} \varepsilon_x \\ \varepsilon_y \\ \gamma_{xy} \end{Bmatrix} \tag{3.86}$$

With the aid of Eq. (3.8) and the parameters of Table 3.1, the following constitutive equations in the x-y-z coordinate system can be obtained:

$$\begin{aligned} \varepsilon_x &= \frac{1}{E}(\sigma_x - \nu\sigma_y - \nu\sigma_z) \quad \gamma_{xy} = \frac{\tau_{xy}}{G} \quad or \quad \varepsilon_{xy} = \frac{\tau_{xy}}{2G} \\ \varepsilon_y &= \frac{1}{E}(\sigma_y - \nu\sigma_x - \nu\sigma_z) \quad \gamma_{xz} = \frac{\tau_{xz}}{G} \quad or \quad \varepsilon_{xz} = \frac{\tau_{xz}}{2G} \\ \varepsilon_z &= \frac{1}{E}(\sigma_z - \nu\sigma_x - \nu\sigma_y) \quad \gamma_{yz} = \frac{\tau_{yz}}{G} \quad or \quad \varepsilon_{yz} = \frac{\tau_{yz}}{2G} \end{aligned} \tag{3.87}$$

The above equations can be written in terms of stresses:

$$\begin{aligned} \sigma_x &= \frac{E}{(1+\nu)(1-2\nu)}[(1-\nu)\varepsilon_x + \nu(\varepsilon_y + \varepsilon_z)] \quad \tau_{xy} = \frac{E}{1+\nu}\varepsilon_{xy} \\ \sigma_y &= \frac{E}{(1+\nu)(1-2\nu)}[(1-\nu)\varepsilon_y + \nu(\varepsilon_x + \varepsilon_z)] \quad \tau_{xz} = \frac{E}{1+\nu}\varepsilon_{xz} \\ \sigma_z &= \frac{E}{(1+\nu)(1-2\nu)}[(1-\nu)\varepsilon_z + \nu(\varepsilon_x + \varepsilon_y)] \quad \tau_{yz} = \frac{E}{1+\nu}\varepsilon_{yz} \end{aligned} \tag{3.88}$$

For the case of plane strain, where $\varepsilon_z = \varepsilon_{xz} = \varepsilon_{yz} = 0$, the constitutive equations can be simplified:

$$\begin{aligned}
\sigma_x &= \frac{E}{(1+\nu)(1-2\nu)}\left[(1-\nu)\varepsilon_x + \nu\varepsilon_y\right] & \tau_{xy} &= \frac{E}{1+\nu}\varepsilon_{xy} \\
\sigma_y &= \frac{E}{(1+\nu)(1-2\nu)}\left[(1-\nu)\varepsilon_y + \nu\varepsilon_x\right] & \tau_{xz} &= 0 \\
\sigma_z &= \frac{\nu E}{(1+\nu)(1-2\nu)}\left(\varepsilon_x + \varepsilon_y\right) & \tau_{yz} &= 0
\end{aligned} \tag{3.89}$$

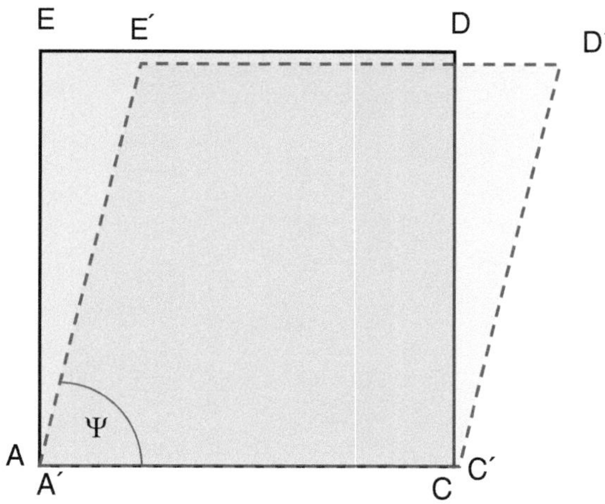

FIG. 3.8

Shear deformation of the plate.

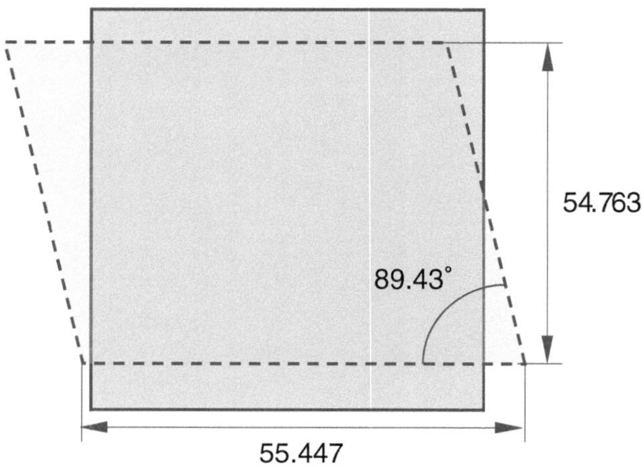

FIG. 3.9

Deformed shape of the plate.

REFERENCES

[1] Pavlou D. Composite materials in piping applications. Destech publications; 2013.

[2] Hyer M. Stress analysis of fiber reinforced composite materials. Destech Publications; 2009.

[3] Reddy JN. Mechanics of laminated composite plates and shelle. CRC press; 2004.

[4] Timoshenko SP, Goodier JN. Theory of elasticity. McGraw-Hill; 1970.

[5] Tsai SW, Wu EM. A general theory of strength for anisotropic materials. Journal of Composite Materials 1971;5:58–80.

[6] Pavlou DG. Essentials of the Finite Elements. Academic Press; 2015.

CHAPTER

Material Testing and Characterization

4

Chapter Outline

The mechanical behavior of materials can be characterized by certain mechanical properties obtained by mechanical tests on specimens. The mechanical properties can be used for design, analysis, and quality control of structure, as well as for investigating the reasons of failures. Typical mechanical tests are the tensile test, impact test, and fatigue test.

4.1 TENSILE TEST

Tensile test is the test we use for obtaining the main mechanical properties of the materials. Although the results are obtained by uniaxial, pseudo-static testing of small specimens, they can be utilized for real-size structural components under uniaxial or multiaxial loading conditions.

In a tensile testing machine (Fig. 4.1), the axial deformation of a uniaxial specimen (cylindrical or prismatic) is correlated with the corresponding uniaxial force. A slow elongation rate is applied in a uniaxial specimen, and the resistant force F of the material and the corresponding elongation ΔL are recorded until the fracture of the specimen. The obtained force-elongation diagram is used for deriving the stress-strain diagram σ-ε. The deformation rate during the test is very slow to simulate static conditions (pseudo-static test).

The cross section of the specimen is uniform, and the main testing parameters are the nominal stress and strain, σ and ε, respectively, given by the following formulae:

$$\sigma = \frac{F}{A_o} \tag{4.1}$$

Advanced Mechanics of Solids and Structures. DOI: http://doi.org/10.1016/B978-0-443-23778-2.00004-4

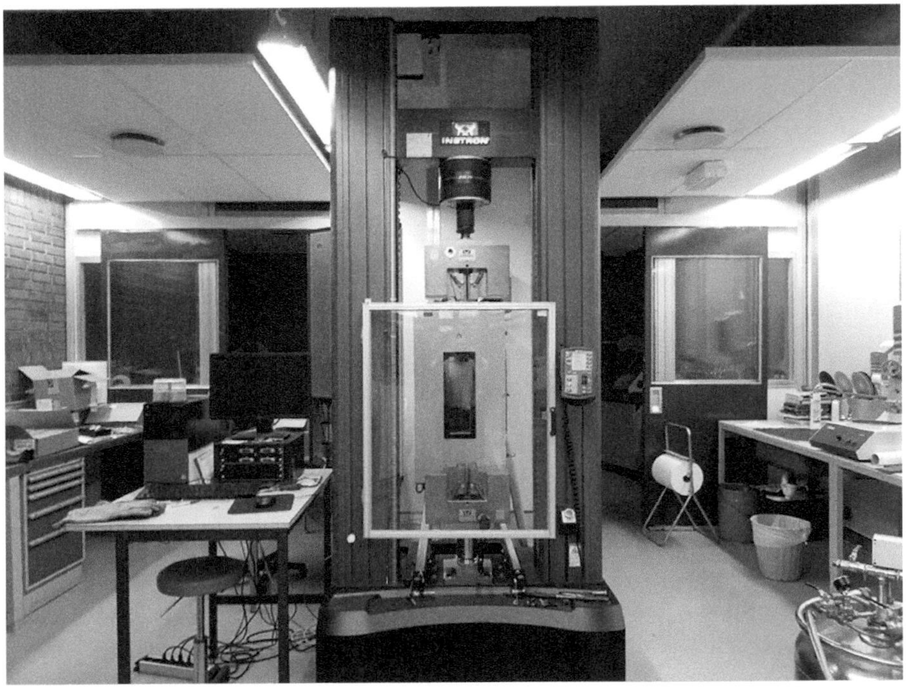

FIG. 4.1

Modern tensile testing machine at the University of Stavanger, Norway.

$$\varepsilon = \frac{\Delta L}{L_o} \tag{4.2}$$

In the above equations, F is the resistant force of the specimen, ΔL is the corresponding elongation of the specimen and A_o, L_o are the cross section and the length of the undeformed specimen (dimensions before the test).

A typical stress-strain curve for a ductile steel is demonstrated in Fig. 4.2. In this schematic demonstration the following material properties can be obtained:

Yield stress Y: The stress corresponding to point A that expresses the maximum stress for elastic behavior of the material and the range of stresses causing elastic deformation corresponding to the straight-line OA that is called Hooke's line (linear elasticity). In some materials (e.g., polymers), Hooke's line can be slightly nonlinear (nonlinear elasticity).

Ultimate stress S_u: The maximum stress the material can afford before failure (material strength).

Yield strain ε_Y: The maximum elastic strain of the material (strain corresponding to the yield stress).

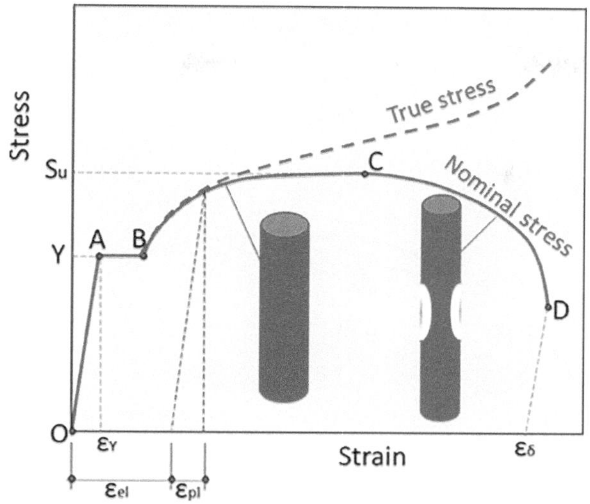

FIG. 4.2

Typical stress-strain curve for a ductile steel.

Modulus of elasticity *E* (or Young's modulus): The material property correlating the elastic stress to the elastic strain $E = Y/\varepsilon_Y$. The elasticity modulus is expressed graphically by the slope of the Hooke's line OA.

Ductility ε_δ: A measure of the permanent strain of a material after the fracture of the specimen. Materials with high value of ε_δ, that is, $20\% < \varepsilon_\delta < 30\%$, are called ductile (e.g., structural steel, copper). Small values of ε_δ have the brittle materials (e.g., ceramics, cast-iron, glass). Schematic demonstration of a ductile and a brittle material is shown in Fig. 4.3.

In definition of the nominal stress in Eq. (4.1), the force F has been divided with the initial cross section A_o. However, during the tensile test a progressive reduction of the cross section of the specimen takes place. The notation A_t is used for the instant value of the cross-section area. Then, the following definition of the real stress σ_t is given by the following formula:

$$\sigma_t = \frac{F}{A_t} \tag{4.3}$$

Following the same concept, the true strain ε_t, that is, the strain where the elongation is normalized with the true length L_t, is given by the following formula:

$$\varepsilon_t = \int_{L_o}^{L_t} \frac{dL_t}{L_t} = \ln\left(\frac{L_t}{L_o}\right) = \ln\left(\frac{L_o + \Delta L}{L_o}\right) = \ln(1+\varepsilon) \tag{4.4}$$

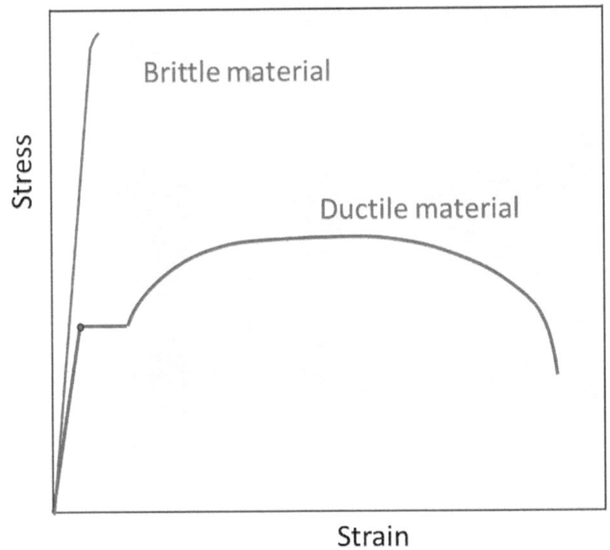

FIG. 4.3

Schematic demonstration of a ductile and a brittle material.

The volume V of the specimen during the tensile test can be considered constant. Therefore:

$$V = A_o L_o = A_t L_t \tag{4.5}$$

The above equation yields

$$\frac{A_o}{A_t} = \frac{L_t}{L_o} = \frac{(L_o + \Delta L)}{L_o} = 1 + \varepsilon \tag{4.6}$$

Then, the true stress can be correlated to the nominal stress by the following formula:

$$\sigma_t = \left(\frac{F}{A_o}\right)\left(\frac{A_o}{A_t}\right) = \sigma(1 + \varepsilon) \tag{4.7}$$

Comparison of the true stress versus nominal stress-strain curve is demonstrated in Fig. 4.2.

4.2 PARAMETERS AFFECTING THE STRESS-STRAIN CURVE

The use of the stress-strain curve instead of the force-displacement one is advantageous for characterizing materials because we eliminate the effect of the size of the specimen (at least in the elastic range). However, the following four parameters can affect the stress-strain curve and the mechanical behavior of the material:

- The loading history
- The strain rate
- The temperature
- The notches on the surface

The loading history is a very important parameter that affects the mechanical behavior of a material. Let's consider a material to be subjected to a stress Y′ beyond the yield stress and be unloaded (path OABHG in Fig. 4a). The unloading path HG is always elastic and parallel to the Hooke's line OA. Reloading from this state will follow the path GHCD where the loading path GH is elastic too. Therefore a material that has been subjected to a plastic deformation in the past has been hardened. For reloading, the material presents a stress-strain curve that has higher yield stress $Y' > Y$ and smaller ductility $\varepsilon'_\delta < \varepsilon_\delta$ (Fig. 4b).

The strain rate and the temperature also affect the mechanical behavior of the material. Increase of strain rate does not give sufficient time to the dislocations to move and fracture without plastic deformation (embrittlement) can happen. Increase of temperature offers thermal energy that helps the mobility of the dislocations, and the fracture is accompanied with plastic deformations. Generally, slow loading rate and high temperature causes ductile behavior. High loading rate (e.g., impact) and low temperature causes considerable embrittlement. The effect of the strain rate and the temperature on the material behavior is demonstrated in Fig. 4.5.

Usually the surface of the tensile test specimens is smooth. However, irregularities in the surface quality can considerably affect the stress-strain curve. Smooth specimen surface without notches yields uniform distribution of the tensile stress and

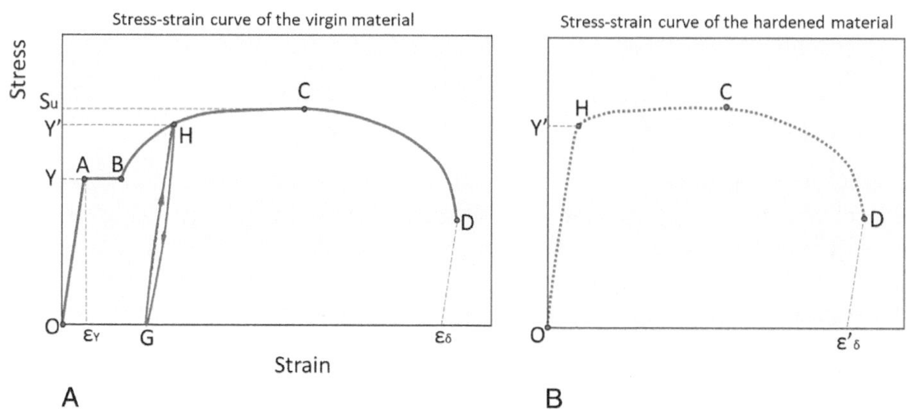

FIG. 4.4

Effect of the loading history to the stress-strain curve. (a) Stress-strain curve of the virgin material. (b) Stress-strain curve of the hardened material.

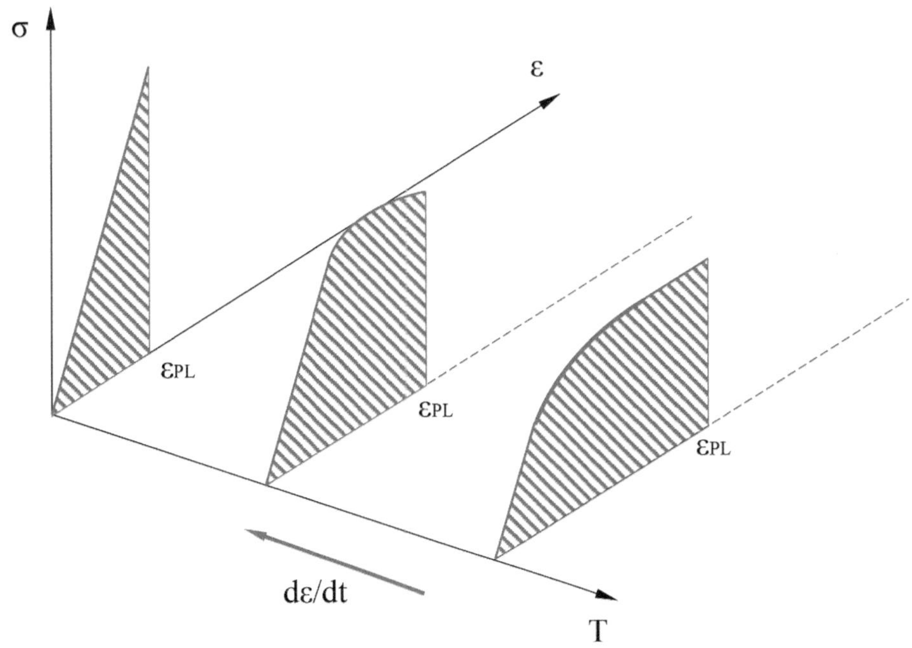

FIG. 4.5

Effect of the strain rate and the temperature on the material behavior.

purely uniaxial stress state. Notches and other local irregularities transform the uniaxial stress state to a multiaxial one where the hydrostatic stress without shear stress is predominant and causes brittle mechanical behavior. The effect of notches on the stress-strain curve is demonstrated in Fig. 4.6.

4.3 IMPACT TEST

Engineering structures in service often suffer from impact damage, for example, a car crash, impact of a ship on the monopile of an offshore wind turbine, and so on. Therefore impact resistance is an important mechanical parameter for material characterization. During impact testing, the energy absorption due to specimen fracture and the resistance of the specimen versus time are mainly measured. For low-velocity impact, two types of testing are usual: drop weight impact testing and Charpy impact testing. Drop weight impact testing is the most common test. The test machine is equipped with a weight that is raised to a prescribed height and released, impacting the specimen. The specimen is a plate that is clamped along its edges. Schematic representation of the values of energy and resistance versus time, obtained by an impact testing, is demonstrated in Fig. 4.7.

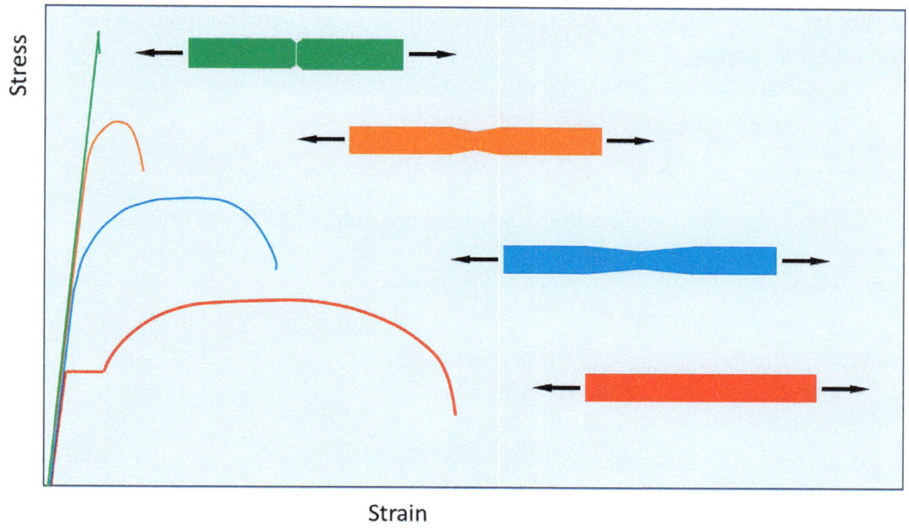

FIG. 4.6

Effect of notches on the stress-strain curve.

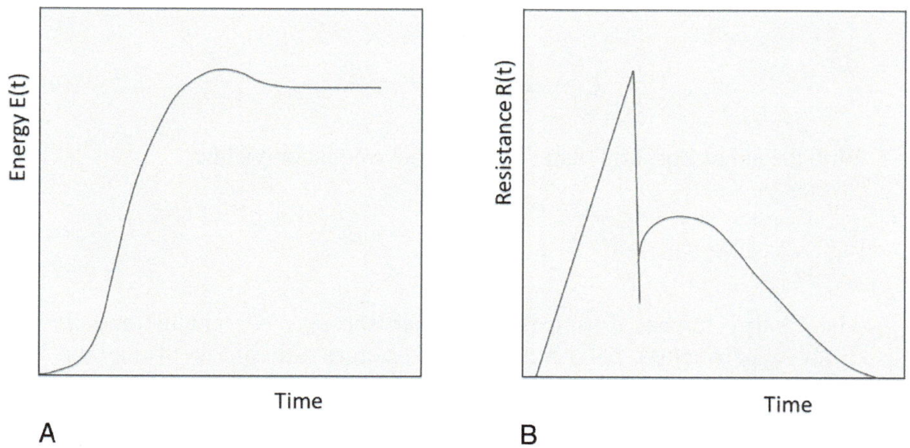

FIG. 4.7

Schematic representation of impact testing results: (a) Absorbed energy vs. time; (b) specimen resistance vs. time.

Once the curve of resistance $R = R(t)$ is measured, the corresponding acceleration (or retardation) $\gamma(t)$ can be obtained by the Newton's law

$$R(t) = m\gamma(t) + mg \tag{4.8}$$

where m is the mass of the impactor and g is the gravitational acceleration ($g = 9.81\,\text{m/sec}^2$).

Therefore velocity $U(t)$ can be obtained by the following integration:

$$U(t) = \int \gamma(t)\,dt \tag{4.9}$$

On the other hand, the energy is given by the following known integral:

$$E(t) = \int R(t)\,dx \tag{4.10}$$

where x is the displacement of the contact point.

Using the definition of the velocity:

$$U(t) = \frac{dx}{dt} \tag{4.11}$$

Eq. (4.10) yields

$$E(t) = \int R(t)\frac{dx}{dt}dt \tag{4.12}$$

or

$$E(t) = \int R(t)\,U(t)\,dt \tag{4.13}$$

With the aid of Eqs. (4.8) and (4.9), the above equation yields:

$$E(t) = \int R(t)\left[\frac{R(t)}{m} - g\right]dt \tag{4.14}$$

The Charpy impact testing is older than the drop weight method. It was used for experimental determination of fracture toughness of metals. The Charpy machine (Fig. 4.8) is equipped with a heavy pendulum that is raised to a known height and released, impacting a prismatic specimen. The difference in the initial and final heights H and h, respectively, is proportional to the amount of the absorbed energy due to damage accumulation in the specimen. Therefore the total absorbed energy can be estimated by the formula:

$$E_{tot} = G(H - h) \tag{4.15}$$

where G is the weight of the impactor.

Modern Charpy machines (Fig. 4.9) are equipped with an electronic impactor that is able to record the specimen's resistance $R(t)$ versus time.

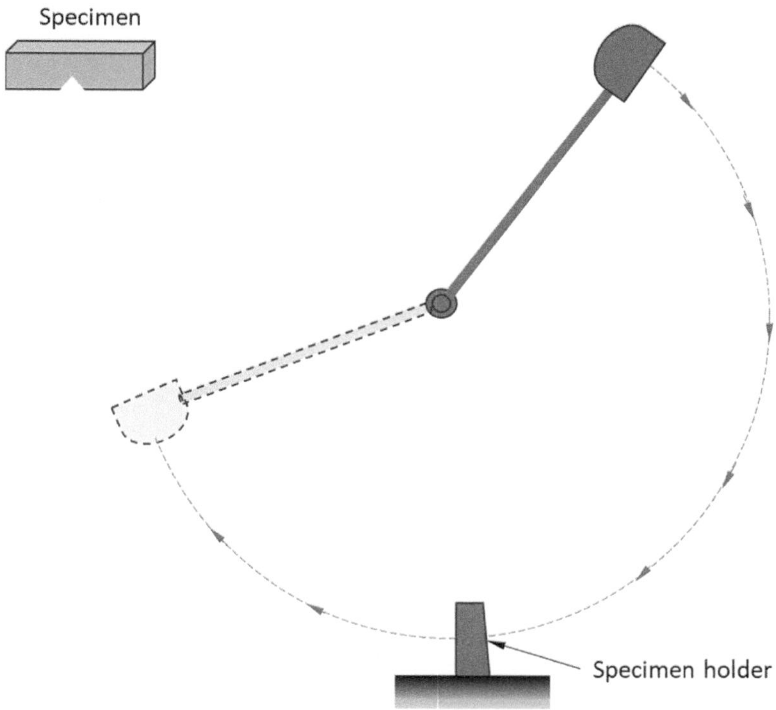

FIG. 4.8

Typical Charpy machine.

4.4 FATIGUE TEST

Variable loading conditions cause damage accumulation within the material microstructure. Macroscopically, the damage accumulation during repetitive loading yields reduction of the strength and the remaining life of the structural part. The above phenomenon is called fatigue. To obtain design data concerning the fatigue behavior of materials and to compare them under standard conditions, fatigue test methods have been developed and standardized. The fatigue loading can be harmonic (i.e., consisting of distinguished loading cycles) or irregular (Fig. 4.10ab).

In the harmonic loading history shown in Fig. 4.10a, σ_a *and* σ_m are the values of stress amplitude and mean stress, respectively, whereas σ_{max} and σ_{min} are the maximum and minimum stress levels. The relation of σ_α and σ_m with σ_{max} and σ_{min} is given by:

$$\sigma_a = \frac{\sigma_{max} - \sigma_{min}}{2} \tag{4.16}$$

$$\sigma_m = \frac{\sigma_{max} + \sigma_{min}}{2} \tag{4.17}$$

FIG. 4.9

Modern Charpy impact testing machine at the University of Stavanger, Norway.

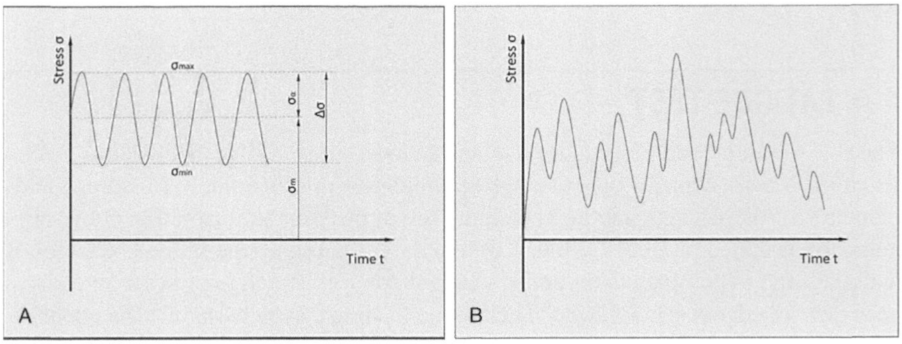

FIG. 4.10

Fatigue loading types: (a) harmonic and (b) irregular.

Materials in engineering practice can often subjected to stepwise constant amplitude loading histories (Fig. 4.11).

The aim of fatigue testing is the derivation of curves providing the fatigue life, that is, the number of loading cycles up to failure versus the stress amplitude of fatigue loading. These curves are called S-N curves or Woehler curves.

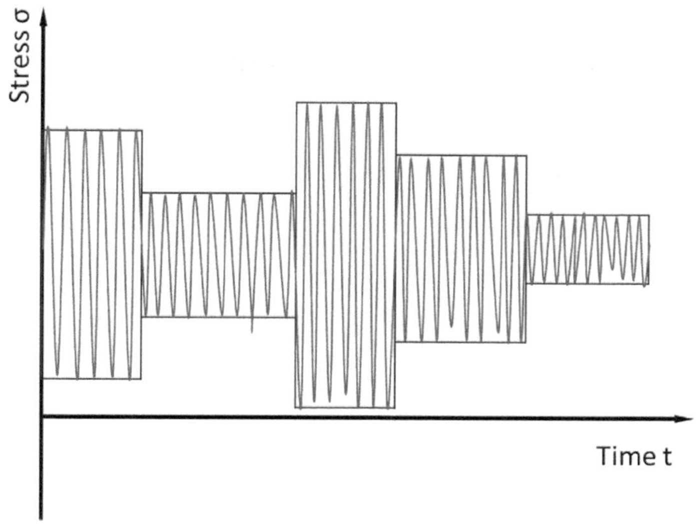

FIG. 4.11

Stepwise constant amplitude fatigue loading.

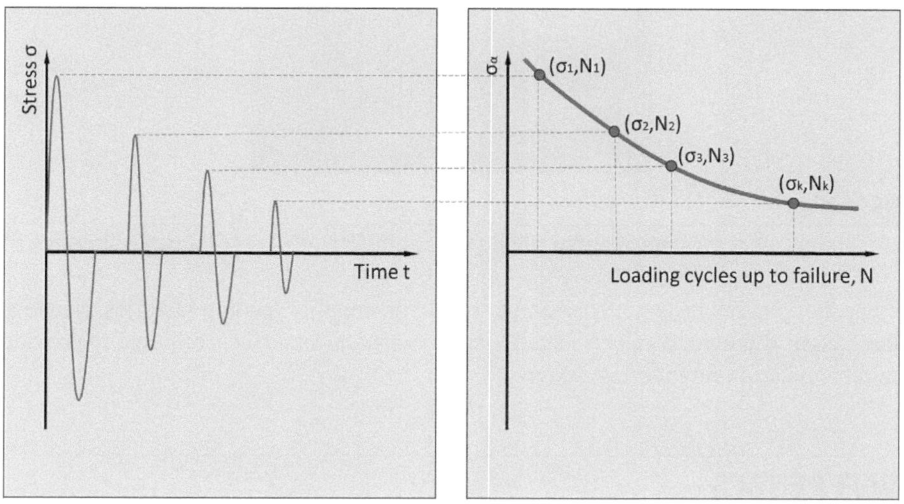

FIG. 4.12

Derivation of an S-N curve from fatigue tests with different stress amplitudes.

Schematic representation of the procedure for the derivation of an S-N curve is shown in Fig. 4.12.

For performing fatigue tests, computer-controlled servo-hydraulic machines are usually used today. These machines offer the ability to perform load or strain

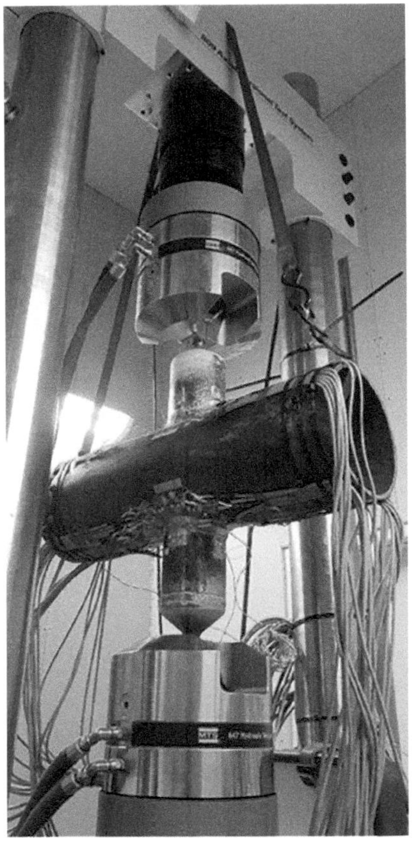

FIG. 4.13

Modern fatigue testing machine for components at the University of Stavanger, Norway.

controlled fatigue tests with harmonic, block, or irregular loading histories. Because the scatter of fatigue results is usually large, statistical analysis is required to provide reliable data for engineering design.

REFERENCES

[1] ASM Metals Handbook, American Society of Metals, Metals Park, Ohio 44073 USA.
[2] Manual of Steel Construction, 8th ed., Chicago, Illinois: American Institute of Steel Construction; 1980.
[3] Pavlou DG. Prediction of fatigue crack growth under real stress histories. Engineering Structures 2000;Volume 22(Issue 12):1707–13.
[4] Pavlou DG. Creep life prediction under stepwise constant uniaxial stress and temperature conditions. Engineering Structures 2001;Volume 23(Issue 6):656–62.
[5] Pavlou D. Composite materials in piping applications. Destech; 2013.

CHAPTER

Inelastic Material Response and Yield Criteria

5

Chapter Outline

5.1 MATERIAL RESPONSE

We can classify the materials according to how they behave under external stress. Generally, the material behavior can be elastic or inelastic. The elastic behavior can be linear or nonlinear or viscoelastic. The inelastic behavior can be plastic or visco-plastic. The stress-strain curve for these categories are represented schematically in Fig. 5.1.

The selection of a material for structural design is mainly based on the stress-strain curve. Stress-strain curves are obtained by a uniaxial tensile test and can provide the most important material properties, that is, yield stress Y, ultimate stress SU, elasticity modulus E, and ductility. In structural analysis it is necessary to use the stress-strain curve of the material to be selected. Metals and alloys are the most used materials for engineering structures. Typical models of the stress-strain curve for metals and alloys are demonstrated in Fig. 5.2.

Advanced Mechanics of Solids and Structures. DOI: http://doi.org/10.1016/B978-0-443-23778-2.00005-6

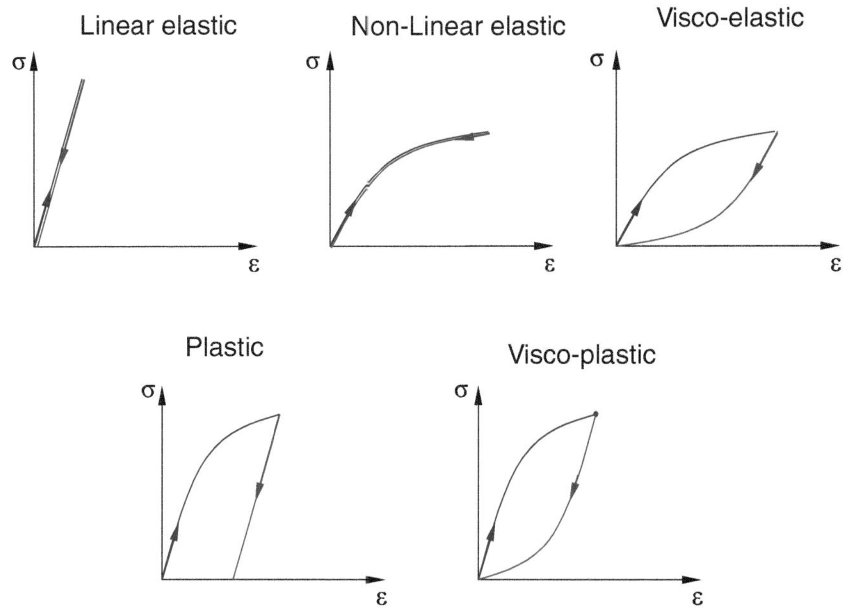

FIG. 5.1

Schematic representation of the stress-strain curve for linear elastic, nonlinear elastic, viscoelastic, plastic, and viscoplastic material behavior.

5.2 YIELD CRITERIA

When a structural element is subjected to uniaxial stress state with a normal stress σ_x (Fig. 5.3), the condition $\sigma_x < Y$ is sufficient to avoid inelastic behavior (yielding), whereas $\sigma_x = Y$ yielding takes place under uniaxial stress. Usually, structural elements are subjected to multiaxial stress state (biaxial or triaxial). Inelastic behavior (yielding) can occur under multiaxial stress, even if none of the individual stress components (Fig. 5.3) exceeds the uniaxial yield stress Y. For this purpose, we need yield criterion to assess whether a material element under multiaxial stress state will yield. To this end, it is necessary to combine the stress components into an effective uniaxial stress that can be compared to the uniaxial yield stress Y:

$$\sigma_{eff}(\sigma_x, \sigma_y, \sigma_z, \tau_{xy}, \tau_{xz}, \tau_{yz}) = Y \tag{5.1}$$

Such a condition like Eq. (5.1) is called yield criterion. Several yield criteria have been proposed for assessing the condition for yielding of materials under multiaxial stress state. Some of them are accurate (like the von Mises and the maximum shear stress criterion), and some have less accuracy.

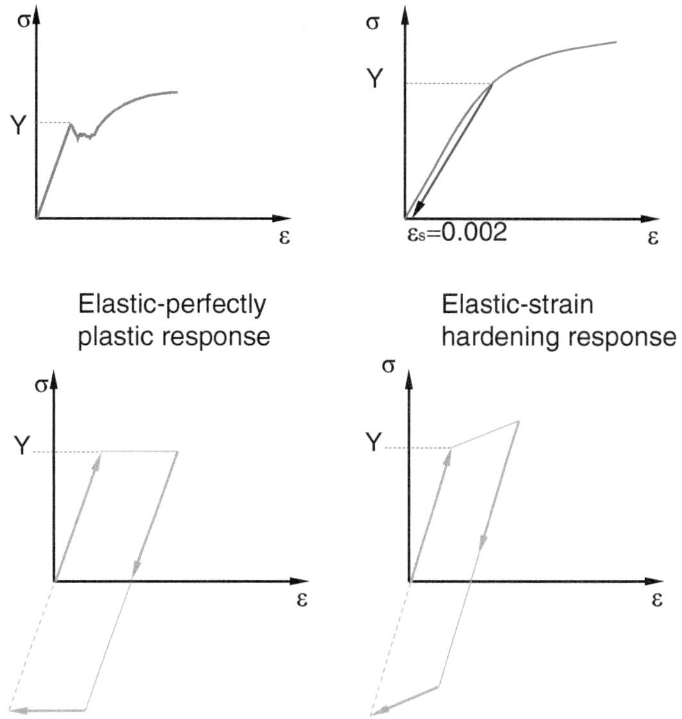

FIG. 5.2

Models of uniaxial stress-strain curves for metals and alloys.

We can classify the yield criteria in three categories: (a) criteria based on stress, (b) criteria based on strain, (c) criteria based on strain energy density (Table 5.1).

5.2.1 MAXIMUM PRINCIPAL STRESS CRITERION

The maximum principal stress criterion provides good accuracy only for brittle materials. According to this criterion, yielding begins when the maximum principal stress at a point reaches the value of uniaxial yield stress Y. For a material element under biaxial stress state, the principal stresses are given by the following equation:

$$\sigma_{1,2} = \frac{\sigma_x + \sigma_y}{2} \pm \sqrt{\frac{1}{4}(\sigma_x - \sigma_y)^2 + \tau_{xy}^2} \tag{5.2}$$

Then the maximum principal stress criterion can be formulated by the following equation:

$$\max\{\sigma_1, \sigma_2\} = Y \tag{5.3}$$

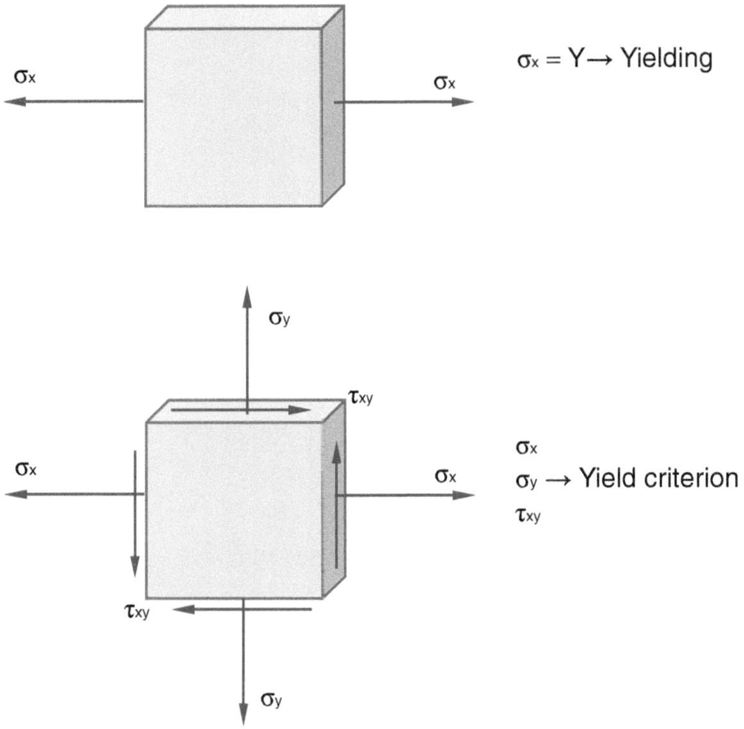

FIG. 5.3

Material element under uniaxial and biaxial stress state.

Table 5.1 Classification of the main yield criteria.

Stress-based criteria	Strain-based criteria	Strain energy density criteria
Maximum principal stress	Maximum p-incipal strain (St. Venant)	Strain energy density
Maximum shear stress (Tresca)		Distortional energy density (von Mises)

Therefore the safe condition against yielding is:

$$\max\{\sigma_1, \sigma_2\} < Y \tag{5.4}$$

In order to apply the maximum principal stress criterion for a triaxial stress state, the principal stresses σ_1, σ_2, and σ_3 should be determined from the solution of the following cubic equation:

$$\sigma^3 - I_1\sigma^2 + I_2\sigma - I_3 = 0 \tag{5.5}$$

where I_1, I_2, and I_3 are the stress invariants (see the Chapter 1: Stress Theory).

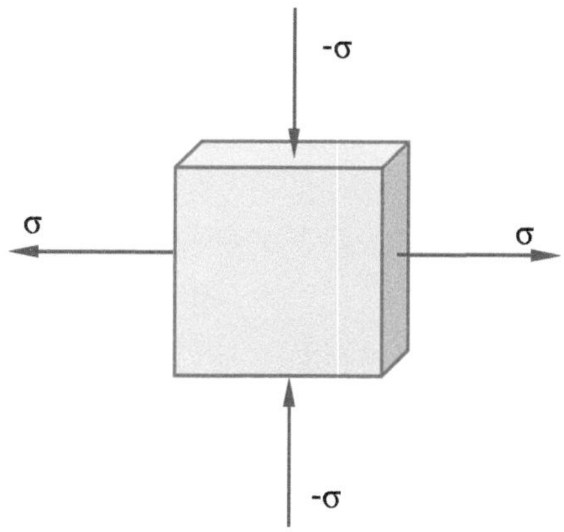

FIG. 5.4

Biaxial stress state of equal normal stress components.

Then, the following equation is the quantitative formulation of the maximum principal stress criterion for general triaxial stress state:

$$\max\{\sigma_1, \sigma_2, \sigma_3\} = Y \tag{5.6}$$

If we apply the maximum principal stress criterion in the biaxial stress state shown in Fig. 5.4, where $\sigma_x = \sigma_1 = \sigma$, $\sigma_y = \sigma_2 = -\sigma$, $\tau_{xy} = 0$, the maximum shear stress is:

$$\tau_{max} = \frac{\sigma_1 - \sigma_2}{2} = \sigma \tag{5.7}$$

Eq. (5.7) implies that in yielding condition the shear yield stress is equal to the normal yield stress $\tau_Y = Y$. Experimental results show that the shear yield stress is considerably smaller (almost half) than the normal yield stress. Therefore the maximum principal stress criterion is not suitable for designing structures made of structural steel and other ductile metals and alloys.

5.2.2 MAXIMUM PRINCIPAL STRAIN CRITERION (ST. VENANT'S CRITERION)

According to the maximum principal strain criterion, yielding begins when the maximum principal strain at a point reaches a value equal to the yield strain. The yield strain is defined by Hooke's law:

$$\varepsilon_Y = \frac{Y}{E} \tag{5.8}$$

For a material element subjected to biaxial stress state, this criterion is given by the following equation:

$$\max\{\varepsilon_1, \varepsilon_2\} = \frac{Y}{E} \tag{5.9}$$

With the aid of the equations for principal strains:

$$\varepsilon_1 = \frac{1}{E}(\sigma_1 - \nu\sigma_2) \tag{5.10}$$

$$\varepsilon_2 = \frac{1}{E}(\sigma_2 - \nu\sigma_1) \tag{5.11}$$

Eq. (5.9) yields

$$\frac{1}{E}(\sigma_1 - \nu\sigma_2) = \frac{Y}{E} \tag{5.12}$$

or

$$\frac{1}{E}(\sigma_2 - \nu\sigma_1) = \frac{Y}{E} \tag{5.13}$$

Graphical presentation of Eqs. (5.12) and (5.13) in coordinate system $\sigma_1 - \sigma_2$ (Fig. 5.5) can provide the area of the safe combination of σ_1, σ_2 against yielding.

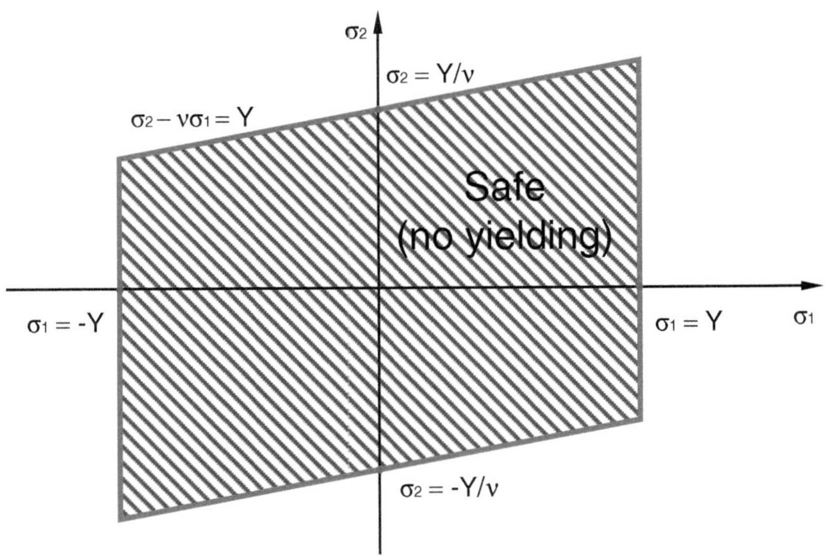

FIG. 5.5

Safe area of σ_1, σ_2 values against yielding according to the maximum principal strain criterion.

With the aid of Eq. (5.2), the maximum principal strain criterion in terms of the components σ_x, σ_x, and τ_{xy} can be provided by the following equation:

$$\underbrace{(1-\nu)\frac{\sigma_x+\sigma_y}{2}+(1+\nu)\sqrt{\frac{1}{4}(\sigma_x-\sigma_y)^2+\tau_{xy}^2}}_{\sigma_{eff}}=Y \tag{5.14}$$

5.2.3 MAXIMUM SHEAR STRESS CRITERION (TRESCA CRITERION)

The maximum shear stress criterion is among the most accurate yield criteria. According to this criterion, yielding begins when the maximum shear stress at a point reaches the value of the maximum shear stress at yield in uniaxial loading.

We recall that the maximum shear stress of a material element subjected to uniaxial stress σ_o is:

$$\tau_{\max}=\frac{\sigma_o}{2} \tag{5.15}$$

Therefore the maximum shear stress at yield for uniaxial loading ($\sigma_o = Y$) is given by the following formula

$$\tau_Y=\frac{Y}{2} \tag{5.16}$$

Then, implementation of the maximum shear stress criterion yields the following quantitative formula:

$$\tau_{\max}=\tau_Y \tag{5.17}$$

or

$$\frac{\sigma_{\max}-\sigma_{\min}}{2}=\frac{Y}{2} \tag{5.18}$$

The graphical presentation of Eq. (5.18) provides the area of the safe combination of σ_1, σ_2 against yielding (Fig. 5.6). In the specific case of biaxial stress state where the principal stresses are given by Eq. (5.2), Eq. (5.18) yields:

$$\underbrace{\sqrt{(\sigma_x-\sigma_y)^2+4\tau_{xy}^2}}_{\sigma_{eff}}=Y \tag{5.19}$$

For triaxial stress state, the three principal stresses σ_1, σ_2, and σ_3 must be obtained by the solution of the cubic equation (Eq. 5.5). Then, the following equation for maximum shear stress criterion can be obtained:

$$\frac{\max\{\sigma_1,\sigma_2,\sigma_3\}-\min\{\sigma_1,\sigma_2,\sigma_3\}}{2}=\frac{Y}{2} \tag{5.20}$$

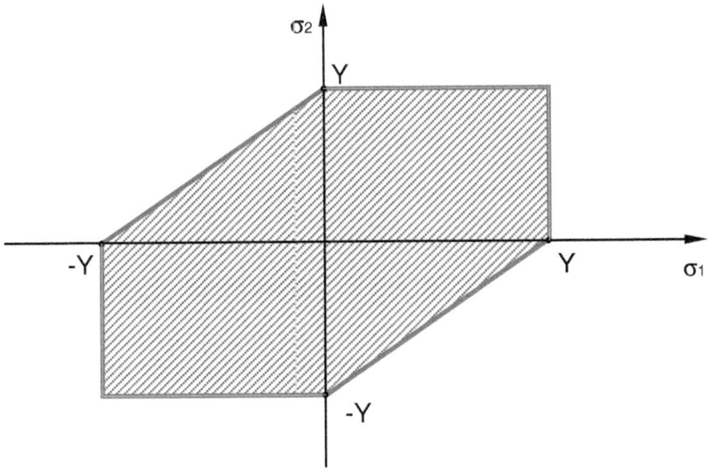

FIG. 5.6

Safe area of σ_1, σ_2 values against yielding according to the maximum shear stress criterion.

5.2.4 STRAIN ENERGY DENSITY CRITERION

Because strain energy density is a product of stress and strain, it accounts for the effect on yielding of both stress and strain. The strain energy density criterion is the first energy-based criterion and assumes that yielding at a point begins when the strain-energy-density in a multiaxial stress state reaches the value of the strain energy density at yield in uniaxial tension or compression.

According to Fig. 5.7, the strain energy density at yield in uniaxial stress state is:

$$U_Y = \frac{Y^2}{2E} \tag{5.21}$$

In a multiaxial stress state, all the stress components contribute to the strain energy density. We consider a material element under the principal stresses σ_1, σ_2, and σ_3 that accumulates strain energy density $U(\sigma_1, \sigma_2, \sigma_3)$. Then, the strain energy density criterion can be expressed by the following equation:

$$U(\sigma_1, \sigma_2, \sigma_3) = U_Y \tag{5.22}$$

or

$$U(\sigma_1, \sigma_2, \sigma_3) = \frac{Y^2}{2E} \tag{5.23}$$

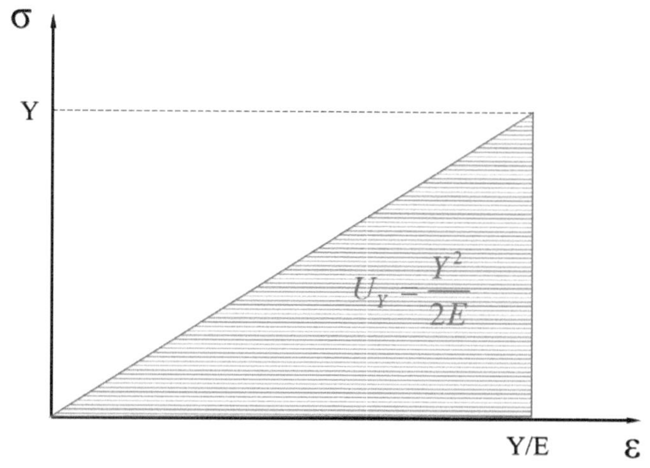

FIG. 5.7

Strain energy density at yield U_Y in uniaxial stress.

For calculating the strain energy density $U(\sigma_1, \sigma_2, \sigma_3)$ at a multiaxial stress state, we consider three steps:

Step 1: A material element is subjected to a stress σ_1 (Fig. 5.8) that accumulates elastic energy density:

$$U_1 = \frac{\sigma_1^2}{2E} \tag{5.24}$$

Step 2: In the stress state of Step 1, we add the stress component σ_2 (Fig. 5.8).

The stress component σ_2 accumulates elastic energy density $\sigma_2^2/2E$, and the stress component σ_1 that has reached its maximum value in Step 1 accumulates energy density $\sigma_1 \varepsilon_1$.

Therefore the strain energy density in Step 2 is:

$$U_2 = \frac{\sigma_2^2}{2E} + \sigma_1 \varepsilon_1 \tag{5.25}$$

As

$$\varepsilon_1 = -\nu \varepsilon_2 \tag{5.26}$$

and

$$\varepsilon_2 = \frac{\sigma_2}{E} \tag{5.27}$$

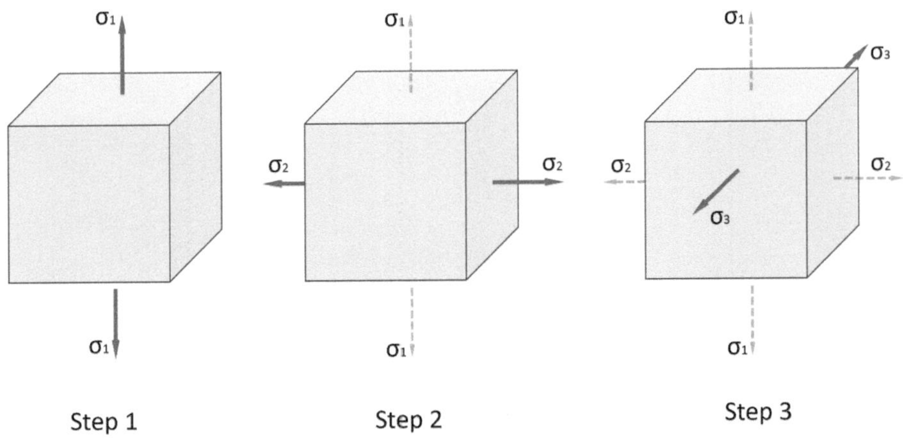

FIG. 5.8

Successive application of stress to a material element in three steps.

Eq. (5.25) yields:

$$U_2 = \frac{\sigma_2^2}{2E} - \frac{\nu}{E}\sigma_1\sigma_2 \tag{5.28}$$

Step 3: In the stress state of Step 2, we add the stress component σ_3 (Fig. 5.8).

The stress component σ_3 accumulates elastic energy density $\sigma_3^2/2E$, and the stress components σ_1 and σ_2 that reached their maximum value in Steps 1 and 2 accumulate energy density $\sigma_1\varepsilon_1$ and $\sigma_2\varepsilon_2$, respectively. Therefore the strain energy density accumulation in Step 3 is:

$$U_3 = \frac{\sigma_3^2}{2E} + \sigma_1\varepsilon_1 + \sigma_2\varepsilon_2 \tag{5.29}$$

Considering the following equations

$$\varepsilon_1 = \varepsilon_2 = -\nu\varepsilon_3 \tag{5.30}$$

$$\varepsilon_3 = \frac{\sigma_3}{E} \tag{5.31}$$

Eq. (5.29) gives:

$$U_3 = \frac{\sigma_3^2}{2E} - \frac{\nu}{E}\sigma_1\sigma_3 - \frac{\nu}{E}\sigma_2\sigma_3 \tag{5.32}$$

The total elastic strain energy density accumulation for a material element subjected to triaxial stress state can be obtained by superposition of the elastic strain energy of the three steps:

$$U = U_1 + U_2 + U_3 \tag{5.33}$$

or

$$U = \frac{1}{2E}\left[\sigma_1^2 + \sigma_2^2 + \sigma_3^2 - 2\nu\left(\sigma_1\sigma_2 + \sigma_1\sigma_3 + \sigma_2\sigma_3\right)\right] \tag{5.34}$$

With the aid of the above equation, Eq. (5.23) yields the general quantitative expression of the strain energy density criterion for triaxial stress state:

$$\underbrace{\sqrt{\sigma_1^2 + \sigma_2^2 + \sigma_3^2 - 2\nu\left(\sigma_1\sigma_2 + \sigma_1\sigma_3 + \sigma_2\sigma_3\right)}}_{\sigma_{eff}} = Y \tag{5.35}$$

For biaxial stress state, the above equation can be simplified:

$$\underbrace{\sqrt{\sigma_1^2 + \sigma_2^2 - 2\nu\sigma_1\sigma_2}}_{\sigma_{eff}} = Y \tag{5.36}$$

Graphical presentation of Eq. (5.36) provides the area of the safe combination of σ_1, σ_2 against yielding according to the strain energy density criterion (Fig. 5.9).

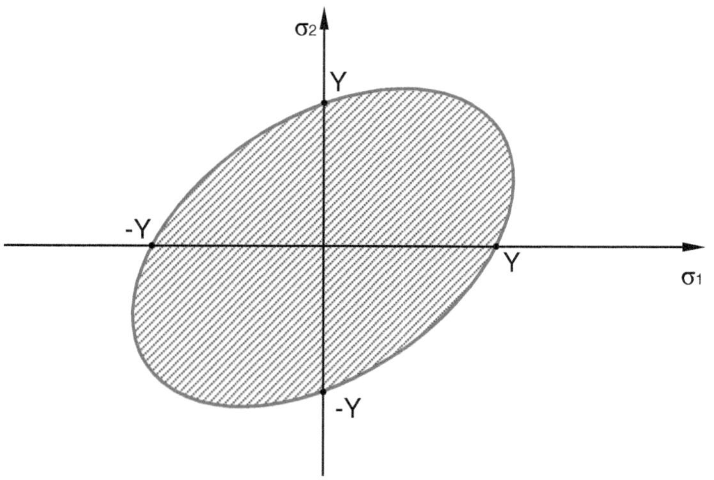

FIG. 5.9

Safe area of σ_1, σ_2 values against yielding according to the strain energy density criterion.

5.2.5 DISTORTIONAL ENERGY DENSITY CRITERION (VON MISES CRITERION)

The strain energy density includes two components: strain energy density for volume change and strain energy density for shape change (distortional strain energy density). The first component is associated with the hydrostatic pressure and has no contribution to material yielding. Therefore the only the distortional energy density contributes to material yielding and can provide an effective yield criterion. We now have to obtain the formula for the distortional energy density from Eq. (5.34). For this purpose, the following algebraic identities can be used:

$$\sigma_1^2 + \sigma_2^2 + \sigma_3^2 = (\sigma_1 + \sigma_2 + \sigma_3)^2 - 2(\sigma_1\sigma_2 + \sigma_1\sigma_3 + \sigma_2\sigma_3) \tag{5.37}$$

$$\begin{aligned} &2(\sigma_1\sigma_2 + \sigma_1\sigma_3 + \sigma_2\sigma_3) \\ &= 2\left(\sigma_1^2 + \sigma_2^2 + \sigma_3^2\right) - \left[(\sigma_1 - \sigma_2)^2 + (\sigma_2 - \sigma_3)^2 + (\sigma_1 - \sigma_3)^2\right] \end{aligned} \tag{5.38}$$

With the aid of Eq. (5.37), Eq. (5.38) yields:

$$\begin{aligned} &6(\sigma_1\sigma_2 + \sigma_1\sigma_3 + \sigma_2\sigma_3) \\ &= 2(\sigma_1 + \sigma_2 + \sigma_3)^2 - \left[(\sigma_1 - \sigma_2)^2 + (\sigma_2 - \sigma_3)^2 + (\sigma_1 - \sigma_3)^2\right] \end{aligned} \tag{5.39}$$

Considering Eq. (5.37), Eq. (5.34) can be written:

$$2EU = (\sigma_1 + \sigma_2 + \sigma_3)^2 - 2(\sigma_1\sigma_2 + \sigma_1\sigma_3 + \sigma_2\sigma_3) - 2\nu(\sigma_1\sigma_2 + \sigma_1\sigma_3 + \sigma_2\sigma_3)$$

or

$$2EU = (\sigma_1 + \sigma_2 + \sigma_3)^2 - 2(1 + \nu)(\sigma_1\sigma_2 + \sigma_1\sigma_3 + \sigma_2\sigma_3) \tag{5.40}$$

With the aid of Eq. (5.39), the above equation yields:

$$\begin{aligned} 2EU &= (\sigma_1 + \sigma_2 + \sigma_3)^2 \\ &-2(1 + \nu)\left[\frac{2(\sigma_1 + \sigma_2 + \sigma_3)^2}{6} - \frac{\left[(\sigma_1 - \sigma_2)^2 + (\sigma_2 - \sigma_3)^2 + (\sigma_1 - \sigma_3)^2\right]}{6}\right] \end{aligned} \tag{5.41}$$

or

$$U = \frac{(\sigma_1 + \sigma_2 + \sigma_3)^2}{\left(\frac{6E}{1 - 2\nu}\right)} + \frac{(\sigma_1 - \sigma_2)^2 + (\sigma_2 - \sigma_3)^2 + (\sigma_1 - \sigma_3)^2}{\left(\frac{6E}{1 + \nu}\right)} \tag{5.42}$$

Using the definitions of the bulk modulus K and the shear modulus G of elasticity:

$$K = \frac{E}{3(1 - 2\nu)} \tag{5.43}$$

$$G = \frac{E}{2(1 + \nu)} \tag{5.44}$$

Eq. (5.42) can be written in the following form:

$$U = \underbrace{\frac{(\sigma_1 + \sigma_2 + \sigma_3)^2}{18K}}_{\text{Volume change}} + \underbrace{\frac{(\sigma_1 - \sigma_2)^2 + (\sigma_2 - \sigma_3)^2 + (\sigma_1 - \sigma_3)^2}{12G}}_{\text{Shape change (Distortional Energy Density)}} \tag{5.45}$$

In the above equation, the term

$$U_V = \frac{(\sigma_1 + \sigma_2 + \sigma_3)^2}{18K} \tag{5.46}$$

includes the effect of normal (hydrostatic) stresses and expresses the stored strain energy density for changing the volume.

The component

$$U_D = \frac{(\sigma_1 - \sigma_2)^2 + (\sigma_2 - \sigma_3)^2 + (\sigma_1 - \sigma_3)^2}{12G} \tag{5.47}$$

includes the terms $(\sigma_1 - \sigma_2)$, $(\sigma_2 - \sigma_3)$, $(\sigma_1 - \sigma_3)$ expressing shear stresses. The component U_D is the strain energy density contribution to shape change, called distortional strain energy density, and is responsible for material yielding. According to the von Mises criterion, yielding begins when the distortional strain energy density at a point reaches the distortional strain energy density at yield in uniaxial stress state.

The distortional strain energy density for a uniaxial state can be obtained by Eq. (5.47) by setting $\sigma_1 = \sigma$, $\sigma_2 = 0$, $\sigma_3 = 0$:

$$U_{D,\text{uniaxial}} = \frac{\sigma^2}{6G} \tag{5.48}$$

At the yielding state where $\sigma = Y$, Eq. (5.48) provides:

$$U_{D,\text{uniaxial}} = \frac{Y^2}{6G}$$

As the distortional energy density at multiaxial stress state should be equal to the distortional strain energy density at uniaxial yielding (von Mises criterion), that is, $U_D = U_{D,uniaxial}$, the following equation can be obtained:

$$\frac{(\sigma_1 - \sigma_2)^2 + (\sigma_2 - \sigma_3)^2 + (\sigma_1 - \sigma_3)^2}{12G} = \frac{Y^2}{6G} \tag{5.49}$$

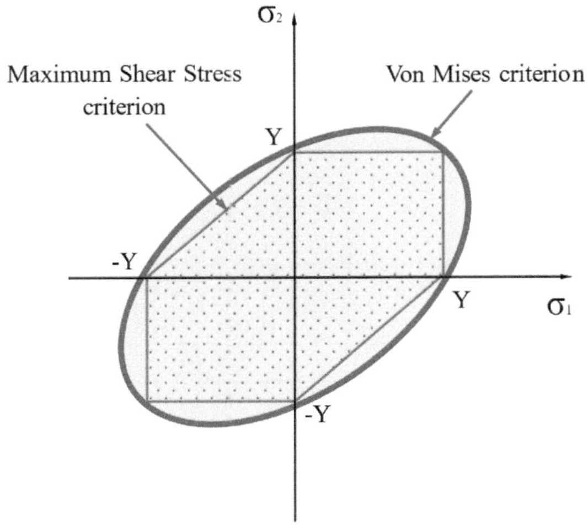

FIG. 5.10

Safe area of σ_1, σ_2 values against yielding according to the von Mises criterion and comparison with the maximum shear stress criterion.

or

$$\underbrace{\sqrt{\frac{1}{2}(\sigma_1-\sigma_2)^2+(\sigma_2-\sigma_3)^2+(\sigma_1-\sigma_3)^2}}_{\sigma_{eff}}=Y \tag{5.50}$$

Eq. (5.50) expresses the von Mises criterion for multiaxial stress state. In terms of the stress components in planes x, y, and z, the above criterion can be written as:

$$\sqrt{\frac{1}{2}\left[(\sigma_x-\sigma_y)^2+(\sigma_y-\sigma_z)^2+(\sigma_x-\sigma_z)^2\right]+3\left(\tau_{xy}^2+\tau_{yz}^2+\tau_{xz}^2\right)}=Y \tag{5.51}$$

Graphical representation of Eq. (5.50) for biaxial stress state is demonstrated in Fig. 5.10. Comparison of von Mises with the maximum shear stress criterion (Fig. 5.10) shows that maximum shear stress criterion is more conservative.

5.2.6 BENDING CAPACITY OF BEAMS BEYOND THE ELASTIC LIMIT

The yield criteria are used for designing structures with elastic behavior. The material behavior is governed by Hooke's law, and the strains are elastic. However, beams can continue to carry bending moments even when the maximum bending stress exceeds the yield stress Y. Fig. 5.11 shows the stress distribution on a beam cross section for bending stress beyond the yield stress. The capacity of beam to withstand bending moment in the fully plastic state can be determined.

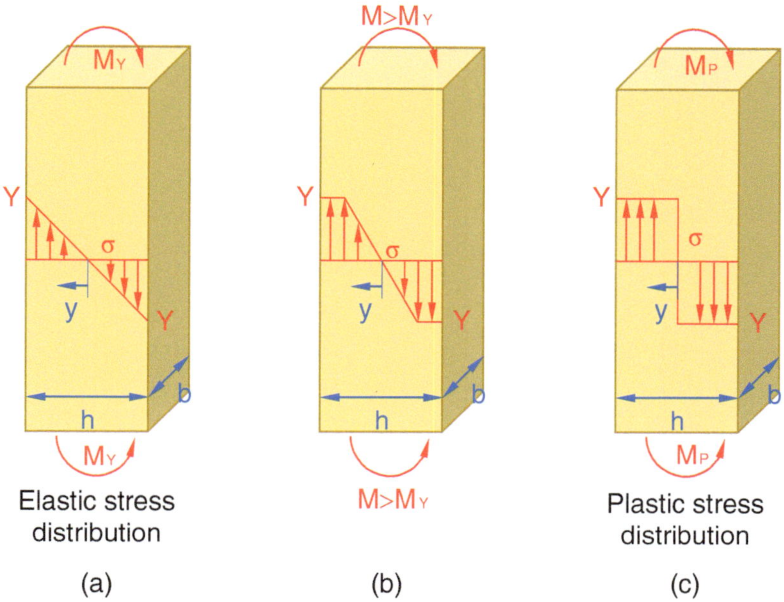

FIG. 5.11

Bending stress distribution on beam cross section for (a) fully elastic stress state, (b) elasto-plastic stress state, and (c) fully plastic stress state.

According to the bending theory, the maximum elastic bending stress on a beam cross section of h (Fig. 5.1) is given by the following equation:

$$\sigma_{\max} = \frac{M}{I}\frac{h}{2} \tag{5.52}$$

When the maximum bending stress reaches the yield stress Y (Fig. 5.11a), the following equation can be obtained:

$$Y = \frac{M_Y}{I}\frac{h}{2} \tag{5.53}$$

where M_Y is the bending moment causing fully elastic stress state where the maximum bending stress reaches the material yield stress Y. For beams with rectangular cross section of dimensions $h \times b$ (Fig. 5.11), the moment of inertia is

$$I = \frac{bh^3}{12} \tag{5.54}$$

Combining Eqs. (5.53) and (5.54) the following result can be obtained:

$$M_Y = Y\frac{bh^2}{6} \tag{5.55}$$

In the fully plastic state (Fig. 5.11c), the maximum bending stress that is equal to the yield stress Y has uniform distribution, and the corresponding bending moment is:

$$M_P = F_P \frac{h}{2} \tag{5.56}$$

In the above equation, F_P is the resultant force of the plastic stress distribution in the cross-section area under tensile or compressive stresses:

$$F_P = Y \frac{h}{2} b \tag{5.57}$$

Combining Eqs. (5.56) and (5.57), the following result for M_P can be obtained:

$$M_P = Y \frac{bh^2}{4} \tag{5.58}$$

Obviously, the bending moment M_P in the fully plastic state is larger than the fully elastic bending moment M_Y. Comparing Eqs. (5.55) and (5.59), the following conclusion can be obtained:

$$M_P = 1.5 M_Y \tag{5.59}$$

Therefore, after reaching the elastic limit, the beam has still capacity to carry loads. For a rectangular cross section, the bending capacity of a beam in the fully plastic state is 50% larger than the bending capacity in the fully elastic state.

EXERCISE 1

For the cross section shown in the following figure, determine:

1. The ratio M_P/M_Y of the fully plastic moment M_P and the maximum elastic moment M_Y.
2. After the fully plastic state, the beam is unloaded elastically. Determine the residual stress distribution in the cross section.
   Consider that during loading the upper layer of the flange is subjected to compressive stresses.

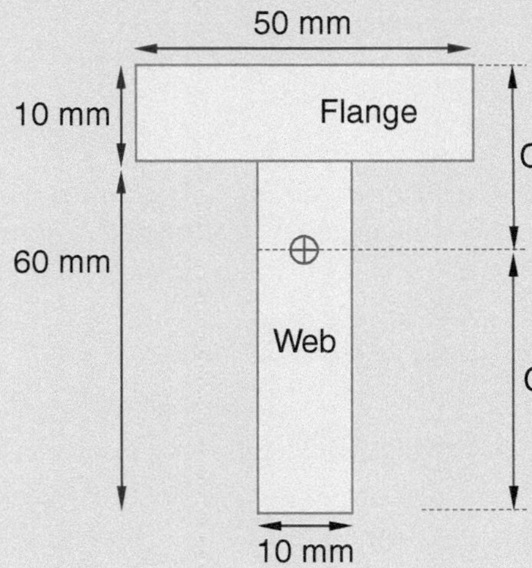

EXERCISE 1—CONT'D

Solution

1. To determine the maximum elastic moment M_Y we determine the location of the centroid and the moment of inertia of the whole section.

$$C_1 = \frac{\sum_{i=1}^{2} A_i y_i}{\sum_{i=1}^{2} A_i} = \frac{(50 \cdot 10)65 + (60 \cdot 10)30}{(50 \cdot 10) + (60 \cdot 10)} = 45.9\ mm$$

and

$$C_2 = 70 - C_1 = 24.1\ mm$$

The moment of inertia can be calculated with the aid of Steiner's theorem.

$$I_x = \underbrace{\left[\frac{50 \cdot 10^3}{12} + (50 \cdot 10)(C_2 - 5)^2\right]}_{flange} + \underbrace{\left[\frac{10 \cdot 60^3}{12} + (10 \cdot 60)(70 - 30 - C_2)^2\right]}_{web}$$

or

$$I_x = 343257\ mm^4$$

The maximum elastic moment can be calculated by the following formula for bending stresses:

$$\frac{M_Y}{I_x} C_1 = Y$$

or

$$M_Y = \frac{I_x Y}{C_1} = 7478\ Y$$

Next, we have to determine the plastic moment M_P. For that we first calculate e, the distance of the neutral axis in plastic stress distribution to the bottom layer of the flange. To this end, the force equilibrium when the cross section is fully plastic should be used. Then, the yield stress times the area under compression must be equal to the yield stress times the area under tension:

$$YA_t = YA_c$$

or

$$(50 \cdot 10) + 10e = 10(60 - e)$$

The above equation yields:

$$e = 5\ mm$$

(Continued)

EXERCISE 1—CONT'D

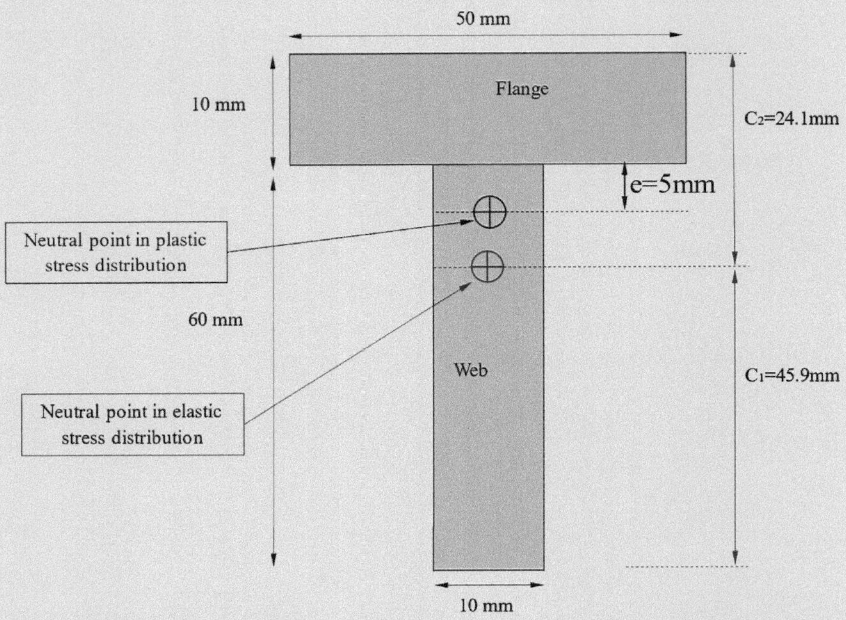

The plastic moment can be calculated by the force resultant of the plastic stress distribution of each subarea times the corresponding distance from the neutral point in plastic stress distribution:

$$M_P = \underbrace{10(60-e)Y}_{force}\underbrace{\left(\frac{60-e}{2}\right)}_{distance} + \underbrace{(10\cdot e)Y}_{force}\underbrace{\frac{e}{2}}_{distance} + \underbrace{50\cdot 10Y}_{force}\underbrace{\left(\frac{10}{2}+e\right)}_{distance} = 20250Y$$

Then, the following ratio of the moments M_P, M_Y can be obtained:

$$\frac{M_P}{M_Y} = \frac{20250\ Y}{7478\ Y} = 2.7$$

2. Determination of the residual stresses
   The bending stresses during loading up to the fully plastic state follow a nonlinear path, but the unloading path is always linear and parallel to the Hooke's loading path (see figure below).

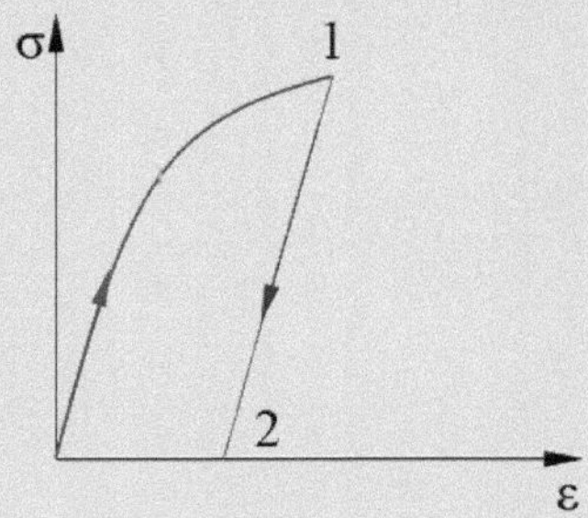

EXERCISE 1—CONT'D

We determine the linear unloading stresses at the top and the bottom layer of the cross section with the aid of the bending stress formulae that have been developed for linear stress-strain behavior. However, the maximum bending moment corresponding to point 1 in the stress-strain curve is the moment M_P:

$$\sigma_{top} = \frac{M_P}{I_x} C_2 = \frac{20250\ Y}{343257}(-24.1) = -1.42\ Y$$

$$\sigma_{bottom} = \frac{M_P}{I_x} C_1 = \frac{20250\ Y}{343257}(45.9) = 2.7\ Y$$

The plastic stress distribution during loading and the elastic stress distribution during unloading are demonstrated in the following figure. The residual stresses after unloading from the fully plastic state can be obtained by subtracting the unloading linear stress distribution by the loading plastic one. In the location where the plastic stress distribution has discontinuity, the corresponding stress σ_e of the linear stress distribution can be obtained by the following relation:

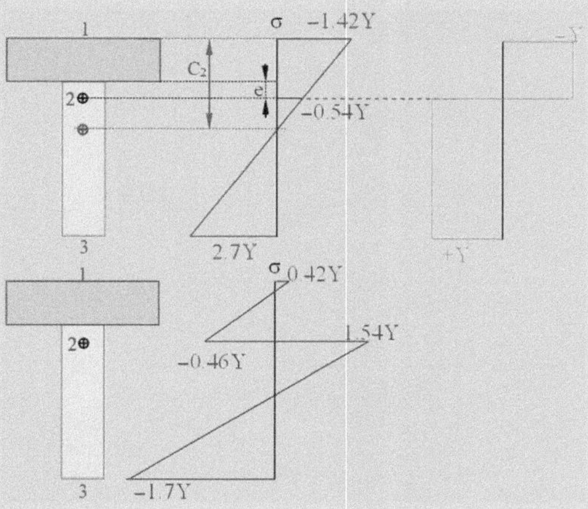

$$\frac{\sigma_e}{C_2 - 10 - e} = \frac{-1.42\ Y}{C_2}$$

or

$$\sigma_e = -0.54\ Y$$

Then, the residual stresses at locations 1, 2, and 3 can be obtained as follows:

$$\sigma_1^r = -Y - (-1.42Y) = 0.42\ Y$$

$$\sigma_{2^+}^r = -Y - \sigma_e = -Y - (-0.54\ Y) = -0.46\ Y$$

$$\sigma_{2^-}^r = +Y - \sigma_e = +Y - (-0.54\ Y) = 1.54\ Y$$

$$\sigma_3^r = +Y - 2.7\ Y = -1.7\ Y$$

APPENDIX

MOMENTS OF INERTIA OF CROSS SECTIONS

$$I_x = \int_A y^2 dA$$
$$I_y = \int_A x^2 dA$$
$$J = \int_A r^2 dA \text{ (Polar moment of inertia)}$$
$$I_{xy} = \int_A xy\, dA \text{ (Product of inertia)}$$

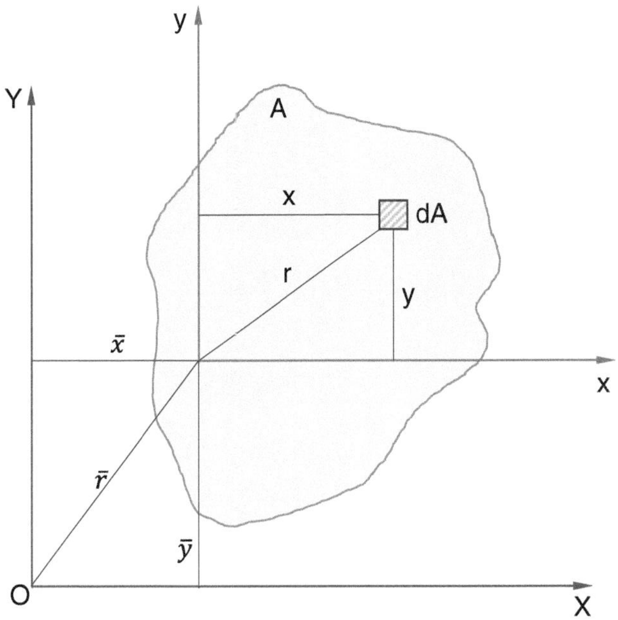

STEINER'S THEOREM FOR MOMENTS OF INERTIA

$$I_X = I_x + A\bar{y}^2$$
$$I_Y = I_y + A\bar{x}^2$$
$$J_O = J_o + A\bar{r}^2$$
$$I_{XY} = I_{xy} + A\bar{x}\bar{y}$$

CENTROID OF CROSS SECTIONS

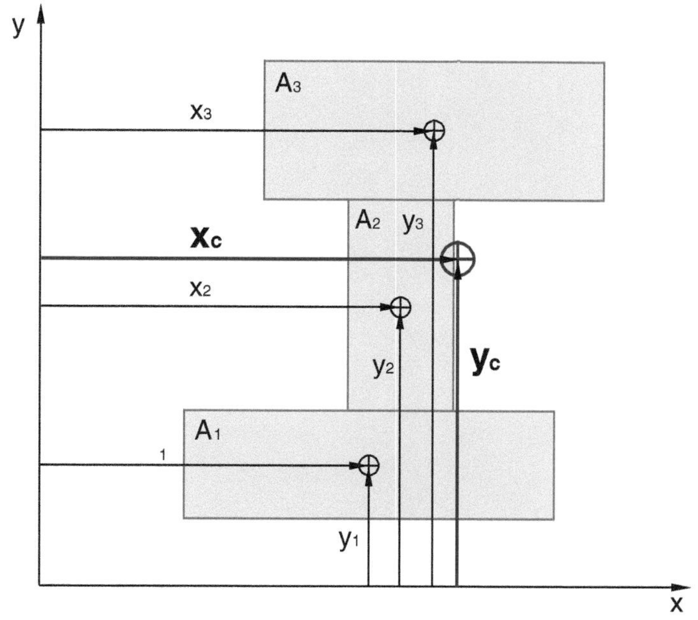

$$y_c = \frac{\sum_{i=1}^{3} A_i y_i}{\sum_{i=1}^{3} A_i}, \quad x_c = \frac{\sum_{i=1}^{3} A_i x_i}{\sum_{i=1}^{3} A_i}$$

REFERENCES

[1] Boresi Arthur P, Schmidt Richard J. Advanced Mechanics of Materials. Wiley; 2002.
[2] Fenner RT. Engineering elasticity. Ellis Horwood ltd; 1986.
[3] Ugural Ansel, Fenster Saul. Advanced Mechanics of Materials and Applied Elasticity. Pearson; 2019.
[4] Solecki Roman, Conant RJay. Advanced Mechanics of Materials. Oxford University Press; 2003.
[5] Cook Robert. Advanced Mechanics of Materials. Pearson; 1998.
[6] Bickford William. Advanced Mechanics of Materials. Longman Pub Group; 1998.
[7] Timoshenko SP, Goodier JN. Theory of elasticity. McGraw-Hill; 1970.
[8] Anand Lallit, Govindjee Sanjay. Continuum Mechanics of Solids. OUP Oxford; 2020.

CHAPTER

Bending Theory of Beams

6

Chapter Outline

6.1 BENDING-INDUCED NORMAL STRESSES

Let's assume a beam is subjected to a bending moment M_x (Fig. 6.1). We select the origin of the coordinate system to be located at the centroid of the beam cross section. The shape of the cross section and the orientation of the y and x axes can be arbitrary, whereas the z axis is directed along the centroid axis of the beam. For the sake of simplicity, we demonstrate in Fig. 6.1 a rectangular shape of the cross section. According to the Bernoulli assumption, the cross section of the beam remains plane after deformation and normal to the deformed axis of the beam. The consequence of this assumption is that the strain in the longitudinal direction z is a linear function of the variable y:

$$\varepsilon_{zz} = by \tag{6.1}$$

In the same way, in a beam subject to a bending moment M_y (Fig. 6.2), the following linear relation for the strain ε_{zz} can be obtained:

$$\varepsilon_{zz} = cx \tag{6.2}$$

The coefficients b and c in Eqs. (6.1) and (6.2) are constants.

Superposition of Eqs. (6.1) and (6.2) yields the strain distribution of a beam subjected to both M_x and M_y (Fig. 6.3):

Advanced Mechanics of Solids and Structures. DOI: http://doi.org/10.1016/B978-0-443-23778-2.00006-8

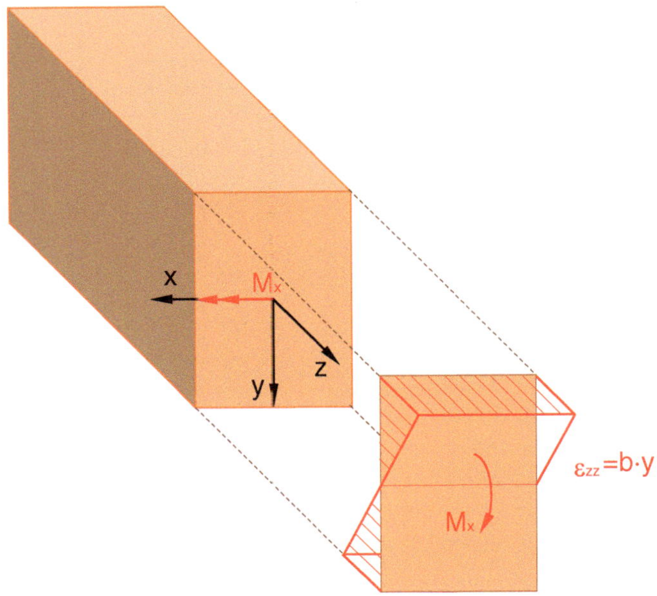

FIG. 6.1

Beam subjected to a bending moment M_x.

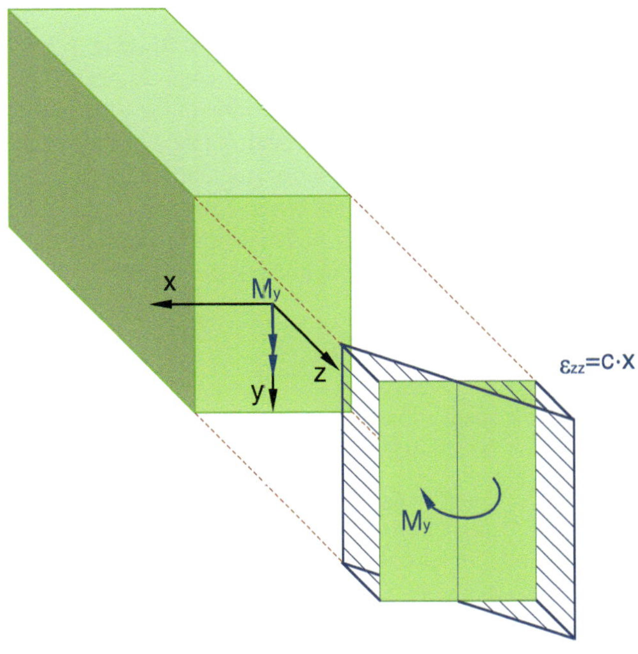

FIG. 6.2

Beam subjected to bending moment M_y.

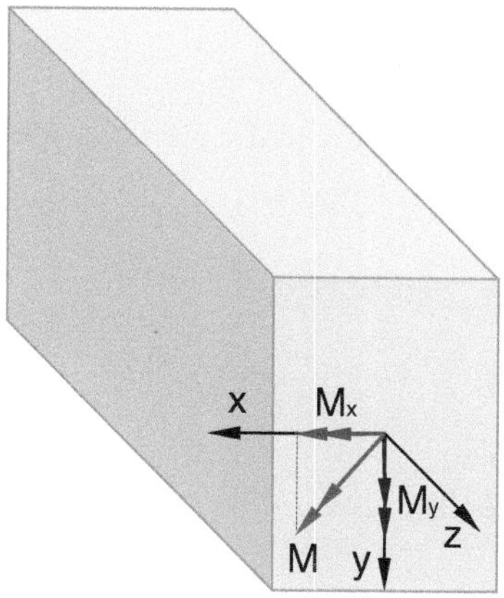

FIG. 6.3

Superposition of bending moments M_x and M_y.

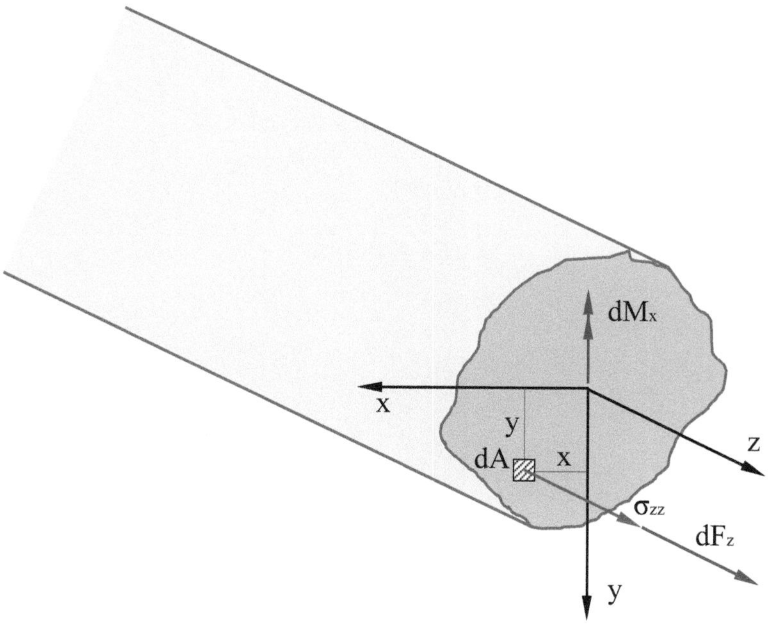

FIG. 6.4

Elementary force dF_Z on an arbitrary point of the cross section.

$$\varepsilon_{zz} = by + cx \tag{6.3}$$

Multiplying the above equation, with the elasticity modulus we get

$$E\varepsilon_{zz} = Eb\ y + Ec\ x \tag{6.4}$$

or

$$\sigma_{zz} = Cx + By \tag{6.5}$$

where B and C are constants.

The force on any elementary area dA located on an arbitrary point (x,y) is given by the following equation:

$$dF_{zz} = \sigma_{zz} dA \tag{6.6}$$

or

$$dF_{zz} = (Cx + By)dA \tag{6.7}$$

The corresponding components of the bending moment are:

$$dM_x = dF_{zz}y = (Cx + By)dA\ y \tag{6.8}$$

$$dM_y = -dF_{zz}x = -(Cx + By)dA\ x \tag{6.9}$$

Integration of Eqs. (6.8) and (6.9) yields:

$$M_x = \int_A (Cx + By)y\ dA \tag{6.10}$$

$$M_y = \int_A -(Cx + By)x\ dA \tag{6.11}$$

or

$$M_x = C\underbrace{\int_A xy\ dA}_{I_{xy}} + B\underbrace{\int_A y^2\ dA}_{I_x} \tag{6.12}$$

$$M_y = -C\underbrace{\int_A x^2\ dA}_{I_z} - B\underbrace{\int_A xy\ dA}_{I_{xy}} \tag{6.13}$$

Because the moments of inertia I_x, I_y and the product of inertia I_{xy} are given by the following well-known formulae

$$\int_A y^2 \, dA = I_x \tag{6.14}$$

$$\int_A x^2 \, dA = I_y \tag{6.15}$$

$$\int_A xy \, dA = I_{xy} \tag{6.16}$$

the system of Eqs. (6.12) and (6.13) can provide the parameters B and C:

$$B = \frac{M_x I_y + M_y I_{xy}}{I_x I_y - I_{xy}^2} \tag{6.17}$$

$$C = -\frac{M_y I_x + M_x I_{xy}}{I_x I_y - I_{xy}^2} \tag{6.18}$$

With the aid of the last equations, Eq. (6.5) yields:

$$\sigma_{zz} = -\left(\frac{M_y I_x + M_x I_{xy}}{I_x I_y - I_{xy}^2} \right) x + \left(\frac{M_x I_y + M_y I_{xy}}{I_x I_y - I_{xy}^2} \right) y \tag{6.19}$$

It should be noted that the product of inertia I_{xy} will be zero for cross sections of symmetrical shape when a coordinate axis is also an axis of symmetry. In this case, Eq. (6.19) can be simplified:

$$\sigma_{zz} = -\frac{M_y}{I_y} x + \frac{M_x}{I_x} y \tag{6.20}$$

We can define a neutral line as a line in the section where the longitudinal stress σ_{zz} is null (Fig. 6.5). By setting $\sigma_{zz} = 0$ in Eq. (6.20), the following function of the neutral line can be obtained:

$$-\left(\frac{M_y I_x + M_x I_{xy}}{I_x I_y - I_{xy}^2} \right) x + \left(\frac{M_x I_y + M_y I_{xy}}{I_x I_y - I_{xy}^2} \right) y = 0 \tag{6.21}$$

or

$$y = \left(\frac{M_y I_x + M_x I_{xy}}{M_x I_y + M_y I_{xy}} \right) x \tag{6.22}$$

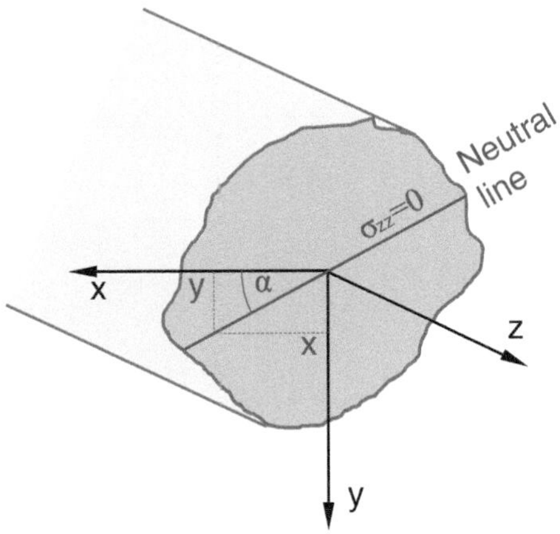

FIG. 6.5

Neutral line is a cross section.

The last equation shows that the neutral line is a straight line (Fig. 6.5) with inclination angle a:

$$\tan\alpha = \frac{y}{x} = \frac{M_y I_x + M_x I_{xy}}{M_x I_y + M_y I_{xy}} \tag{6.23}$$

6.2 BENDING-INDUCED DISPLACEMENTS

For a beam subjected to bending (Fig. 6.6), the vertical displacement w_y can be determined by the similarity of the triangles ABC and KDC (Fig. 6.7):

$$\frac{AB}{BC} = \frac{CD}{DK} \tag{6.24}$$

or

$$\frac{\varepsilon_{zz}\Delta z}{y} = \frac{\Delta z}{R_y} \tag{6.25}$$

With the aid of the geometric definition of curvature

$$\frac{1}{R_y} = \frac{\frac{d^2w_y}{dz^2}}{\left[1+\left(\frac{d^2w_y}{dz^2}\right)^2\right]^{3/2}} \approx \frac{d^2w_y}{dz^2} \tag{6.26}$$

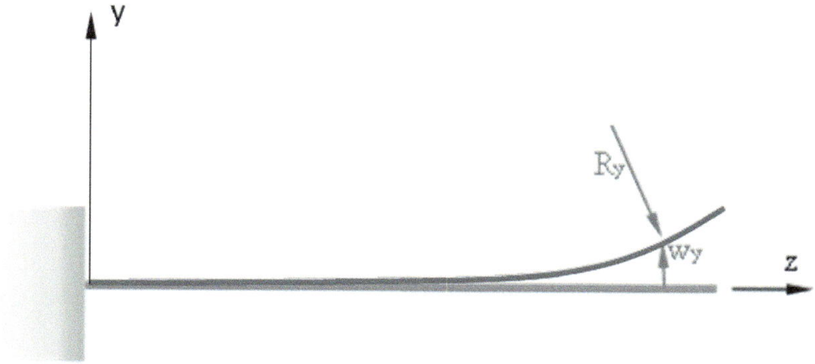

FIG. 6.6

Bending-induced displacement.

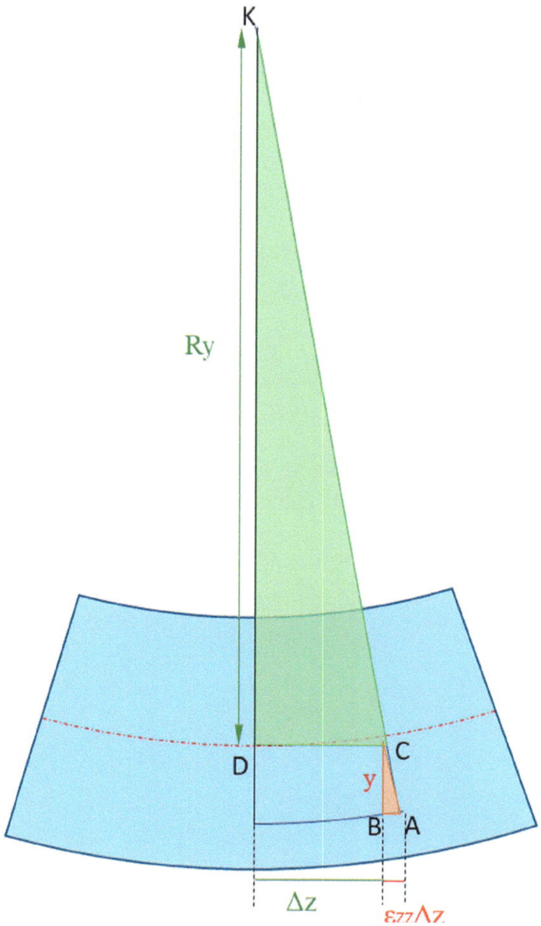

FIG. 6.7

Radius of curvature of a beam under bending.

Eq. (6.25) yields:

$$\frac{\varepsilon_{zz}}{y} = \frac{d^2 w_y}{dz^2} \tag{6.27}$$

Multiplying the above equation with the elasticity modulus, the following formula can be obtained:

$$\frac{\sigma_{zz}}{y} = E\frac{d^2 w_y}{dz^2} \tag{6.28}$$

Combining Eqs. (6.19) and (6.28) and setting $x = 0$, the following differential equation for calculating the vertical displacement of the centroid of the cross section can be determined:

$$\frac{d^2 w_y}{dz^2} = \frac{1}{E}\left(\frac{M_x I_y + M_y I_{xy}}{I_x I_y - I_{xy}^2}\right) \tag{6.29}$$

Because the displacement w of the centroid is always perpendicular to the neutral axis, the horizontal displacement w_x of the centroid (Fig. 6.8) can be calculated by the following equation:

$$w_x = w_y \tan(a) \tag{6.30}$$

or

$$w_x = w_y \frac{M_y I_x + M_x I_{xy}}{M_x I_y + M_y I_{xy}} \tag{6.31}$$

For symmetrical cross sections where a coordinate axis is an axis of symmetry, that is, $I_{xy} = 0$, Eqs. (6.29) and (6.31) can be simplified:

$$\frac{d^2 w_y}{dz^2} = \frac{M_x}{EI_x} \tag{6.32}$$

$$w_x = w_y \frac{M_y\, I_x}{M_x\, I_y} \tag{6.33}$$

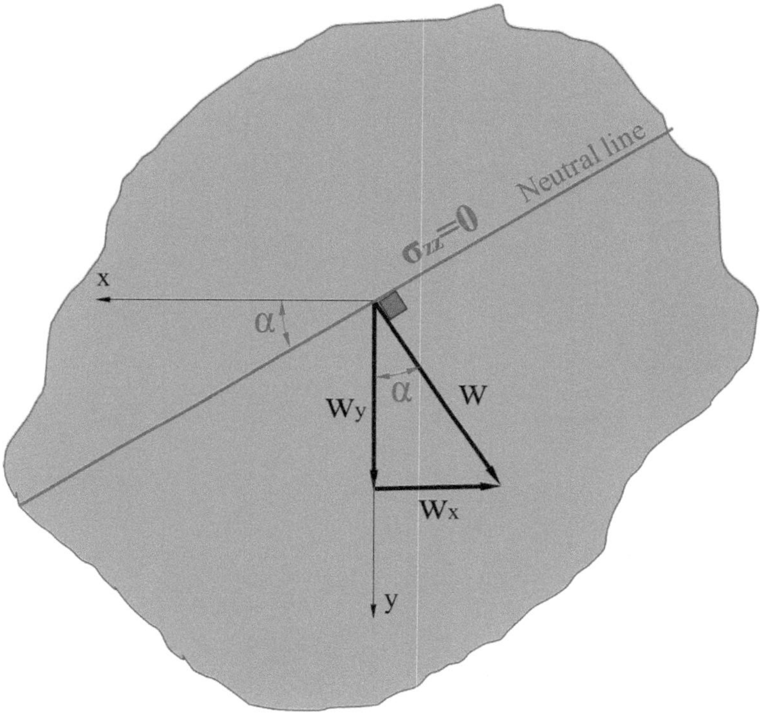

FIG. 6.8

Vertical and horizontal displacement of the centroid.

6.3 BENDING-INDUCED SHEAR STRESSES

Let's assume a cantilever beam subjected to a point load, as shown in Fig. 6.9. For the sake of simplicity, a rectangular cross section is demonstrated in the figure. The load F causes a nonuniform bending moment distribution. Because of the gradient of the bending moment distribution, the magnitude of the normal stresses on two cross sections is different. Because of the difference of the normal stresses, the corresponding forces $F(x)$ and $F(x + \Delta x)$ on the cross sections of a layer lying between a distance y from the centroid and the bottom surface of the beam (Fig. 6.9) are different. This difference is equilibrated by the existence of shear stresses $\tau(y)$ acting on the layer surface $B\ \Delta x$. For $\Delta x \to dx$, the following equilibrium equation can be formulated for the layer in Fig. 6.9:

$$\tau(y)Bdx = F(x + dx) - F(x) = (F(x) + dF(x)) - F(x) \qquad (6.34)$$

or

$$\tau(y)Bdx = dF(x) \qquad (6.35)$$

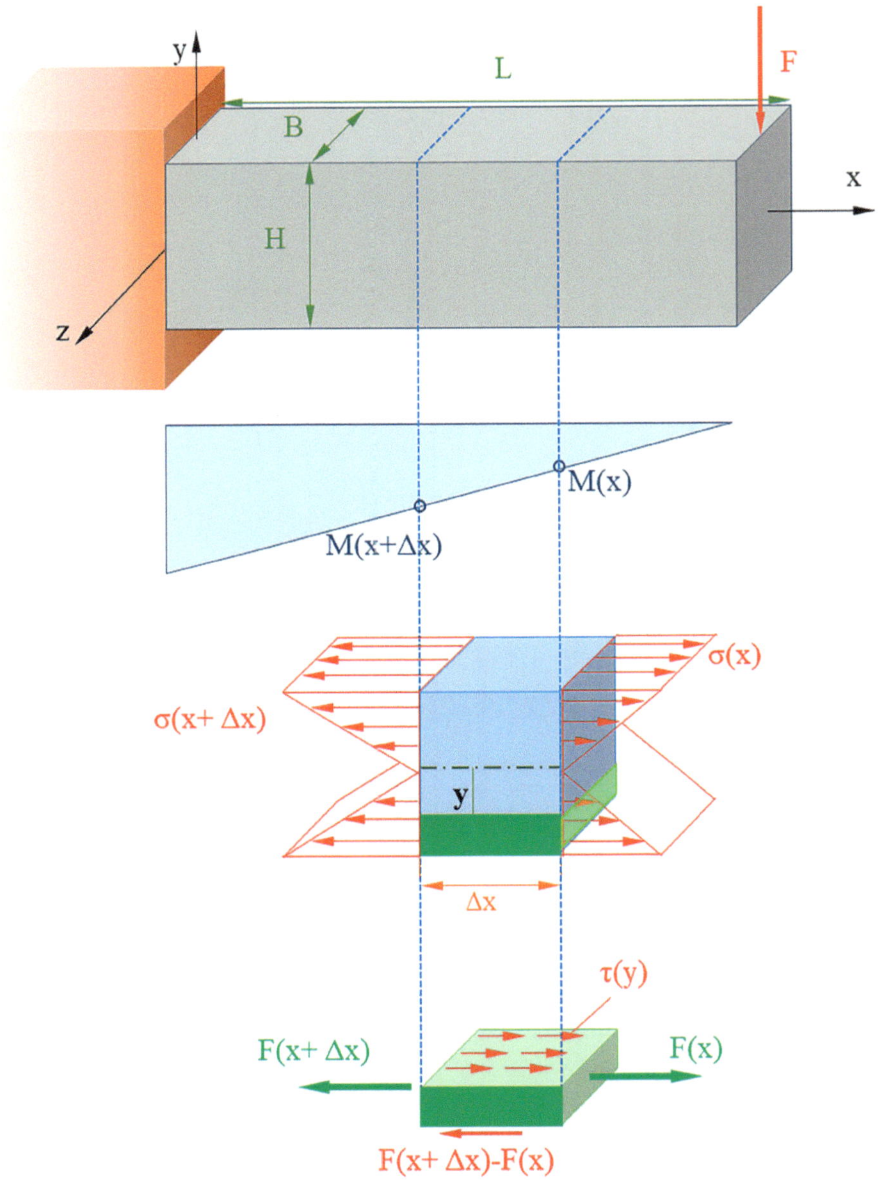

FIG. 6.9

Bending-induced shear stresses.

Because

$$dF(x) = d(\sigma(x, y)A(y)) = \frac{dM(x)}{I_z} \int_y^{H/2} y\, dA \tag{6.36}$$

Eq. (6.35) yields:

$$\tau(y)B = \frac{dM/dx}{I_z} \int_y^{H/2} y \ dA \tag{6.37}$$

Obviously, the reason of the shear stresses $\tau(y)$ is the gradient $dM(x)/dx$ that expresses the shear force $Q(x)$. It should be noted that Eq. (6.37) yields $\tau(y) = 0$ for pure shear where M = const. The term $\tau(y)B$ is called shear flow, and the integral $\int_y^{H/2} y \ dA$ expresses the static moment of area $S_z(y)$ of the cross section area lying between y and the bottom (or top) surface of the beam:

$$\underbrace{\tau(y)B}_{\text{Shear flow}} = \frac{Q(x)}{I_z} S_z(y) \tag{6.38}$$

EXERCISE 1

Calculate the shear stress distribution on a beam of rectangular surface under (nonpure) bending.

Solution

Let's assume the rectangular cross section of Fig. 6.10. An elementary area can be specified by the formula $dA = B \ dy$. With the aid of Eq. (6.37), the following equation for the shear flow can be obtained:

$$\tau(y)B = \frac{Q}{I} \int_y^{H/2} y \ Bdy \tag{6.39}$$

or

$$\tau(y) = \frac{Q}{I} \int_y^{H/2} y \ dy \tag{6.40}$$

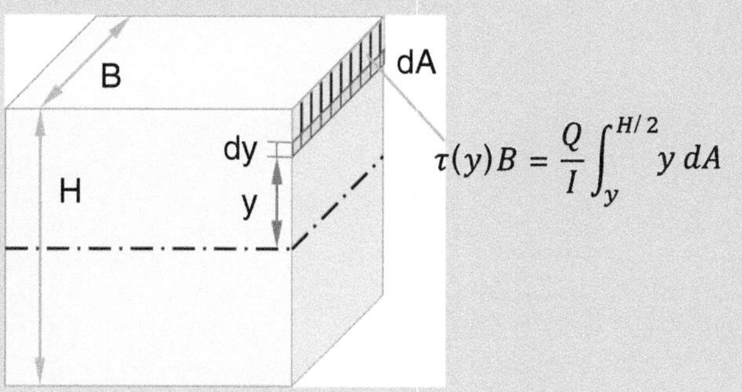

FIG. 6.10

Bending-induced shear stresses on a rectangular cross section.

(Continued)

EXERCISE 1—CONT'D

or

$$\tau(y) = \frac{Q}{2I}\left[\frac{H^2}{4} - y^2\right] \tag{6.41}$$

Graphical representation of the above equation is demonstrated in Fig. 6.11. The resulting distribution shows that, unlike the distribution of the bending-induced normal stresses, the shear stress has null magnitude on the beam's external surfaces and maximum value on the centroid. For the rectangular cross section, the maximum shear stress is:

$$\tau_{\max} = \tau(0) = \frac{Q}{2I_z}\left[\frac{H^2}{4}\right] = \frac{QH^2}{8I_z} \tag{6.42}$$

We can determine the mean shear stress τ_m for a rectangular cross section using the equation:

$$\tau_m = \frac{Q}{BH} \tag{6.43}$$

Then, the maximum shear stress in a rectangular cross section can be correlated to the mean shear stress by the following formula:

$$\tau_{\max} = \frac{QH^2}{8I_z} = \frac{QH^2}{8\left(\frac{BH^3}{12}\right)} = \frac{3Q}{2BH} = \frac{3}{2}\tau_m \tag{6.44}$$

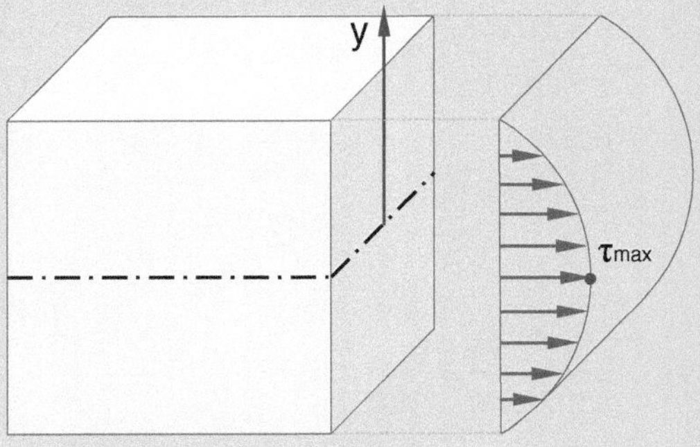

FIG. 6.11

Bending-induced shear stress distribution on a rectangular cross section.

6.4 THIN-WALL OPEN CROSS SECTIONS

Eq. (6.38) is based on the bending stress distribution on a cross section subjected to a bending moment vector that acts on the centroid. Therefore the shear force $Q = dM(x)/dx$ acts on the centroid too. Fig. 6.12 demonstrates schematically the bending-induced shear flow on a thin wall open cross section. Obviously, the shear flow in Fig. 6.12 causes a torsion of the cross section. To avoid torsion in open cross sections, the force Q should not be applied on the centroid. The point C, through which, if the force Q is applied, does not produce any torsion is called shear center (Fig. 6.12). The aim of this subchapter is the determination of the distance e in order to locate the shear center of the cross section. The methodology will be demonstrated for the semicircular, thin-wall cross section in Fig. 6.12.

With the aid of the following formulae

$$y = R\cos\theta \tag{6.45}$$

$$x = R\sin\theta \tag{6.46}$$

$$dA = tR\,d\theta \tag{6.47}$$

Eq. (6.38) yields:

$$\tau(\theta)t = \frac{Q}{I_x}\int (R\cos\theta)tR\;d\theta \tag{6.48}$$

or

$$\tau(\theta) = -\frac{QR^2}{I_x}\sin\theta \tag{6.49}$$

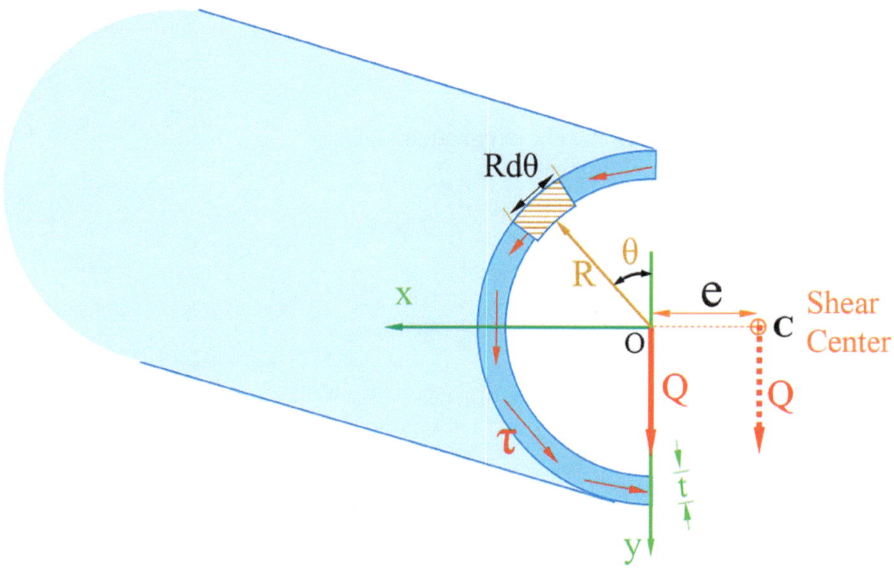

FIG. 6.12

Shear flow on an open, thin-wall cross section.

For the thin-wall cross section of Fig. 6.12, the moment of inertia I_x is

$$I_x = \int_0^{\pi} y^2 dA = \int_0^{\pi} (R\cos\theta)^2\, tR\, d\theta = \frac{1}{2}R^3\pi t \tag{6.50}$$

Considering the last result, Eq. (6.49) yields:

$$\tau(\theta) = -\frac{2Q}{R\pi t}\sin\theta \tag{6.51}$$

The above shear stress distribution causes a torsional moment with respect to the centroid O.

The contribution of an elementary surface $dA = tRd\theta$ on the torsional moment is:

$$dM_t = RdF_t = R(\tau(\theta)dA) = R\tau(\theta)tRd\theta \tag{6.52}$$

With the aid of Eq. (6.51), the above equation yields:

$$dM_t = \left[-\frac{2QR\sin\theta}{\pi}\right]d\theta \tag{6.53}$$

or

$$M_t = -\frac{2QR}{\pi}\int_0^{\pi}\sin\theta d\theta = -\frac{4QR}{\pi} \tag{6.54}$$

To eliminate the torsional moment M_t, the shear force Q should be applied in a distance e (Fig. 6.12). The distance e determines the location of the shear center and can be obtained by the following equilibrium equation:

$$Qe = M_t \tag{6.55}$$

Considering Eq. (6.54), the above equation yields:

$$e = -\frac{4R}{\pi} \tag{6.56}$$

EXERCISE 2

For a beam of length 2 m, with the cross section shown in Fig. 6.13, determine the bending-induced normal stress in point A. The unit of the dimensions in the cross section is mm.

Solution

The bending moment

$$M = P \cdot L = 14000 \times 2 = 28000\ Nm$$

can be analyzed into the following components:

$$M_x = M\cos\varphi = 28000\cos(72) = 8652\ Nm$$

$$M_y = M\sin\varphi = 28000\sin(72) = 26630\ Nm$$

EXERCISE 2—CONT'D

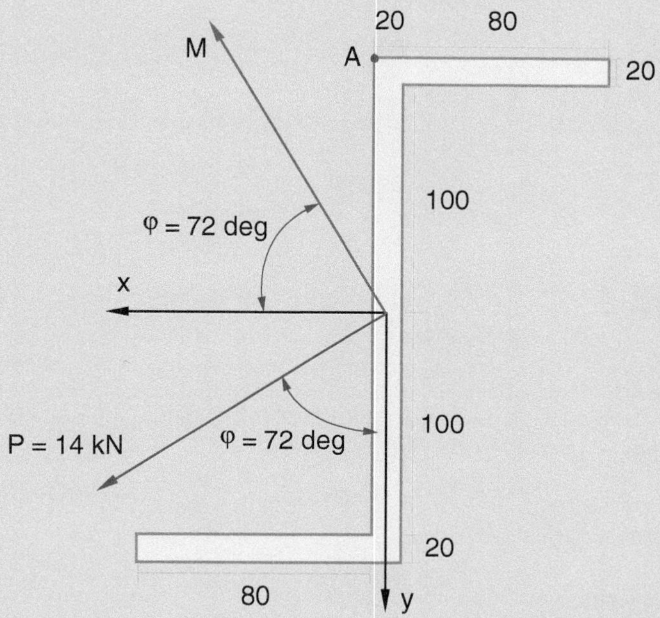

FIG. 6.13

Beam of nonsymmetric cross section under bending.

For calculating the moments of inertia I_x, I_y we divide the cross section into three areas (top flange, bottom flange, and web area):

$$I_x = \frac{80(20)^3}{12} + (80 \cdot 20)90^2 + \frac{20(200)^3}{12} + \frac{80(20)^3}{12} + (80 \cdot 20)(-90)^2 = 39360000\ mm^4$$

$$I_y = \frac{20(80)^3}{12} + (20 \cdot 80)50^2 + \frac{200(20)^3}{12} + \frac{20(80)^3}{12} + (20 \cdot 80)(-50)^2 = 9840000\ mm^4$$

$$I_{xy} = 20 \cdot 80 \cdot 50 \cdot 90 + 20 \cdot 80 \cdot (-50)(-90) = 14400000\ mm^4$$

In Eq. (6.19)

$$\sigma_{zz} = -\left(\frac{M_y I_x + M_x I_{xy}}{I_x I_y - I_{xy}^2}\right)x + \left(\frac{M_x I_y + M_y I_{xy}}{I_x I_y - I_{xy}^2}\right)y$$

we set the coordinates of $A(x_A, y_A)$

$$x_A = 10 \times 10^{-3}\ m$$
$$y_A = 120 \times 10^{-3}\ m$$

(Continued)

EXERCISE 2—CONT'D

Then, the following result can be obtained:

$$\sigma_{\max}(x_A, y_A) = -\left(\frac{26630 \times 3.936 \times 10^{-5} + 8652 \times 1.44 \times 10^{-5}}{3.936 \times 10^{-5} \times 9.84 \times 10^{-6} - (1.44 \times 10^{-5})^2}\right) 10 \times 10^{-3} +$$

$$\left(\frac{8652 \times 9.84 \times 10^{-6} + 26630 \times 1.44 \times 10^{-5}}{3.936 \times 10^{-5} \times 9.84 \times 10^{-6} - (1.44 \times 10^{-5})^2}\right) 120 \times 10^{-3} = 247 \times 10^{6}\ Pa$$

EXERCISE 3

A beam DG of rectangular cross section is simply supported on its ends D and G (Fig. 6.14a). The centroid of the cross section on the point B is subjected to forces Fx, Fy, and Fz oriented in the directions x, y, z, respectively. The details of the cross section on point A are shown in Fig. 6.14b, and the material element of point C (a, b, h) of the cross section A is shown in Fig. 6.14c. The plane MHJ (Fig. 6.14c) has the following stresses:

$$\begin{Bmatrix} \sigma_{cx} \\ \sigma_{cy} \\ \sigma_{cz} \end{Bmatrix} = \begin{Bmatrix} 133 \\ 42 \\ 27 \end{Bmatrix} MPa$$

The stress components σ_{xx}, σ_{yy}, and σ_{xy} on the point C (a, b, h) are null ($\sigma_{xx} = 0$, $\sigma_{yy} = 0$, $\sigma_{xy} = 0$). Calculate the forces Fx, Fy, and Fz. *(Tip: Assume that the forces Fx, Fy, and Fz are equally shared on the supports.)*

Given data: Beam length $a = 1.0\,\text{m}$, cross section dimensions $b = 5\,\text{mm}$, $h = 2\,\text{mm}$.

Useful formulae:

$$I_y = \frac{4b(4h)^3}{12},\ I_z = \frac{4h(4b)^3}{12},\ I_{yz} = 0,\ \sigma_{xz}(z) = \frac{Q_z}{2I_y}(4h^2 - z^2),\ \sigma_{xy}(z) = \frac{Q_z}{2I_z}(4b^2 - y^2)$$

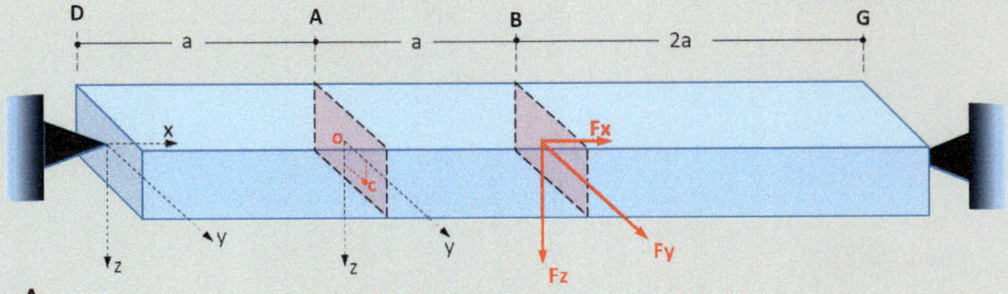

A

FIG. 6.14A

Solution

Orientation of the plane HJM:

Vector $\overrightarrow{HM} = (0-0)\vec{i} + (0-3)\vec{j} + (1-0)\vec{k} = -3\vec{j} + \vec{k}$

Vector $\overrightarrow{HJ} = (1-0)\vec{i} + (0-3)\vec{j} + (0-0)\vec{k} = \vec{i} - 3\vec{j}$

EXERCISE 3—CONT'D

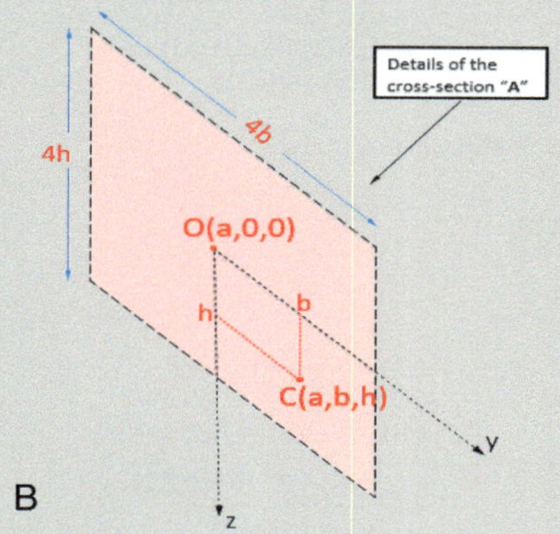

FIG. 6.14B

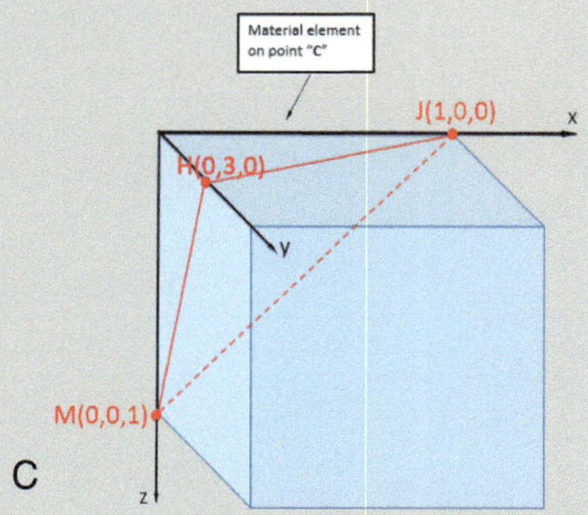

FIG. 6.14C

$$\overrightarrow{HM}\times\overrightarrow{HJ}=\begin{vmatrix}\vec{i} & \vec{j} & \vec{k}\\ 0 & -3 & 1\\ 1 & -3 & 0\end{vmatrix}=\begin{vmatrix}-3 & 1\\ -3 & 0\end{vmatrix}\vec{i}-\begin{vmatrix}0 & 1\\ 1 & 0\end{vmatrix}\vec{j}+\begin{vmatrix}0 & -3\\ 1 & -3\end{vmatrix}\vec{k}=$$

$$3\vec{i}+\vec{j}+3\vec{k}$$

$$l=\frac{3}{\sqrt{3^2+1^2+3^2}}=0.688$$

$$m=\frac{1}{\sqrt{3^2+1^2+3^2}}=0.229$$

(Continued)

EXERCISE 3—CONT'D

$$n = \frac{3}{\sqrt{3^2 + 1^2 + 3^2}} = 0.688$$

Calculation of the stress components σ_{xz}, σ_{yz}, σ_{zz}

$$\begin{Bmatrix} \sigma_{cx} \\ \sigma_{cy} \\ \sigma_{cz} \end{Bmatrix} = \begin{bmatrix} \sigma_{xx} & \sigma_{xy} & \sigma_{xz} \\ \sigma_{xy} & \sigma_{yy} & \sigma_{yz} \\ \sigma_{xz} & \sigma_{yz} & \sigma_{zz} \end{bmatrix} \begin{Bmatrix} l \\ m \\ n \end{Bmatrix} \Rightarrow$$

$$\begin{Bmatrix} 133 \\ 42 \\ 27 \end{Bmatrix} = \begin{bmatrix} 0 & 0 & \sigma_{xz} \\ 0 & 0 & \sigma_{yz} \\ \sigma_{xz} & \sigma_{yz} & \sigma_{zz} \end{bmatrix} \begin{Bmatrix} 0.688 \\ 0.229 \\ 0.688 \end{Bmatrix} \Rightarrow$$

$$\left.\begin{aligned} 133 &= 0.688\sigma_{xz} \\ 42 &= 0.688\sigma_{yz} \\ 27 &= 0.688\sigma_{xz} + 0.229\sigma_{yz} + 0.688\sigma_{zz} \end{aligned}\right\} \Rightarrow \left\{\begin{aligned} \sigma_{xz} &= 193.3MPa \\ \sigma_{yz} &= 61.0MPa \\ \sigma_{zz} &= -174.4MPa \end{aligned}\right.$$

Calculation of forces F_x, F_y, and F_z.

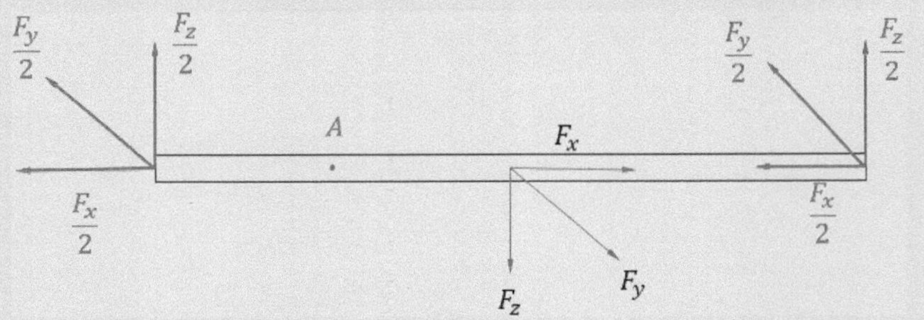

Due to symmetry, the support reactions are half of the F_x, F_y, and F_z forces. Bending moments and shear forces on cross section A:

$$M_y^A = \frac{F_z}{2}3a - F_z a = \frac{F_z}{2}a$$

$$M_z^A = F_y a - \frac{F_y}{2}3a = -\frac{F_y}{2}a$$

$$Q_y^A = F_y - \frac{F_y}{2} = \frac{F_z}{2}$$

$$Q_z^A = F_z - \frac{F_z}{2} = \frac{F_z}{2}$$

EXERCISE 3—CONT'D

The forces F_x, F_y, and F_z can be calculated with the aid of the given stresses on point C:

$$\sigma_{xx}(a, b, h) = \frac{F_x - \frac{F_x}{2}}{4b \times 4h} + \frac{M_y^A I_z + M_z^A I_{yz}}{I_y I_z - I_{yz}^2} h - \frac{M_z^A I_y + M_y^A I_{yz}}{I_y I_z - I_{yz}^2} b = 0$$

$$\sigma_{xz} = \frac{Q_z^A}{2I_y}(4h^2 - h^2)$$

$$\sigma_{xy} = \frac{Q_y^A}{2I_z}(4b^2 - b^2)$$

The last three equations yield:

$$\frac{F_x}{32bh} + \frac{\left(\frac{F_z a}{2}\right)}{I_y} h + \frac{\left(\frac{F_y a}{2}\right)}{I_z} b = 0$$

$$193.3 = \frac{\left(\frac{F_z}{2}\right)}{2I_y} 3h^2$$

$$0 = \frac{\left(\frac{F_y}{2}\right)}{2I_z} 3b^2$$

Solution of the system of the last three equations results:

$$F_x = -4123.2ab$$

$$F_y = 0$$

$$F_z = 257.7 \frac{I_y}{h^2}$$

6.5 BEAM ON ELASTIC FOUNDATION

Foundations, rail tracks, cylindrical silos, and so on are some applications of the theory of the beam on elastic foundation. The mechanical model consists of a beam lying on an elastic substrate (Fig. 6.15), which reacts with a pressure proportional to the deflection of the beam at each point. Therefore the pressure on an elementary segment of the beam (Fig. 6.15) is:

$$p(x) = q(x) - k\ w(x) \tag{6.57}$$

The equilibrium equations of the beam segment of length dx in Fig. 6.15 are:

$$\text{Vertical forces: } -q(x)dx + kw(x)dx + Q - (Q + dQ) = 0 \tag{6.58}$$

$$\text{Moments: } M_z + Qdx + kw(x)dx\frac{dx}{2} - q(x)dx\frac{dx}{2} - (M_z + dM_z) = 0 \tag{6.59}$$

Neglecting the terms $(dx)^2 \approx 0$, the above equations yield:

$$\frac{dQ}{dx} = kw(x) - q(x) \tag{6.60}$$

and

$$Q = \frac{dM_z}{dx} \tag{6.61}$$

Combining the last two equations, we get

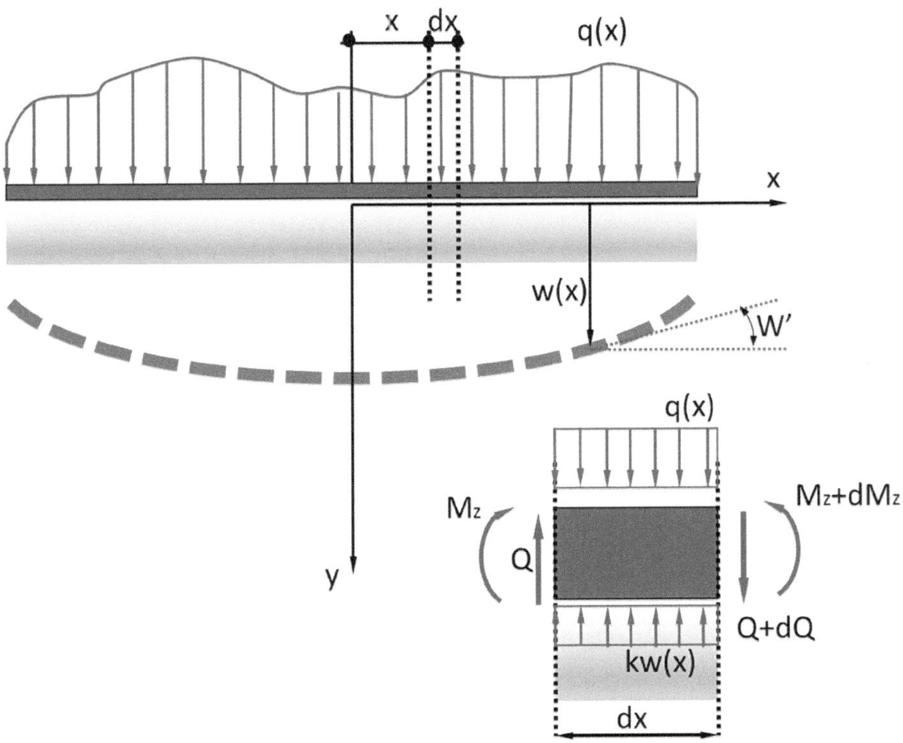

FIG. 6.15

Beam on elastic foundation.

$$\frac{d^2M_z}{dx^2} = kw(x) - q(x) \tag{6.62}$$

Taking into account that coordinate system of Fig. 6.15, Eq. (6.32) can be written as:

$$-\frac{M_z}{EI_z} = \frac{d^2w(x)}{dx^2} \tag{6.63}$$

With the aid of Eq. (6.63), Eq. (6.62) yields:

$$EI_z\frac{d^4w(x)}{dx^4} + kw(x) = q(x) \tag{6.64}$$

Eq. (6.64) is the governing equation of an infinite beam on elastic foundation. Eq. (6.64) is usually written in the following form:

$$w''''(x) + \lambda w(x) = \frac{q(x)}{EI_z} \tag{6.65}$$

where

$$\lambda = \frac{k}{EI_z} \tag{6.66}$$

6.6 INFINITE BEAM ON ELASTIC FOUNDATION UNDER A POINT FORCE

Fig. 6.16 shows an infinite beam on elastic foundation under a point force.

Then, Eq. (6.65) can be written in the following form:

$$w''''(x) + \lambda w(x) = \frac{P}{EI_z}\delta(x) \tag{6.67}$$

where $\delta(x)$ is the Dirac-Delta function.

The solution of the differential equation (Eq. 6.67) can be benefited by the properties of integral transforms. Taking a suitable integral transform in the differential Eq. (6.67), we can obtain an algebraic equation of the transformed variable that can be solved easily. Then, we can inverse the integral transform of the transformed variable in order to get the solution of the differential equation. This procedure is explained schematically in Fig. 6.17.

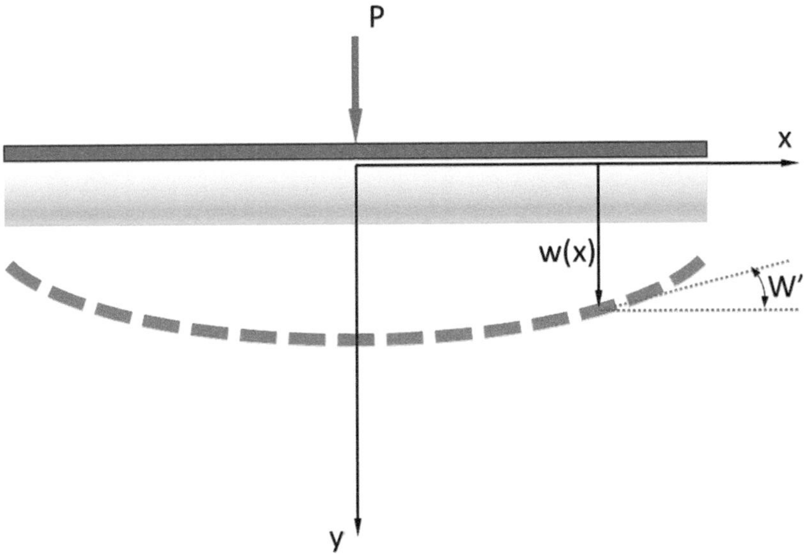

FIG. 6.16

Infinite beam on elastic foundation under a point force.

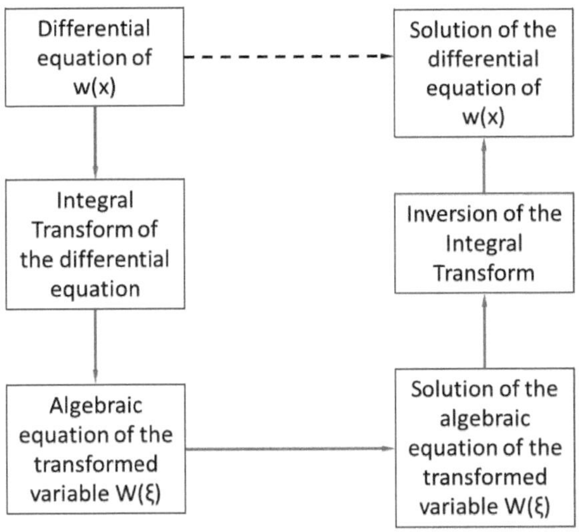

FIG. 6.17

Solution of differential equations with integral transforms.

Because of the symmetry of the geometry and the loading, the following conditions are valid:

$$w'(0) = 0 \quad \text{(the slope of the deformed shape at x = 0 is null)} \tag{6.68}$$

$$w'''(0) = \frac{Q(0)}{EI_z} = 0 \text{ (the shear force at x} = 0 \text{ is null)} \tag{6.69}$$

At $x \to \infty$ we assume that the deflection, slope, bending moment, and shear force are null:

$$w(\infty) = w'(\infty) = w''(\infty) = w'''(\infty) = 0 \tag{6.70}$$

Because of the conditions given in Eqs. (6.68)–(6.70), the Fourier Cosine Integral Transform

$$F_c\{w(x);\ \xi\} = W_c(\xi) = \int_0^{\infty} w(x)\cos(\xi x)dx \tag{6.71}$$

is suitable to be applied in Eq. (6.67).

Taking into account the following properties of the Fourier Cosine Integral Transform

$$F_c\{w''''(x);\ \xi\} = \xi^4 W_c(\xi) \tag{6.72}$$

$$F_c\{kw(x);\ \xi\} = kW_c(\xi) \tag{6.73}$$

$$F_c\{P\delta(x);\ \xi\} = \frac{P}{\sqrt{2\pi}} \tag{6.74}$$

the differential equation (Eq. 6.67) can be transformed to the following algebraic equation:

$$\xi^4 W_c(\xi) + \lambda W_c(\xi) = \frac{P}{EI_z\sqrt{2\pi}} \tag{6.75}$$

or

$$W_c(\xi) = \frac{P}{EI_z\sqrt{2\pi}}\frac{1}{\xi^4 + \lambda} \tag{6.76}$$

The inversed Fourier Cosine Transform has the form:

$$w(x) = F_c^{-1}\{W_c(\xi);\ x\} = \frac{2}{\pi}\int_0^{\infty} W_c(\xi)\cos(x\xi)d\xi \tag{6.77}$$

or

$$w(x) = F_c^{-1}\{W_c(\xi);\ x\} = \frac{2}{\pi}\frac{P}{EI_z\sqrt{2\pi}}\int_0^{\infty}\frac{\cos(x\xi)}{\xi^4 + \lambda}d\xi \tag{6.78}$$

The inversion of the Fourier Cosine Transform can be carried out by the standard software Mathematica by formulae provided by specialized books. The integral in Eq. (6.78) has an analytic solution that yields the following result:

$$w(x) = \frac{P}{EI_z\sqrt{2\pi}}\left(-\frac{(-1)^{3/4}e^{-(-\lambda)^{1/4}x}\left(i + e^{i\lambda^{1/4}\sqrt{2}x}\right)\sqrt{\frac{\pi}{2}}}{2\lambda^{3/4}}\right) \tag{6.79}$$

EXERCISE 4

An infinite beam of flexural stiffness $EI_z = 150 \times 10^3\ Pam^3$ is lying on an elastic foundation (Fig. 6.18) with $k = 10^4\ kN/m^3$. The beam is subjected to a point force $P = 10\ kN$. Determine the deflection distribution within a distance $0 < x < 3.0m$.

Solution

The value of the parameter λ is:

$$\lambda = \frac{k}{EI_z} = 66.6$$

With the aid of Eq. (6.79), the following deformed shape of the beam is obtained:

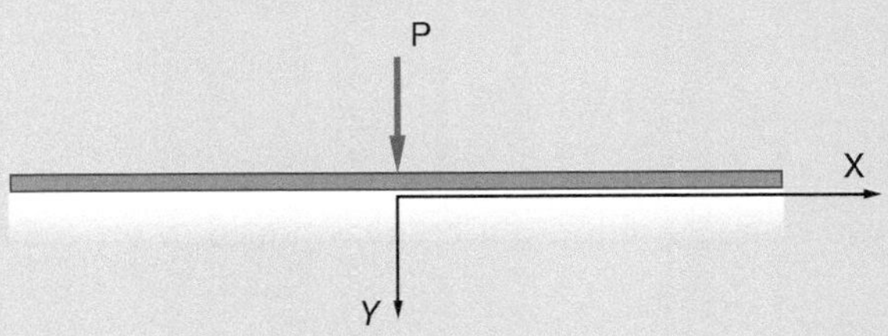

FIG. 6.18

Infinite beam on an elastic foundaticn under point force.

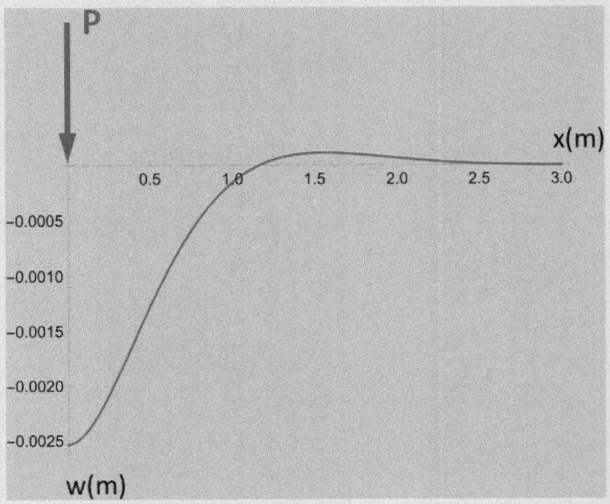

6.7 INFINITE BEAM ON ELASTIC FOUNDATION UNDER A UNIFORM LOAD DISTRIBUTION

Fig. 6.19 shows an infinite beam on elastic foundation subjected to a uniformly distributed load in the area $-a < x < a$. With the aid of the Heaviside Step function

$$H(x) = \begin{cases} 1 & for\ x \geq 0 \\ 0 & for\ x < 0 \end{cases} \tag{6.80}$$

the distributed load in the area $-a < x < a$ can be expressed by the following formula

$$q(x) = q_o [H(x+a) - H(x-a)] \tag{6.81}$$

and Eq. (6.65) takes the form

$$w''''(x) + \lambda w(x) = \frac{q_o}{EI_z}[H(x+a) - H(x-a)] \tag{6.82}$$

Considering the following property of Fourier Cosine Transform

$$F_c\{H(x+a) - H(x-a);\ \xi\} = \sqrt{\frac{2}{\pi}}\frac{\sin(a\xi)}{\xi} \tag{6.83}$$

the Fourier Cosine Transform of Eq. (6.82) yields:

$$\xi^4 W_c(\xi) + \lambda W_c(\xi) = \frac{q_o}{EI_z}\sqrt{\frac{2}{\pi}}\frac{\sin(a\xi)}{\xi} \tag{6.84}$$

or

$$W_c(\xi) = \frac{q_o}{EI_z}\sqrt{\frac{2}{\pi}}\frac{\sin(a\xi)}{\xi(\xi^4 + \lambda)} \tag{6.85}$$

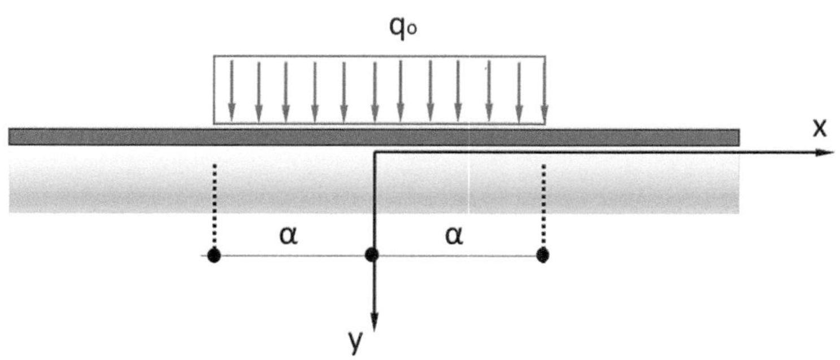

FIG. 6.19

Infinite beam on elastic foundation under uniformly distributed load.

Inversion of the above Fourier Cosine Transform provides the solution of Eq. (6.82):

$$w(x) = F_c^{-1}\{W_c(\xi); x\} = \left(\frac{2}{\pi}\right)^{3/2} \frac{q_o}{EI_z} \int_0^{\infty} \frac{\sin(a\xi)}{\xi(\xi^4 + \lambda)} \cos(x\xi) d\xi \tag{6.86}$$

Inversion of the above Fourier Cosine Transform was carried out by the software Mathematica. Then, the following result was obtained:

$$w(x) = \frac{q_o}{EI_z} \frac{1}{2\lambda} \left(\frac{(a-x)\left(1 - e^{-\frac{\sqrt{(a-x)^2}}{\sqrt{2}\left(\frac{1}{\lambda}\right)^{1/4}}} \cos\left[\frac{\sqrt{(a-x)^2}}{\sqrt{2}\left(\frac{1}{\lambda}\right)^{1/4}}\right]\right)}{\sqrt{(a-x)^2}} + \frac{(a+x)\left(1 - e^{-\frac{\sqrt{(a+x)^2}}{\sqrt{2}\left(\frac{1}{\lambda}\right)^{1/4}}} \cos\left[\frac{\sqrt{(a+x)^2}}{\sqrt{2}\left(\frac{1}{\lambda}\right)^{1/4}}\right]\right)}{\sqrt{(a+x)^2}} \right) \tag{6.87}$$

EXERCISE 5

An infinite beam of flexural stiffness $EI_z = 150 \times 10^3$ Pam^3 is lying on an elastic foundation (Fig. 6.20) with $k = 10^4$ kN/m^3. The beam is subjected to a uniform load $q_o = 10$ kN/m acting on the area $-3.0 < x < 3.0m$. Determine the deflection and the bending moment distribution within a distance $-6.0 < x < 6.0m$.

Solution

The value of the parameter λ is:

$$\lambda = \frac{k}{EI_z} = 66.6$$

With the aid of Eq. (6.87), the following deformed shape of the beam is obtained: The bending moment distribution can be determined with the following equation:

$$\frac{M_z(x)}{EI_z} = w''(x)$$

EXERCISE 5—CONT'D

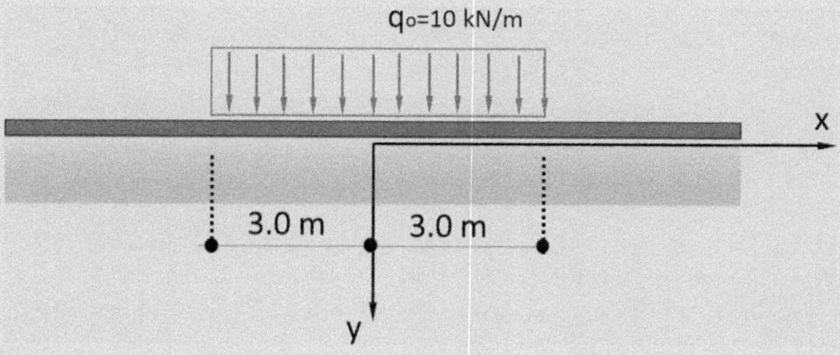

FIG. 6.20

Infinite beam on elastic foundation under uniform load in the area $-3.0 < x < 3.0m$.

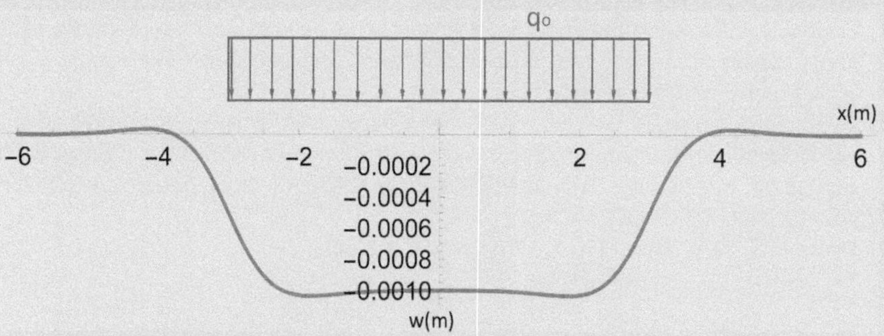

Taking the second derivative of Eq. (6.87), the following result can be obtained:

$$\frac{M_z(x)}{EI_z} = -\frac{q_o}{EI_z}\frac{1}{2\sqrt{\lambda}}\, e^{-\frac{\sqrt{(a-x)^2}+\sqrt{(a+x)^2}}{\sqrt{2}\left(\frac{1}{\lambda}\right)^{1/4}}}\left(e^{\frac{\sqrt{(a+x)^2}}{\sqrt{2}\left(\frac{1}{\lambda}\right)^{1/4}}} \sin\left[\frac{a-x}{\sqrt{2}\left(\frac{1}{\lambda}\right)^{1/4}}\right] + e^{\frac{\sqrt{(a-x)^2}}{\sqrt{2}\left(\frac{1}{\lambda}\right)^{1/4}}} \sin\left[\frac{a+x}{\sqrt{2}\left(\frac{1}{\lambda}\right)^{1/4}}\right]\right)$$

Graphical representation of the above distribution is shown in the following figure.

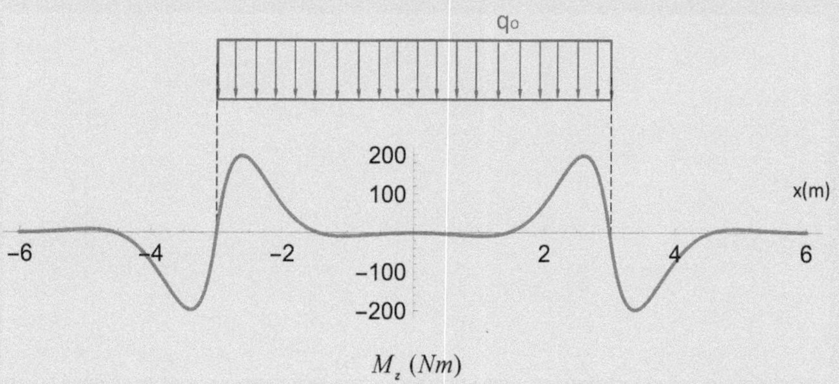

REFERENCES

[1] Fenner RT. Engineering Elasticity. Ellis Horwood Ltd; 1986.
[2] Boresi AP, Schmidt RJ. Advanced Mechanics of Materials. Wiley; 2003.
[3] Timoshenko SP, Goodier JN. Theory of Elasticity. McGraw-Hill; 1951.
[4] Timoshenko SP, Woinowsky-Krieger S. Theory of plates and shells. New York: McGraw-Hill; 1959.
[5] Hetenyi M. Beams on Elastic Foundation. Ann Arbor: The University of Michigan Press; 1961.
[6] Sneddon IN. The use of Integral Transforms. New York: McGraw-Hill; 1972.
[7] Sneddon IN. Fourier Transforms. New York: Dover Publications; 1995.
[8] Fryba L. Vibration of solids and structures under moving loads. Thomas Telford; 1999.
[9] Pavlou DG. Essentials of the finite element method. Academic Press; 2015.
[10] Pavlou DG. Green's function for a pre-stressed thin plate on an elastic foundation under axisymmetric loading. Engineering Analysis with Boundary Elements 2005;29(5).
[11] Pavlou D, et al. An analytical solution of the annular plate on elastic foundation. Structural Engineering and Mechanics 2005;Volume 20(Number 2):209–23. https://doi.org/10.12989/sem.2005.20.2.209
[12] Pavlou D. Elastodynamic analysis of a thin layer bonded on a visco-elastic medium under combined in-plane and pulse lateral loads. Mechanics research communications 2011;Volume 38(Issue 8):546–52.
[13] Pavlou DG. Boundary element analysis of a plate on an elastic foundation loaded by a twisted annular rigid stamp. Proceedings of the Institution of Mechanical Engineers, Part J: Journal of Engineering Tribology 2004;218(6):549–56. https://journals.sagepub.com/doi/10.1243/1350650042794761
[14] Wolfram S. MATHEMATICA. USA: Wolfram Media; 2003.

CHAPTER

Torsion Theory of Beams

7

Chapter Outline

Unlike bending, the torsion of prismatic bars causes warping of the cross sections (Fig. 7.1) that makes the determination of stresses more complex. Let's assume a bar of an arbitrary cross section under torsion (Fig. 7.2a).

A reference point A with coordinates (x,y) within the cross section is moved to the location A' (Fig. 7.2b). If we denote by θ the angle of rotation per unit length, the angle of rotation of the segment OA to the new position OA' is θz, and the coordinates of the point A' are $[r\cos(\varphi+\theta z), r\sin(\varphi+\theta z)]$. Therefore the corresponding displacements u, v in the x, and y directions are:

$$u = r\cos(\phi + \theta z) - x \tag{7.1}$$

$$v = r\sin(\phi + \theta z) - y \tag{7.2}$$

The trigonometric terms of Eqs. (7.1) and (7.2) can be transformed with the aid of the following trigonometric identities. Then, Eqs. (7.1) and (7.2) yield:

$$\cos(\phi + \theta z) = \cos(\phi)\cos(\theta z) - \sin(\phi)\sin(\theta z) \tag{7.3}$$

Advanced Mechanics of Solids and Structures. DOI: http://doi.org/10.1016/B978-0-443-23778-2.00011-1

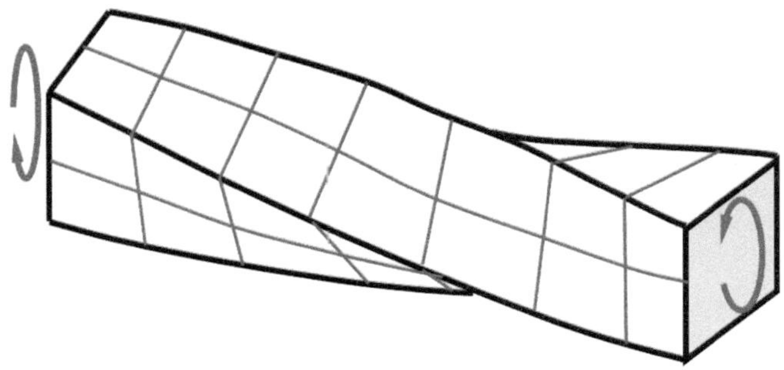

FIG. 7.1

Warping of the cross sections of prismatic bars under torsion.

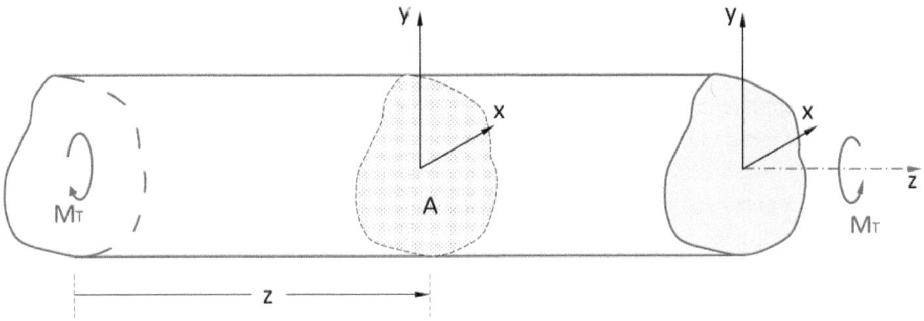

FIG. 7.2A

Beam of arbitrary cross section under torsion.

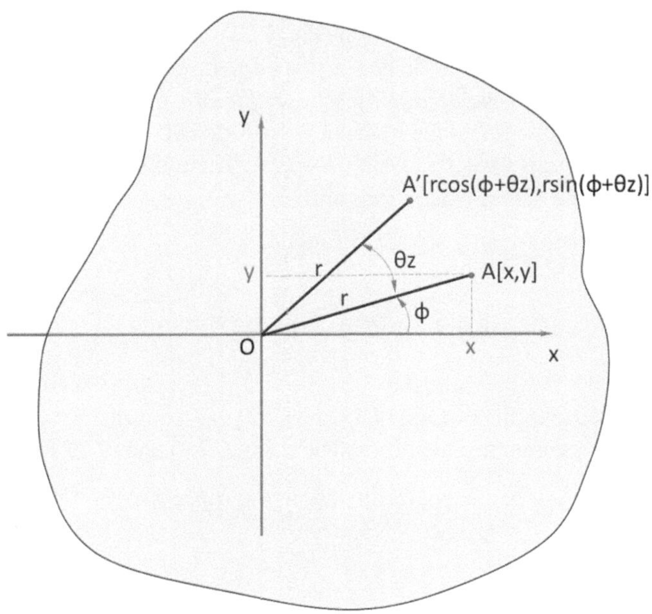

FIG. 7.2B

Deformations on an arbitrary cross section of a beam under torsion.

$$\sin(\phi + \theta z) = \sin(\phi)\cos(\theta z) + \cos(\phi)\sin(\theta z) \tag{7.4}$$

Because θz is small, the simplifications $\cos(\theta z) = 1$ and $\sin(\theta z) = \theta z$ can be used. Then, Eqs. (7.3) and (7.4) can be transformed into

$$\cos(\phi + \theta z) = \cos(\phi) - \theta z \sin(\phi) \tag{7.5}$$

$$\sin(\phi + \theta z) = \sin(\phi) + \theta z \cos(\phi) \tag{7.6}$$

Because $\cos(\phi) = x/r$ and $\sin(\phi) = y/r$ (see Fig. 7.2b), the above equations can be simplified:

$$\cos(\phi + \theta z) = \frac{1}{r}(x - \theta z y) \tag{7.7}$$

$$\sin(\phi + \theta z) = \frac{1}{r}(y + \theta z x) \tag{7.8}$$

With the aid of the above equations, Eqs. (7.1) and (7.2) yield:

$$u = -\theta z y \tag{7.9}$$

$$v = \theta z x \tag{7.10}$$

The warping w of the cross section can be defined as

$$w = \theta\, \Omega(x, y) \tag{7.11}$$

where the unknown function $\Omega(x, y)$ should be determined.

Let's now recall the strain-displacement relationships:

$$\varepsilon_{xx} = \frac{\partial u}{\partial x} \tag{7.12}$$

$$\varepsilon_{yy} = \frac{\partial v}{\partial y} \tag{7.13}$$

$$\varepsilon_{xx} = \frac{\partial w}{\partial z} \tag{7.14}$$

$$\gamma_{xy} = \frac{\partial u}{\partial y} + \frac{\partial v}{\partial x} \tag{7.15}$$

$$\gamma_{yz} = \frac{\partial v}{\partial z} + \frac{\partial w}{\partial y} \tag{7.16}$$

$$\gamma_{xz} = \frac{\partial u}{\partial z} + \frac{\partial w}{\partial x} \tag{7.17}$$

Replacing the displacements given in Eqs. (7.9)–(7.11), the above equations yield:

$$\varepsilon_{xx} = \varepsilon_{yy} = \varepsilon_{zz} = \gamma_{xy} = 0 \tag{7.18}$$

$$\gamma_{xz} = \theta\left(\frac{\partial \Omega}{\partial x} - y\right) \tag{7.19}$$

$$\gamma_{yz} = \theta\left(\frac{\partial \Omega}{\partial y} + x\right) \tag{7.20}$$

Then, the corresponding stress components can be provided by the following equations:

$$\begin{aligned}
\sigma_x &= \frac{E}{(1+\nu)(1-2\nu)}\left[(1-\nu)\varepsilon_x + \nu\left(\varepsilon_y + \varepsilon_z\right)\right] = 0 \\
\sigma_y &= \frac{E}{(1+\nu)(1-2\nu)}\left[(1-\nu)\varepsilon_y + \nu\left(\varepsilon_x + \varepsilon_z\right)\right] = 0 \\
\sigma_z &= \frac{E}{(1+\nu)(1-2\nu)}\left[(1-\nu)\varepsilon_z + \nu\left(\varepsilon_x + \varepsilon_y\right)\right] = 0
\end{aligned} \tag{7.21}$$

$$\begin{aligned}
\tau_{xy} &= G\gamma_{xy} = 0 \\
\tau_{xz} &= G\gamma_{xz} = \theta\left(\frac{\partial \Omega}{\partial x} - y\right) \\
\tau_{yz} &= G\gamma_{yz} = \theta\left(\frac{\partial \Omega}{\partial y} + x\right)
\end{aligned} \tag{7.22}$$

The above stress components should satisfy the following equilibrium equations of stresses

$$\frac{\partial \sigma_{xx}}{\partial x} + \frac{\partial \tau_{xy}}{\partial y} + \frac{\partial \tau_{xz}}{\partial z} + B_x = 0 \tag{7.23}$$

$$\frac{\partial \tau_{yx}}{\partial x} + \frac{\partial \sigma_{yy}}{\partial y} + \frac{\partial \tau_{yz}}{\partial z} + B_y = 0 \tag{7.24}$$

$$\frac{\partial \tau_{zx}}{\partial x} + \frac{\partial \tau_{zy}}{\partial y} + \frac{\partial \sigma_{zz}}{\partial z} + B_z = 0 \tag{7.25}$$

Replacing the results $\sigma_{xx} = \sigma_{yy} = \sigma_{zz} = \tau_{xy} = 0$ into the above equations, and neglecting the body forces B_x, B_y, and B_z, the following equations can be obtained:

$$\frac{\partial \tau_{zx}}{\partial z} = 0 \tag{7.26}$$

$$\frac{\partial \tau_{zy}}{\partial z} = 0 \tag{7.27}$$

$$\frac{\partial \tau_{zx}}{\partial x} + \frac{\partial \tau_{zy}}{\partial y} = 0 \tag{7.28}$$

We assume a function $\Phi(y, y)$ such that

$$\tau_{xz} = \frac{\partial \Phi}{\partial y} \tag{7.29}$$

Then, with the aid of Eq. (7.28), the following formula for the stress component τ_{yz} can be obtained:

$$\tau_{yz} = -\frac{\partial \Phi}{\partial x} \tag{7.30}$$

With the aid of Eqs. (7.29) and (7.30), Eq. (7.22) yields:

$$\frac{\partial \Phi}{\partial y} = G\theta\left(\frac{\partial \Omega}{\partial x} - y\right) \tag{7.31}$$

$$\frac{\partial \Phi}{\partial x} = -G\theta\left(\frac{\partial \Omega}{\partial y} + x\right) \tag{7.32}$$

Taking the derivatives of the above equations, the first with respect to y and the second with respect to x, and eliminating the term $\partial^2\Omega/\partial x\partial y$, the following equation can be obtained:

$$\frac{\partial^2\Phi}{\partial x^2} + \frac{\partial^2\Phi}{\partial y^2} = -2G\theta \tag{7.33}$$

Let's now recall the stress components σ_{Px}, σ_{Py}, and σ_{Pz} to any plane ABC with orientation defined by the direction cosines l, m, and n (Fig. 7.3).

$$\begin{Bmatrix} \sigma_{Px} \\ \sigma_{Py} \\ \sigma_{Pz} \end{Bmatrix} = \begin{bmatrix} \sigma_{xx} & \tau_{xy} & \tau_{xz} \\ \tau_{yx} & \sigma_{yy} & \tau_{yz} \\ \tau_{zx} & \tau_{zy} & \sigma_{zz} \end{bmatrix} \begin{Bmatrix} l \\ m \\ n \end{Bmatrix} \tag{7.34}$$

If we choose the plane ABC to be an element of the lateral surface of a bar under torsion (Fig. 7.4a), where n cos $(\pi/2) = 0$ and $\sigma_{Pz} = 0$, the following result can be obtained by Eq. (7.34):

$$\tau_{zx} l + \tau_{zy} m = 0 \tag{7.35}$$

With the aid of Fig. 7.4b, the following formulae can be derived:

$$l = \cos(\theta_x) = \frac{-dy}{ds} \tag{7.36}$$

$$l = \cos(\theta_y) = \frac{dx}{ds} \tag{7.37}$$

With the aid of Eqs. (7.29), (7.30), (7.36), and (7.37), Eq. (7.35) yields:

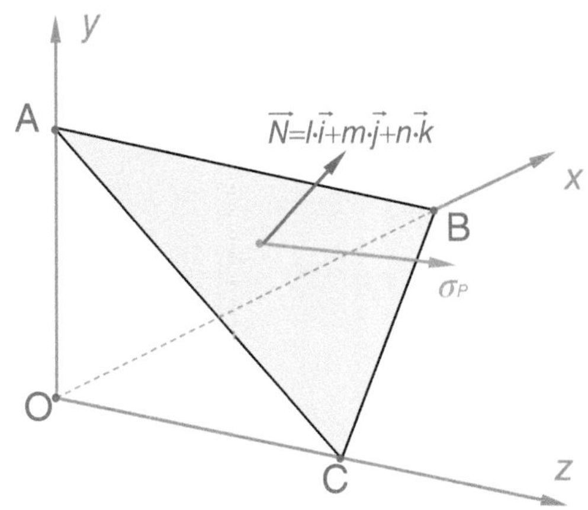

FIG. 7.3

Stress components σ_{Px}, σ_{Py}, and σ_{Pz} to any plane ABC with orientation defined by the direction cosines *l*, *m*, and *n*.

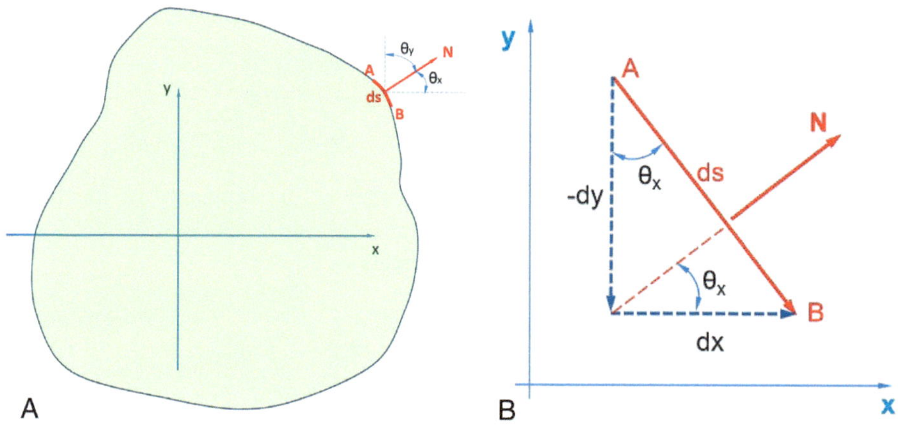

FIG. 7.4

(a) The unit vector $\overrightarrow{N}$ expresses the orientation of any infinitesimal element AB in the lateral surface. (b) Details of the infinitesimal element AB.

$$-\frac{\partial \Phi}{\partial y}\frac{dy}{ds} - \frac{\partial \Phi}{\partial x}\frac{dx}{ds} = 0 \tag{7.38}$$

or

$$\frac{d\Phi}{ds} = 0 \tag{7.39}$$

The above equation means that the function Φ is constant on the lateral surface of the beam

$$\Phi = const \tag{7.40}$$

that is, $\tau_{zx} = \partial\Phi/\partial y = 0$, $\tau_{zy} = -\partial\Phi/\partial x = 0$, we can consider

$$\Phi = 0 \ \ on \ \ the \ \ lateral \ \ surface \tag{7.41}$$

The torsional moment (torque) M_t on a cross section is given by the following equation:

$$M_t = \int x\,\tau_{zy} dA - \int y\,\tau_{zx} dA \tag{7.42}$$

With the aid of Eqs. (7.29) and (7.30), the above equation yields:

$$M_t = -\int x \frac{\partial \Phi}{\partial x} dA - \int y \frac{\partial \Phi}{\partial y} dA = -\iint x \frac{\partial \Phi}{\partial x} dxdy - \iint y \frac{\partial \Phi}{\partial y} dxdy \tag{7.43}$$

For any strip with infinitesimal thickness dy, the following formula can be obtained:

$$\iint x \frac{\partial \Phi}{\partial x} dxdy = dy \int x \frac{\partial \Phi}{\partial x} dx = dy \int_{\Phi(x_1)}^{\Phi(x_2)} x d\Phi \tag{7.44}$$

where x_1 and x_2 are the x-coordinates of the strip at the boundary of the beam (Fig. 7.5).

Integration by parts of the above integral yields:

$$\iint x \frac{\partial \Phi}{\partial x} dxdy = dy \int_{\Phi(x_1)}^{\Phi(x_2)} x d\Phi = dy \left[x \left. \Phi \right|_{x_1}^{x_2} - \int_{x_1}^{x_2} \Phi dx \right] \tag{7.45}$$

Taking into account Eq. (7.41), that means $\Phi(x)$ is equal to zero on the boundary, that is, $\Phi(x_1) = \Phi(x_2) = 0$ the term $\left. \Phi \right|_{x_1}^{x_2}$ is

$$\left. \Phi \right|_{x_1}^{x_2} = \Phi(x_2) - \Phi(x_1) = 0$$

and Eq. (7.45) can be written:

$$\iint x \frac{\partial \Phi}{\partial x} dxdy = -dy \int_{x_1}^{x_2} \Phi dx \tag{7.46}$$

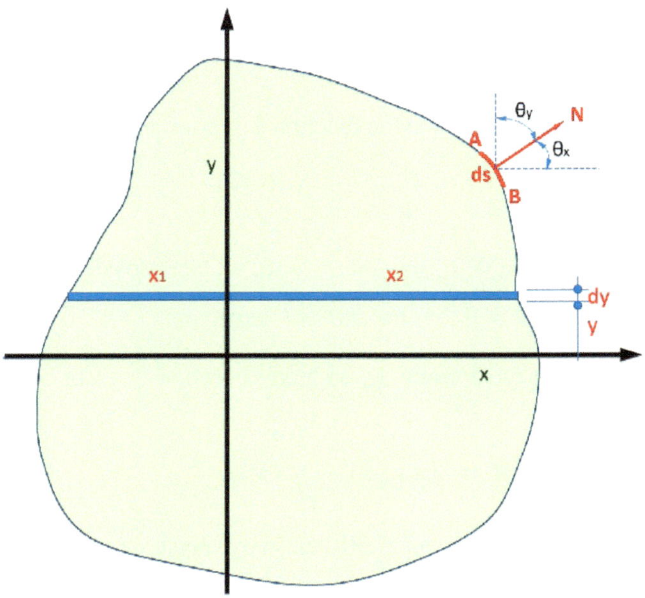

FIG. 7.5

Strip of infinitesimal thickness *dy* on a cross section.

Following a similar procedure, the following result can be obtained:

$$\iint y \frac{\partial \Phi}{\partial y} dxdy = -dx \int_{y_1}^{y_2} \Phi dy \tag{7.47}$$

Summing the integrals for all the strips, and considering Eqs. (7.46) and (7.47), Eq. (7.43) yields:

$$M_t = 2 \iint \Phi dxdy \tag{7.48}$$

7.1 TORSION OF BEAMS WITH ELLIPTICAL CROSS SECTION

Let's consider a beam with elliptical cross section (Fig. 7.6) under torsion with a torque M_t. Our aim is the calculation of the torsion-induced stresses on the cross sections with

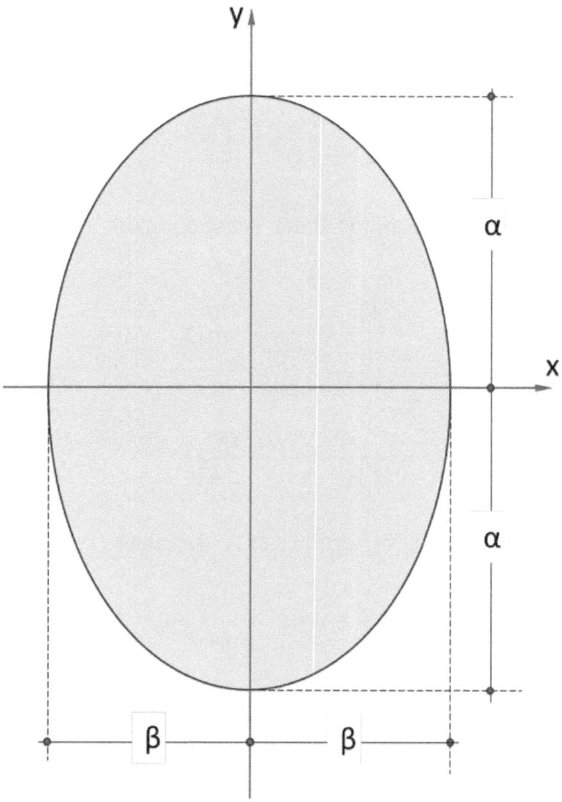

FIG. 7.6

Elliptic cross section.

the aid of Eqs. (7.29) and (7.30). To this end, a formula for the stress function Φ is necessary. According to Eq. (7.41), Φ should be null on the lateral surface of the beam.

Because the equation of the ellipsoidal contour surrounding the elliptical cross section is

$$\frac{x^2}{b^2} + \frac{y^2}{a^2} = 1 \tag{7.49}$$

any function

$$\Phi = \lambda\left(\frac{x^2}{\beta^2} + \frac{y^2}{a^2} - 1\right) \tag{7.50}$$

yields $\Phi = 0$ in the surrounding contour and thus satisfies Eq. (7.41).

Replacing Eq. (7.50) into Eq. (7.33), the following result can be obtained:

$$\lambda = -\frac{a^2\beta^2 G\theta}{a^2 + \beta^2} \tag{7.51}$$

Then,

$$\Phi = -\frac{a^2\beta^2 G\theta}{a^2 + \beta^2}\left(\frac{x^2}{\beta^2} + \frac{y^2}{a^2} - 1\right) \tag{7.52}$$

Utilizing the above equation, the shear stress components on the elliptical cross section are

$$\tau_{zx} = \frac{\partial\Phi}{\partial y} = -\frac{2\beta^2 G\theta}{a^2 + \beta^2}y \tag{7.53}$$

$$\tau_{zy} = -\frac{\partial\Phi}{\partial x} = \frac{2a^2 G\theta}{a^2 + \beta^2}x \tag{7.54}$$

With the aid of Eq. (7.52), the torque can be calculated by Eq. (7.48):

$$\begin{aligned} M_t &= -\frac{a^2\beta^2 G\theta}{a^2 + \beta^2}\left(\frac{2}{\beta^2}\int x^2 dA + \frac{2}{a^2}\int y^2 dA - 2\int dA\right) \\ &= -\frac{a^2\beta^2 G\theta}{a^2 + \beta^2}\left(\frac{2}{\beta^2}I_y + \frac{2}{a^2}I_x - 2A\right) \end{aligned} \tag{7.55}$$

where I_y and I_x are the second moments of inertia for the elliptical cross section, given by the following formulae:

$$I_x = \int y^2 dA = \frac{\pi}{4}\beta a^3 \tag{7.56}$$

$$I_y = \int x^2 dA = \frac{\pi}{4}a\beta^3 \tag{7.57}$$

Then, Eq. (7.55) yields:

$$M_t = -\frac{\pi a^3 \beta^3 G\theta}{a^2 + \beta^2} \tag{7.58}$$

For engineering applications, it is convenient to express the stresses τ_{xz} and τ_{yz} in terms of the applied torque M_t. To this end, the rotation angle θ can be calculated by Eq. (7.58),

$$\theta = -\frac{a^2 + \beta^2}{\pi a^3 \beta^3 G} M_t \tag{7.59}$$

and be replaced in Eqs. (7.53) and (7.54), yielding:

$$\tau_{zx} = \frac{2M_t}{\pi a^3 \beta} y \tag{7.60}$$

$$\tau_{zy} = -\frac{2M_t}{\pi a \beta^3} x \tag{7.61}$$

By using the notation J for the geometric constant

$$J = \frac{\pi a^3 \beta^3}{a^2 + \beta^2} \tag{7.62}$$

Eq. (7.58) can be written in the following form:

$$M_t = -GJ\theta \tag{7.63}$$

The term GJ is called torsional rigidity.

7.2 TORSION OF BEAMS WITH CIRCULAR CROSS SECTION

For a beam of circular cross section (Fig. 7.7) under torsion, the formulae for stresses and rotation angle can be easily obtained by the solution of the beam of elliptical cross section.

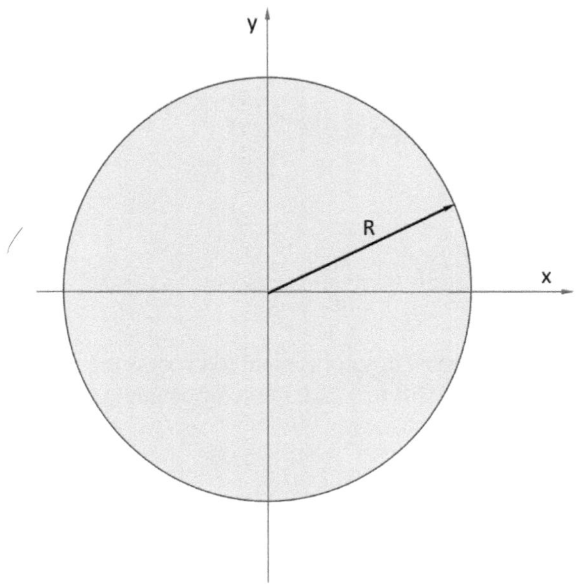

FIG. 7.7

Circular cross section.

If we set $a = \beta = R$, Eqs. (7.59)–(7.61) yield:

$$\theta = -\frac{M_t}{GJ_p} \tag{7.64}$$

$$\tau_{zx} = \frac{M_t}{J_p} y \tag{7.65}$$

$$\tau_{zy} = -\frac{M_t}{J_p} x \tag{7.66}$$

where

$$J_p = \frac{\pi R^4}{2} \tag{7.67}$$

is the polar moment of inertia for a circular cross section of radius R.

Obviously, the shear stress vector lies on the plane-z, that is the surface of the cross section. The shear stress vector can be specified with the aid if the shear stress components:

$$\vec{\tau} = \tau_{zx}\vec{i} + \tau_{zy}\vec{j} \tag{7.68}$$

or

$$\vec{\tau} = \frac{M_t}{J_p} y\, \vec{i} - \frac{M_t}{J_p} x\, \vec{j} \tag{7.69}$$

The magnitude of the vector $\vec{\tau}$ is:

$$\tau = \sqrt{\left(\frac{M_t}{J_p} y\right)^2 + \left(\frac{M_t}{J_p} x\right)^2} = \frac{M_t}{J_p}\sqrt{y^2 + x^2} = \frac{M_t}{J_p} r \tag{7.70}$$

The shear stress for a bar of circular cross section takes the maximum value on $r = R$:

$$\tau_{\max} = \frac{M_t}{J_p} R \tag{7.71}$$

7.3 THE PRANDTL ANALOGY

In the previous sections we demonstrated the determination of torsion-induced stresses by the solution of Eq. (7.33). This equation is analogous with the partial differential equation of the displacements $u_z(x, y)$ of an elastic membrane of the same shape as the cross section, under lateral pressure (Prandtl 1903).

The vertical component of the force per unit length N on the side AB of an elementary surface of the membrane (Fig. 7.8) is:

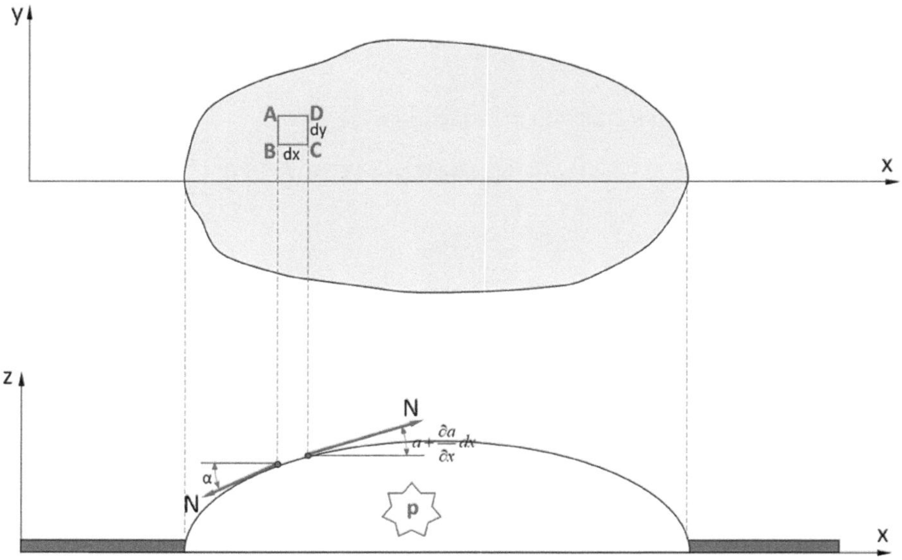

FIG. 7.8

Deformation of elastic membrane under lateral pressure p.

$$V_{AB} = -N\ dy\ \sin(a) \tag{7.72}$$

Because the slope a is small, the following simplification can be used:

$$\sin(a) = \tan(a) = \frac{\partial u_z}{\partial x} \tag{7.73}$$

and

$$\sin\left(a + \frac{\partial a}{\partial x}dx\right) = \tan\left(a + \frac{\partial a}{\partial x}dx\right) = \frac{\partial}{\partial x}\left(u_z + \frac{\partial u_z}{\partial x}dx\right) \tag{7.74}$$

Therefore:

$$V_{AB} = -N\ dy\frac{du_z}{dx} \tag{7.75}$$

Following the same procedure, the following results can be obtained:

$$V_{DC} = N\ dy\frac{d}{dx}\left(u_z + \frac{\partial u_z}{\partial x}dx\right) \tag{7.76}$$

$$V_{AD} = -N\ dx\frac{du_z}{dy} \tag{7.77}$$

$$V_{BC} = N\ dx\frac{d}{dy}\left(u_z + \frac{\partial u_z}{\partial y}dy\right) \tag{7.78}$$

Then, the equilibrium equation in z-direction

$$\sum V_z = V_{AB} + V_{BC} + V_{CD} + V_{AD} + p\ dx\ dy = 0 \tag{7.79}$$

yields

$$\frac{\partial^2 u_z}{\partial x^2} + \frac{\partial^2 u_z}{\partial y^2} = -\frac{p}{N} \tag{7.80}$$

This equation has analogy with Eq. (7.33), with the following analogous quantities:

$$u_z \to \Phi \tag{7.81}$$

$$\frac{p}{N} \to 2G\theta \tag{7.82}$$

7.4 TORSION OF A BEAM OF THIN RECTANGULAR CROSS SECTION

With the aid of the Prandtl analogy, an approximate solution for the torsion of a beam of thin rectangular cross section (Fig. 7.9) can easily be obtained. Fig. 7.10 shows the deformed shape of a membrane with the dimensions of the thin rectangular cross section of Fig. 7.9. Assuming that the displacement u_z is constant along the axis x, and parabolic along the axis y, the following function can be adopted:

$$u_z = u_{z0}\left(1 - \frac{y^2}{b^2}\right) \tag{7.83}$$

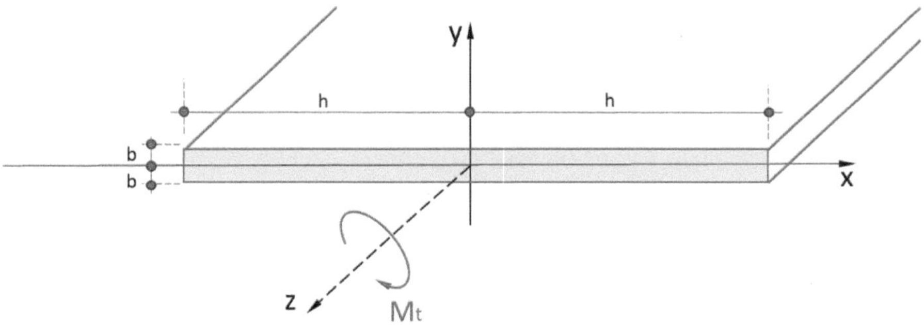

FIG. 7.9

Beam of a thin rectangular cross section under torsion.

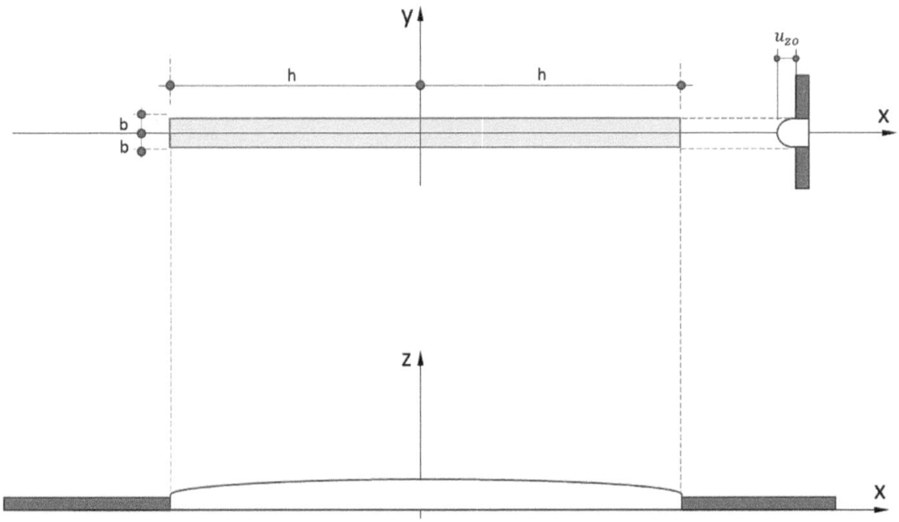

FIG. 7.10

Deformed membrane for a narrow rectangular cross section.

where u_{zo} is the maximum displacement at $y = 0$.

If we try the solution of Eq. (7.83) into Eq. (7.80), the following result can be obtained:

$$-\frac{2u_{zo}}{b^2} = -\frac{p}{N} \tag{7.84}$$

or

$$u_{zo} = \frac{b^2}{2}\frac{p}{N} \tag{7.85}$$

Then, Eq. (7.83) yields:

$$u_z = \frac{b^2}{2}\frac{p}{N}\left(1 - \frac{y^2}{b^2}\right) \tag{7.86}$$

With the aid of Eqs. (7.81) and (7.82), the following analogous solution can be used:

$$\Phi = b^2 G\theta\left(1 - \frac{y^2}{b^2}\right) \tag{7.87}$$

With the aid of the above equation, the stress components can now be obtained by Eqs. (7.29) and (7.30):

$$\tau_{zx} = \frac{\partial\Phi}{\partial y} = -2G\theta y \tag{7.88}$$

$$\tau_{zy} = -\frac{\partial\Phi}{\partial x} = 0 \tag{7.89}$$

The angle of twist θ can be correlated with the applied torque M_t with the aid of Eq. (7.48). With the aid of Eq. (7.87), Eq. (7.48) yields:

$$M_t = 2\int_{-h}^{h}\int_{-b}^{b}\Phi dxdy = \frac{1}{3}G\theta(2h)(2b)^3 \tag{7.90}$$

Using the following notation for the geometric parameters of the cross section

$$I_o = \frac{1}{3}(2h)(2b)^3 \tag{7.91}$$

the following formula can provide the angle of twist:

$$\theta = \frac{M_t}{GI_o} \tag{7.92}$$

Then, the stress component τ_{zx} can be correlated to the torque M_t by the following equation:

$$\tau_{zx} = -2\frac{M_t}{I_o}y \tag{7.93}$$

Obviously, the maximum value of the shear stress take place on $y = \pm b$. Therefore

$$\tau_{\max} = \pm 2\frac{M_t}{I_o}b \tag{7.94}$$

It should be reminded that the torsion analysis of thin rectangular cross section is based on the Prandtl membrane analogy under the assumption that the displacement u_z is constant along the axis x and parabolic along the axis y (Fig. 7.10). Therefore the derived formulae are approximate. Eqs. (7.92)–(7.94) provide accurate results for $h > 10b$.

7.5 TORSION OF BEAM OF COMPOSITE CROSS SECTION (CROSS SECTIONS CONSISTING OF THIN RECTANGLES) AND THIN-WALL, CURVED OPEN CROSS SECTION

The results of the analysis of the previous section can be utilized for the torsion analysis of bars of composite cross section (cross sections consisting of thin rectangles), as shown in Fig. 7.11. If a cross section consists of n-thin rectangles, the parameter of the entire cross section can be calculated by summing the corresponding I_i of the individual rectangles:

$$I_o = \sum_{i=1}^{n} I_i = \frac{1}{3}\sum_{i=1}^{n}(2h_i)(2b_i)^3 \tag{7.95}$$

With the aid of the above equation, the angle of twist and the maximum shear stress can be calculated by Eqs. (7.92) and (7.94). As it has been mentioned, these results provide sufficient accuracy for $h_i > 10b_i$. If one or more thin rectangles composing the cross section does not satisfy the last condition, that is, $h_i < 10b_i$, the following formula for I_o is recommended:

$$I_o = \sum_{i=1}^{n} I_i = 0.9\frac{1}{3}\sum_{i=1}^{n}(2h_i)(2b_i)^3 \tag{7.96}$$

Eqs. (7.91), (7.92), and (7.94) can also be used for calculating the angle of twist and the maximum shear stress in thin-wall, curved, open, thin cross section. In this case, $2h$ is the length of the circumference of the cross section and $2b$ is the thickness (Fig. 7.12).

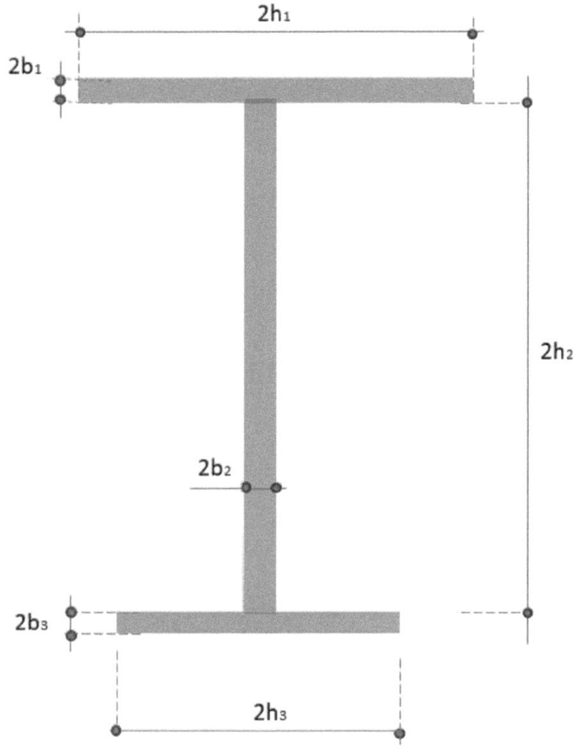

FIG. 7.11

Composite cross section.

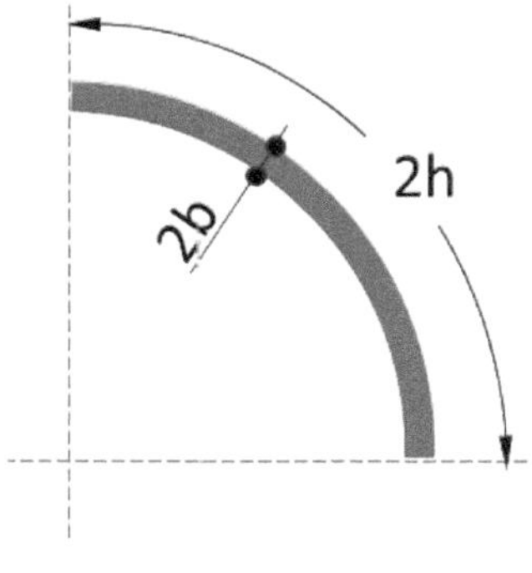

FIG. 7.12

Curved, open, thin, cross section.

7.6 TORSION OF BEAMS OF RECTANGULAR CROSS SECTION

The solution of thin rectangular cross section is based on the Prandtl membrane analogy under the assumption that the displacement u_z is constant along the axis x and parabolic along the axis y (Fig. 7.10). Eqs. (7.92)–(7.94) are valid for slender rectangular cross sections where $h > 10b$. If we discard the last assumption, the solution of Eq. (7.33) for nonslender bars of rectangular cross section (Fig. 7.13) is quite complex. Timoshenko provided in 1934 (*Theory of Elasticity*, McGraw Hill, 1970) the following solution that is valid for any h/b ratio:

$$\Phi = \frac{32G\theta b^2}{\pi^3} \sum_{n=1,3,5,\ldots}^{\infty} \frac{1}{n^3}(-1)^{(n-1)/2}\left(1 - \frac{\cosh(n\pi y/2b)}{\cosh(n\pi h/2b)}\right)\sin\frac{n\pi x}{2b} \tag{7.97}$$

Based on this solution, the following formulae for the angle of twist and the maximum shear stress have been obtained:

$$\theta = \frac{M_t}{\mu G(2b)^3(2h)} \tag{7.98}$$

$$\tau_{\max} = \frac{M_t}{\zeta(2b)^2(2h)} \tag{7.99}$$

The coefficients μ, ζ can be approximated for $1 \leq h/b \leq 10$ by the following formulae:

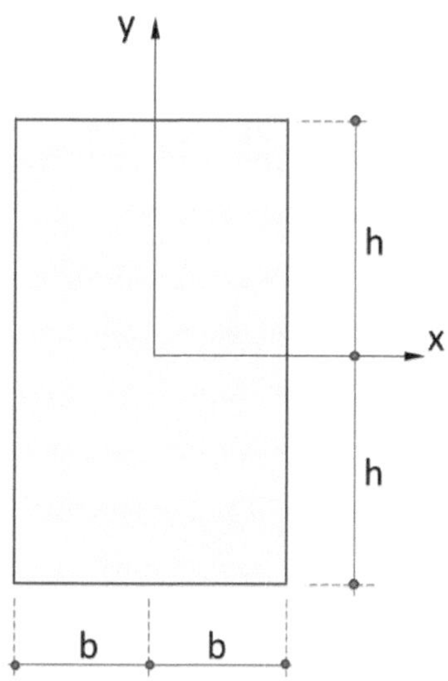

FIG. 7.13

Nonslender beam of rectangular cross section under torsion.

$$\mu = 0.10517 + 0.06166\left(\frac{h}{b}\right) - 0.00412\left(\frac{h}{b}\right)^2 \tag{7.100}$$

$$\zeta = 0.18445 + 0.03248\left(\frac{h}{b}\right) - 0.00198\left(\frac{h}{b}\right)^2 \tag{7.101}$$

7.7 HOLLOW THIN-WALL BEAMS UNDER TORSION

Let's assume a beam of a thin-wall, hollow cross section. The shear stress and the angle of twist versus the applied torque can be determined with the aid of Prandtl membrane theory (Fig. 7.14).

As it has already been mentioned, the displacement u_z is constant along the thin contour of the membrane and varies only in the direction perpendicular to the contour (Figs. 7.10 and 7.14).

If we replace the symbols y and x (Fig. 7.10) by the symbols n and ξ in the local coordinate system (Fig. 7.14), the following equation can be obtained

$$\tau_{z\xi} = \frac{\partial \Phi}{\partial n} \tag{7.102}$$

Let's denote by N the force per unit length. With the aid of Eqs. (7.81) and (7.82), the stress function Φ can be correlated with the membrane displacement u_z:

$$\frac{u_z}{(p/N)} = \frac{\Phi}{2G\theta} \tag{7.103}$$

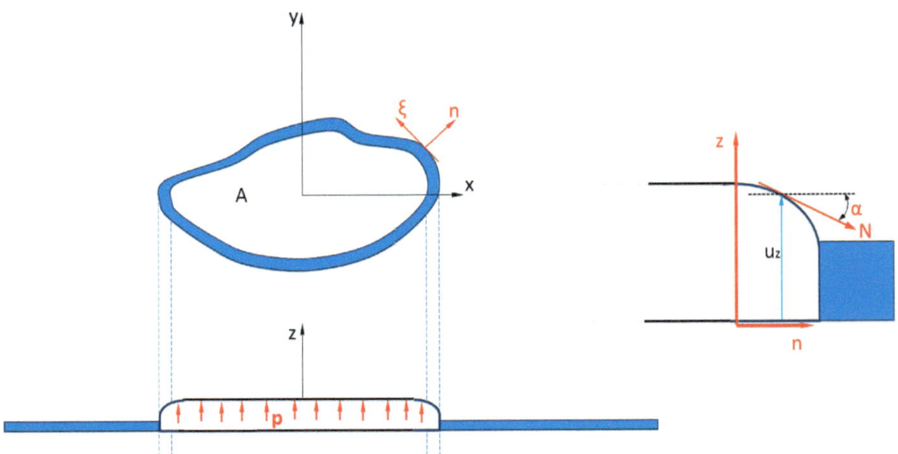

FIG. 7.14

Membrane for hollow, thin-wall cross section under torsion.

or

$$\Phi = \frac{2G\theta N}{p} u_z \tag{7.104}$$

With the aid of the last equation, Eq. (7.102) yields:

$$\tau_{z\xi} = \frac{\partial \Phi}{\partial n} = \frac{2G\theta N}{p} \frac{\partial u_z}{\partial n} \tag{7.105}$$

According to Fig. 7.14, the derivative $\partial u_z / \partial n$ expresses the slope of the membrane in the direction n. Then,

$$\tau_{z\xi} = \frac{\partial \Phi}{\partial n} = \frac{2G\theta N}{p} \tan(\alpha) \tag{7.106}$$

Because the slope a is small, the above equation can be transformed to:

$$\tau_{z\xi} = \frac{2G\theta N}{p} \sin(\alpha) \tag{7.107}$$

or

$$N \sin(\alpha) = \frac{p}{2G\theta} \tau_{z\xi} \tag{7.108}$$

The resultant of all the vertical components $(N\, dl)\sin(\alpha)$ should be equal to the pressure-induced force on the membrane:

$$\oint N \sin(\alpha) d\xi = pA \tag{7.109}$$

Considering Eq. (7.108), the above equation yields:

$$\frac{p}{2G\theta} \oint \tau_{z\xi} d\xi = pA \tag{7.110}$$

or

$$\theta = \frac{1}{2GA} \oint \tau_{z\xi} d\xi \tag{7.111}$$

It has been mentioned that the displacement u_z is constant along the contour of a thin-wall membrane (Fig. 7.10). Considering Eq. (7.81), the stress function Φ is also constant and independent of ξ. If we use the term of the shear flow $q = \tau_{z\xi} t$, Eq. (7.111) can be written as:

$$\theta = \frac{q}{2GA} \oint \frac{d\xi}{t(\xi)} \tag{7.112}$$

If the thickness is stepwise constant, the above equation can be written in the following form:

$$\theta = \frac{q}{2GA}\left(\frac{\xi_1}{t_1} + \frac{\xi_2}{t_2} + \cdots\right) \tag{7.113}$$

The contribution in the torque from an elementary length $d\xi$ of the contour of the cross section (Fig. 7.15) is:

$$dM_t = \left(\tau_{z\xi} t d\xi \cos\gamma\right)x + \left(\tau_{z\xi} t d\xi \sin\gamma\right)y \tag{7.114}$$

Integration of the above equation along the contour yields:

$$M_t = \tau_{z\xi} t \oint (x\cos\gamma + y\sin\gamma)d\xi \tag{7.115}$$

Because

$$\begin{aligned} d\xi\cos\gamma &= dy \\ d\xi\sin\gamma &= -dx \end{aligned} \tag{7.116}$$

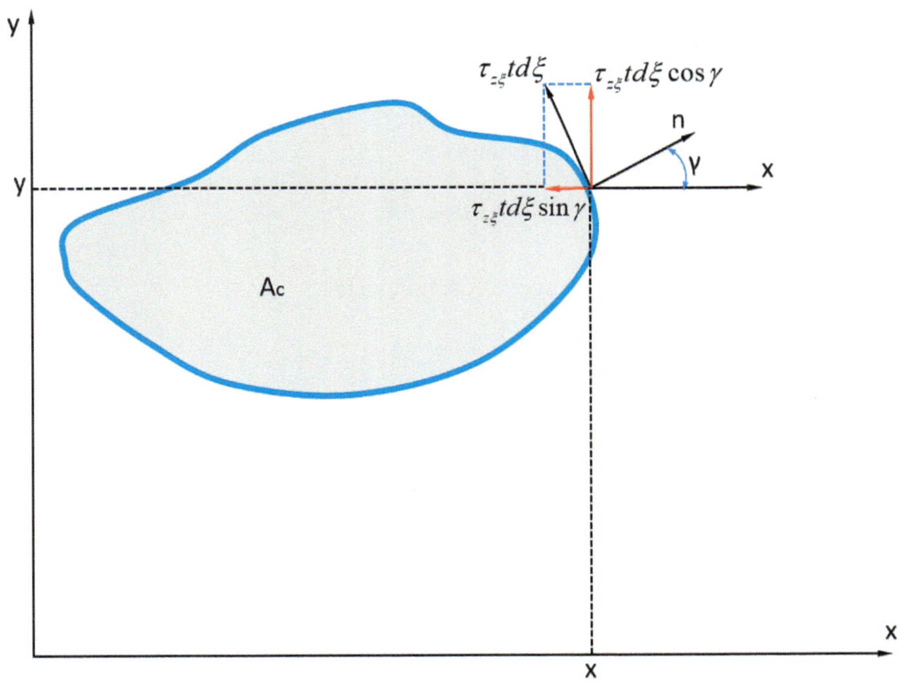

FIG. 7.15

Contribution of an elementary length $d\xi$ of the contour in the torque.

the following result can be obtained by Eq. (7.115):

$$M_t = \tau_{z\xi} t \oint (xdy - ydx) = \tau_{z\xi} t [A - (-A)] = 2A\tau_{z\xi} t \tag{7.117}$$

where A is the area surrounded by the contour of the cross section.

In terms of the shear flow, the above equation can be written as:

$$M_t = 2Aq \tag{7.118}$$

or

$$q = \tau_{z\xi} t = \frac{M_t}{2A} \tag{7.119}$$

With the aid of Eq. (7.112), the above equation can be correlated with θ:

$$M_t = 2A \frac{2GA}{\oint \frac{d\xi}{t(\xi)}} \theta = GJ\theta \tag{7.120}$$

where

$$J = \frac{4A^2}{\oint \frac{d\xi}{t(\xi)}} \tag{7.121}$$

7.8 BEAM OF A CROSS SECTION OF SEVERAL HOLLOW THIN-WALL COMPARTMENTS

The above equations can be utilized to calculate the torsion-induced shear flows q_i and the angle of twist per unit length θ in multicompartment thin-wall cross sections (e.g., Fig. 7.16). For a cross section of N compartments, Eqs. (7.112) and (7.118) can be adjusted accordingly:

$$\theta = \frac{1}{2GA_i} \oint (q_i - q_n) \frac{d\xi}{t} = \frac{1}{2GA_i} \sum_{i=1}^{N} (q_i - q_n) \frac{\xi_n}{t_n} \quad i = 1,2,\ldots, N \tag{7.122}$$

$$T = 2\sum_{i=1}^{N} A_i q_i \tag{7.123}$$

where: q_i $i = 1,2,\ldots, N$ is the shear flow in each compartment of area A_i. q_n is the shear flow of the compartment adjusted to the $i - th$ compartment where the common wall of length ξ_n is located, and t_n is the corresponding thickens. For the walls

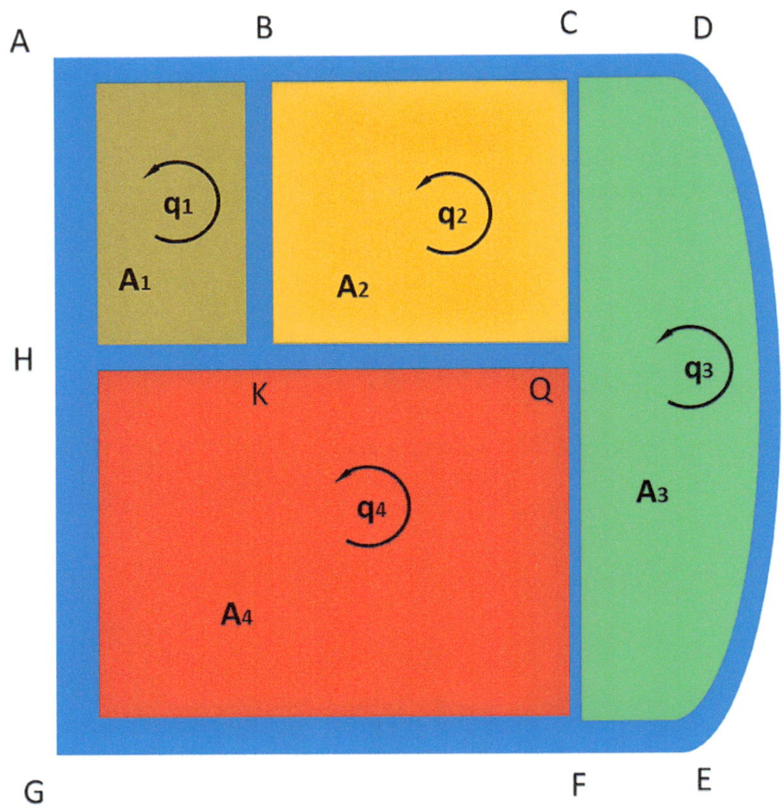

FIG. 7.16

Multicompartment thin-wall cross section under torsion.

on the exterior boundary the value $q_\eta = 0$ should be used. For better understanding, Eqs. (7.122) and (7.123) can be implemented in the cross section of Fig. 7.16, consisting of four compartments. The shear flow on the walls of each compartment is constant, that is, q_1, q_2, q_3, and q_4. Usually, the torque M_t is given and the five quantities θ, q_1, q_2, q_3, and q_4 are the unknowns to be calculated.

With the aid of Eqs. (7.122) and (7.123), the following system of five equations can be formulated:

$$M_t = 2\left(A_1 q_1 + A_2 q_2 + A_3 q_3 + A_4 q_4\right) \tag{7.124}$$

$$\theta = \frac{1}{2GA_1}\left(q_1 \frac{AH}{t_{AH}} + q_1 \frac{AB}{t_{AB}} + (q_1 - q_2)\frac{BK}{t_{BK}} + (q_1 - q_4)\frac{HK}{t_{HK}}\right) \tag{7.125}$$

$$\theta = \frac{1}{2GA_2}\left((q_2 - q_1)\frac{BK}{t_{BK}} + q_2\frac{BC}{t_{BC}} + (q_1 - q_3)\frac{CQ}{t_{CQ}} + (q_2 - q_4)\frac{KQ}{t_{KQ}}\right) \quad (7.126)$$

$$\theta = \frac{1}{2GA_3}\left(q_3\frac{CDEF}{t_{CDEF}} + (q_3 - q_2)\frac{CQ}{t_{CQ}} + (q_3 - q_4)\frac{QF}{t_{QF}}\right) \quad (7.127)$$

$$\theta = \frac{1}{2GA_4}\left(q_4\frac{HG}{t_{HG}} + (q_4 - q_1)\frac{HK}{t_{HK}} + (q_4 - q_2)\frac{KQ}{t_{KQ}} + (q_4 - q_3)\frac{QF}{t_{QF}} + q_4\frac{GF}{t_{GF}}\right) \quad (7.128)$$

Solution of the system of Eqs. (7.124)–(7.128) provides the values of θ, q_1, q_2, q_3, and q_4.

EXERCISE 1

An electric motor transfers power to a productive machine (Fig. 7.17). The maximum transferred power in the machine is *P* (Watt) under rotation frequency *RPM* (rotations per minute). The power in the machine is transferred via a compact shaft of circular cross section of radius *R*. The yield stress of the shaft material is *Sy* and the allowable shear stress is $\tau y = Sy/2$. The safety factor against yielding of the shaft is *SF*. Assuming that the bending of the shaft is negligible, calculate the minimum radius of the circular cross section of the shaft against yielding.

Solution

The maximum shear stress of a shaft under torsion can be calculated by Eq. (7.71):

$$\tau_{\max} = \frac{M_t}{J_p}R$$

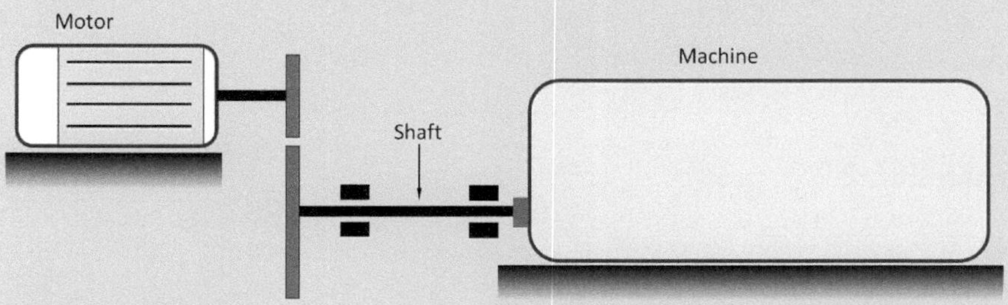

FIG. 7.17

Mechanical installation.

(Continued)

EXERCISE 1—CONT'D

where J_p is the polar moment of inertia of the circular cross section (Eq. 7.67):

$$J_p = \frac{\pi R^4}{2}$$

Therefore:

$$\tau_{\max} = \frac{2M_t}{\pi R^3}$$

The condition for avoiding yielding is $\tau_{\max} < \tau_y$, or

$$\frac{2M_t}{\pi R^3} < \frac{S_y/2}{SF}$$

The torque M_t can be calculated from the power P and the rotation frequency f by the formula

$$P(Watt) = M_t(Nm) \times 2\pi f\,(s^{-1})$$

or, in terms of RPM:

$$P(Watt) = M_t(Nm) \times \frac{2\pi}{60} RPM$$

Thus:

$$M_t(Nm) = \frac{60P(Watt)}{2\pi RPM}$$

Then, the following result can be obtained by the condition for avoiding yielding:

$$\frac{2\left(\frac{60P}{2\pi RPM}\right)}{\pi R^3} < \frac{S_y/2}{SF}$$

or

$$R > \sqrt[3]{\frac{120 \cdot P \cdot SF}{\pi^2 \cdot S_y \cdot RPM}}$$

EXERCISE 2

A hollow, thin-wall shaft of circular cross section (Fig. 7.18) is subjected to a torque M_t. Calculate the shear stress on the cross section and the angle of twist per unit length θ.

Solution

The shear flow q can be calculated by Eq. (7.119):

$$q = \tau_{z\xi} t = \frac{M_t}{2A}$$

EXERCISE 2—CONT'D

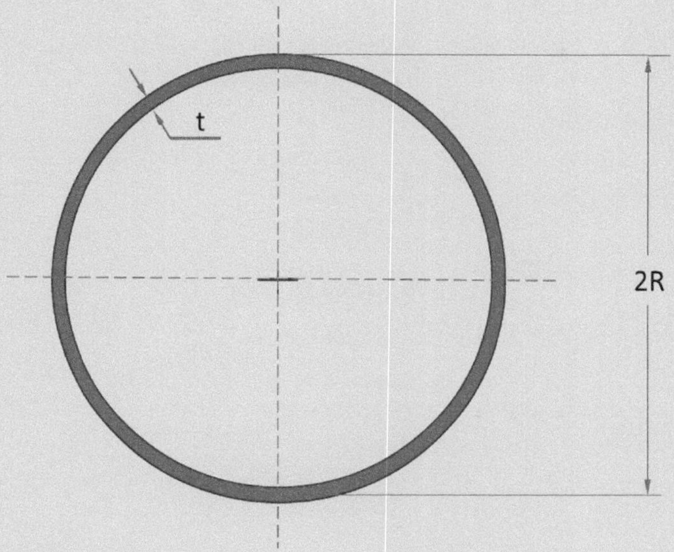

FIG. 7.18

Hollow bar of thin-wall circular cross section under torsion.

$$\text{where } A = \pi R^2$$

Then,

$$\tau_{z\xi} = \frac{M_t}{2\pi R^2 t}$$

The angle of twist per unit length θ can be calculated with the aid of Eq. (7.120):

$$M_t = GJ\theta$$

where (Eq. 7.121)

$$J = \frac{4A^2}{\oint \frac{d\xi}{t(\xi)}} = \frac{4(\pi R^2)^2}{\frac{2\pi R}{t}} = 2t\pi R^3$$

Then, Eq. (7.120) yields:

$$\theta = \frac{M_t}{2t\pi G R^3}$$

EXERCISE 3

A hollow, thin-wall beam of rectangular cross section (Fig. 7.19) is subjected to a torque M_t. Calculate the shear stress on the cross section and the angle of twist per unit length θ.

Solution

The shear flow q on the wall of the cross section can be calculated by Eq. (7.119):

$$q = \tau_{z\xi} t = \frac{M_t}{2A}$$

$$\text{where } A = (2B)(2H) = 4BH$$

Then,

$$\tau_{z\xi} = \frac{M_t}{8BH}$$

The angle of twist per unit length θ can be calculated with the aid of Eq. (7.120):

$$M_t = GJ\theta$$

where (Eq. 7.121)

$$J = \frac{4A^2}{\oint \frac{d\xi}{t(\xi)}} = \frac{4(4BH)^2}{\frac{2B+2H+2B+2H}{t}} = \frac{16B^2H^2t}{B+H}$$

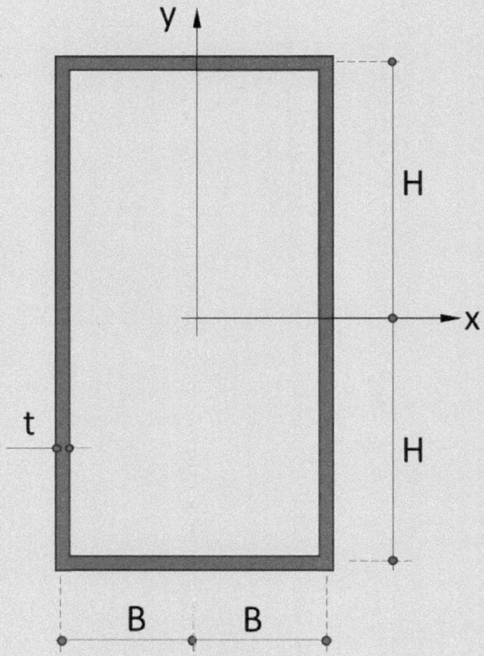

FIG. 7.19

Hollow thin-wall rectangular cross section under torsion.

EXERCISE 3—CONT'D

Then, Eq. (7.120) yields:

$$\theta = \frac{M_t(B+H)}{16GB^2H^2t}$$

EXERCISE 4

The cross section of a thin-wall web-flange beam (Fig. 7.20) is subjected to a torque M_t. Both the web and the flanges are slender, that is, $L_W > 10\ t_W$ and$L_f > 10\ t_f$. Calculate the maximum shear stress on the cross section and the angle of twist per unit length θ.

Solution

The angle of twist per unit length θ and the maximum shear stress on the cross section can be calculated by Eqs. (7.92) and (7.94):

$$\theta = \frac{M_t}{GI_o}$$

$$\tau_{\max} = \pm 2\frac{M_t}{I_o}b$$

where I_o is given by Eq. (7.95)

$$I_o = \sum_{i=1}^{n} I_i = \frac{1}{3}\sum_{i=1}^{n}(2h_i)(2b_i)^3$$

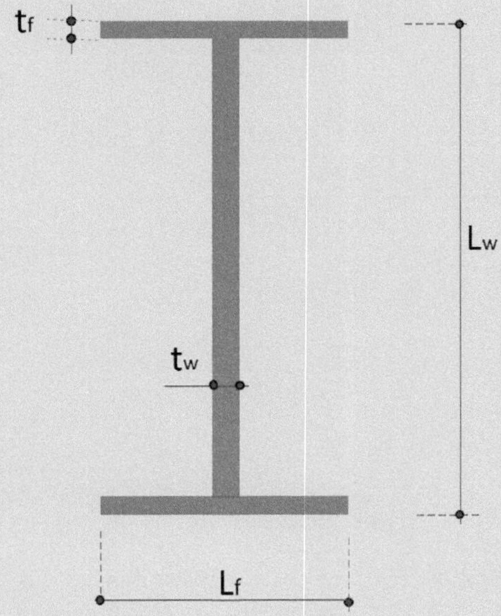

FIG. 7.20

A thin-wall web-flange cross section under torsion.

(Continued)

EXERCISE 4—CONT'D

With the aid of the above equations, the following results can be obtained for the cross section of Fig. 7.20:

$$I_o = \sum_{i=1}^{2} I_i = \frac{1}{3}\left(L_W t_W{}^3 + L_f t_f{}^3\right)$$

$$\theta = \frac{3M_t}{G\left(L_W t_W{}^3 + L_f t_f{}^3\right)}$$

$$\tau_{\max} = \max\left\{\frac{5M_t}{\left(L_W t_W{}^3 + L_f t_f{}^3\right)} t_W, \frac{6M_t}{\left(L_W t_W{}^3 + L_f t_f{}^3\right)} t_f\right\}$$

EXERCISE 5

A cantilever beam with the two-compartment cross section shown in Fig. 7.21 is subjected to torsion. The length of the beam is $L = 2.0\,\text{m}$. All the walls of the cross section have the same thickness, $t = 2\,\text{mm}$. Determine: (a) the maximum torque against yielding and (b) the angle of twist at the free end of the beam. The material has shear modulus $G = 25\,\text{GPa}$ and yield stress $Y = 240\,\text{MPa}$.

Solution

With the aid of Eqs. (7.122) and (7.123), the following system of five equations can be formulated:

$$M_t = 2\left(A_1 q_1 + A_2 q_2\right) = 2\left[(40 \times 70)q_1 + \frac{\pi 30^2}{2} q_2\right]10^{-6} = \left(5.6q_1 + 2.83q_2\right)10^{-3}$$

$$\theta = \frac{1}{2GA_1}\left(q_1 \frac{AB + ED + AE}{t} + (q_1 - q_2)\frac{BD}{t}\right) =$$

$$\frac{1}{2 \times 25 \times 10^9 \times (70 \times 40)10^{-6}}\left(q_1 \frac{40 + 40 + 70}{2} + (q_1 - q_2)\frac{70}{2}\right)$$

$$= \left(7.857q_1 - 2.5q_2\right) \times 10^{-7}$$

$$\theta = \frac{1}{2GA_2} q_2\left(\frac{BCD + DB}{t}\right)$$

$$= \frac{1}{2 \times 25 \times 10^9 \times \left(\frac{30^2\pi}{2}\right)10^{-6}}$$

$$\left(q_2 \frac{30\pi}{2} + \left(q_2 - q_1\right)\frac{70}{2}\right)$$

$$= \left(11.162q_2 - 4.951q_1\right) \times 10^{-7}$$

Dividing by parts the last two equations we can eliminate θ. Then we get:

$$7.857\, q_1 - 2.5\, q_2 = 11.162\, q_2 - 4.951\, q_1$$

EXERCISE 5—CONT'D

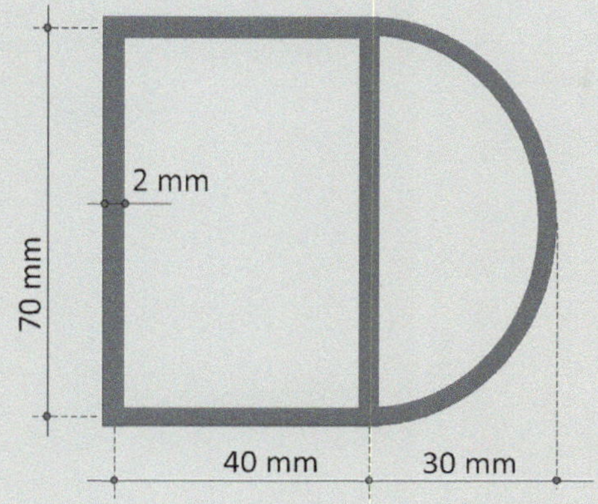

FIG. 7.21

Two-compartment cross section under torsion.

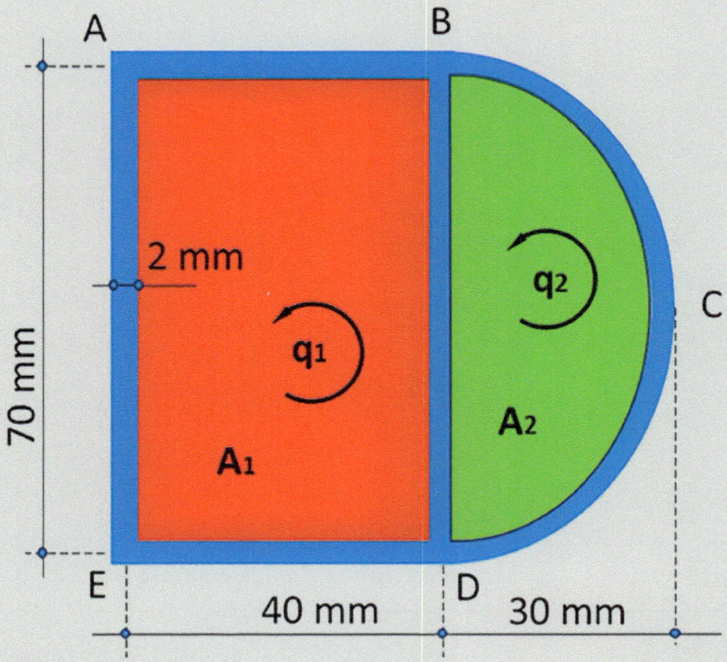

FIG. 7.22

Shear flows in the two compartments.

(Continued)

EXERCISE 5—CONT'D

or

$$q_1 = 0.676\, q_2$$

Then, the first equation yields:

$$M_t = 6.6156 \times 10^{-3} q_2$$

or

$$q_2 = 151.158\, M_t$$

Then,

$$q_1 = 0.676\, q_2 = 102.18\, M_t$$

The corresponding shear stresses in the two compartments are:

$$\tau_1 = \frac{q_1}{t} = 51090 M_t$$

$$\tau_2 = \frac{q_2}{t} = 75579 M_t$$

Obviously,

$$\tau_{max} = \max\{\tau_1, \tau_2\} = 75579 M_t$$

The condition for avoiding yielding is

$$\tau_{max} < \frac{Y}{2}$$

or

$$M_t < 1587 \;\; Nm$$

The angle of twist $\varphi = \theta L$ of the end cross section can be calculated by the following equation:

$$\theta = \frac{\varphi}{L} = \left(7.857 q_1 - 2.5 q_2\right) \times 10^{-7} = (7.857 \times 102.18 - 2.5 \times 151.158) \times 1587 \times 10^{-7} = 0.0674$$

or

$$\varphi = \theta L = 0.0674 \times 2 = 0.135 \;\; rad$$

REFERENCES

[1] Timoshenko SP, Goodier JN. Theory of Elasticity. McGraw-Hill; 1951.
[2] Fenner RT. Engineering Elasticity. Ellis Horwood Ltd; 1986.
[3] Boresi AP, Schmidt RJ. Advanced Mechanics of Materials. Wiley; 2003.
[4] Pavlou DG. Essentials of the finite element method. Academic Press; 2015.

[5] Pavlou DG. Boundary element analysis of a plate on an elastic foundation loaded by a twisted annular rigid stamp. Proceedings of the Institution of Mechanical Engineers, Part J: Journal of Engineering Tribology 2004;218(6):549–56. https://journals.sagepub.com/doi/10.1243/1350650042794761

[6] Pavlou D. Boundary-integral equation analysis of twisted internally cracked axisymmetric bimaterial elastic solids. Computational Mechanics 2002;29:254–64. https://doi.org/10.1007/s00466-002-0338-7

[7] Pavlou, DG. An exact solution for transient wave propagation in an infinite layer under dynamic torsion. Mechanics Research Communications 2010; Volume 37(Issue 4):372–6. https://doi.org/10.1016/j.mechrescom.2010.04.008

[8] Ugural Ansel, Fenster Saul. Advanced Mechanics of Materials and Applied Elasticity. Pearson; 2019.

[9] Solecki Roman, Conant RJay. Advanced Mechanics of Materials. Oxford University Press; 2003.

CHAPTER

8 Energy Methods

Chapter Outline

In statically indeterminate structures, the equilibrium equations are not adequate for determining the unknown forces, that is, the support reactions or/and the forces of the structural members. The set of the equilibrium equations should be completed with more equations. The compatibility equations of displacements and the equilibrium equations of forces are sufficient to analyze statically indetermined structures. However, in complex structures the compatibility equations of displacements are usually not obvious. Because displacements are vectors, it is almost impossible to predict the direction of the displacements of the joints of complex trusses or frames (e.g., offshore structures, buildings, aerospace structures). Unlike the compatibility equations of displacements, the energy methods are advantageous because energy is a scalar quantity and includes both forces and displacements. In the present chapter, attention will be paid to the principle of virtual work, minimum potential energy, and Castigliano's theorem. To this end, necessary background about the law of conservation of energy and its consequences in mechanics will first be provided.

Advanced Mechanics of Solids and Structures. DOI: http://doi.org/10.1016/B978-0-443-23778-2.00008-1

8.1 LAW OF CONSERVATION OF ENERGY (FIRST LAW OF THERMODYNAMICS)

According to thermodynamics, the law of conservation of energy, also known as the first law of thermodynamics, states that the energy input in a closed system is equal to the energy output. According to this law, the work performed by external forces δW and the heat flow δH into a closed system is equal to the increase of the internal energy δU and the increase of the kinetic energy δK. This energy balance can be expressed by the following equation:

$$\delta W + \delta H = \delta U + \delta K \tag{8.1}$$

If we assume adiabatic conditions, that is, no heat flow ($\delta H = 0$) and static equilibrium ($\delta K = 0$), which are valid assumptions for most engineering problems, then the first law of thermodynamics can be simplified:

$$\delta W = \delta U \tag{8.2}$$

To calculate the work performed by external forces, we separate it into the work of surface forces W_S and the work of the body forces W_B:

$$\delta W = \delta W_S + \delta W_B \tag{8.3}$$

We can define the increase of internal energy as the integral of the elastic energy density δU_o over the volume:

$$\delta U = \int_V \delta U_o dV \tag{8.4}$$

Then, Eq. (8.2) can be written as:

$$\delta W_S + \delta W_B = \int_V \delta U_o dV \tag{8.5}$$

8.1.2 CORRELATION OF STRESS AND DISPLACEMENT COMPONENTS TO THE ELASTIC ENERGY DENSITY

With the aid of the above equation, the stress and the strain components in an elastic medium can be correlated to the elastic energy density U_o. The corresponding formulae can be used for deriving quantified expressions of the energy theorems.

For a material element such as in Fig. 8.1 under a stress $\vec{\sigma_P}$, the work of the surface forces W_S and the work of the body forces W_B can be provided by the following equations:

$$\delta W_S = \int_S (\sigma_{P_x} \delta u) dS + \int_S (\sigma_{P_y} \delta v) dS + \int_S (\sigma_{P_z} \delta w) dS \tag{8.6}$$

$$\delta W_B = \int_V (B_x \delta u + B_y \delta v + B_z \delta w) dV \tag{8.7}$$

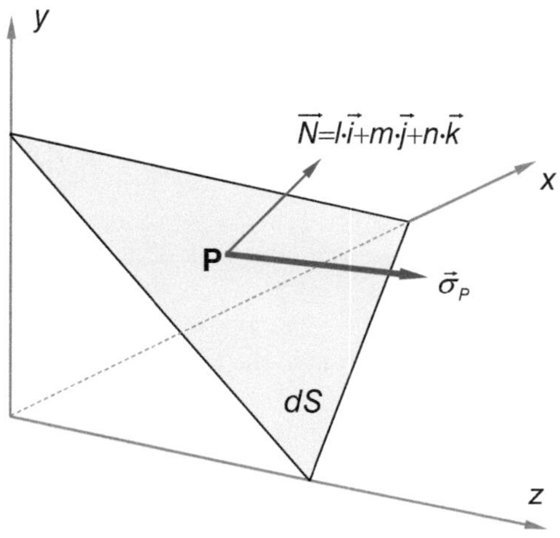

FIG. 8.1

Stress vector $\vec{\sigma}_P$ on an elementary surface dS with arbitrary orientation $\overrightarrow{N} = l\,\vec{i} + m\,\vec{j} + n\,\vec{k}$.

On the other hand, the increase of the elastic energy density can be expressed as follows:

$$\delta U_o = \frac{\partial U_o}{\partial \varepsilon_x}\delta\varepsilon_x + \frac{\partial U_o}{\partial \varepsilon_y}\delta\varepsilon_y + \frac{\partial U_o}{\partial \varepsilon_z}\delta\varepsilon_z + \frac{\partial U_o}{\partial \varepsilon_{xy}}\delta\varepsilon_{xy} + \frac{\partial U_o}{\partial \varepsilon_{xz}}\delta\varepsilon_{xz} + \frac{\partial U_o}{\partial \varepsilon_{yz}}\delta\varepsilon_{yz} \tag{8.8}$$

With the aid of Eqs. (8.6)–(8.8), Eq. (8.5) can be written as:

$$\begin{aligned}&\int_S (\sigma_{Px}\delta u)dS + \int_S (\sigma_{Py}\delta v)dS + \int_S (\sigma_{Pz}\delta w)dS + \int_V (B_x\delta u + B_y\delta v + B_z\delta w)dV = \\ &= \int_V \left(\frac{\partial U_o}{\partial \varepsilon_x}\delta\varepsilon_x + \frac{\partial U_o}{\partial \varepsilon_y}\delta\varepsilon_y + \frac{\partial U_o}{\partial \varepsilon_z}\delta\varepsilon_z + \frac{\partial U_o}{\partial \varepsilon_{xy}}\delta\varepsilon_{xy} + \frac{\partial U_o}{\partial \varepsilon_{xz}}\delta\varepsilon_{xz} + \frac{\partial U_o}{\partial \varepsilon_{yz}}\delta\varepsilon_{yz} \right) dV\end{aligned} \tag{8.9}$$

We recall from the stress theory (Chapter 1) the following equations:

$$\sigma_{Px} = \sigma_{xx}\cdot l + \tau_{xy}\cdot m + \tau_{xz}\cdot n \tag{8.10}$$

$$\sigma_{Py} = \tau_{xy}\cdot l + \sigma_{yy}\cdot m + \tau_{yz}\cdot n \tag{8.11}$$

$$\sigma_{Pz} = \tau_{xz}\cdot l + \tau_{yz}\cdot m + \sigma_{zz}\cdot n \tag{8.12}$$

Utilizing the above equations, Eq. (8.9) yields:

$$\begin{aligned}&\int_S (\sigma_{xx} l + \tau_{xy} m + \tau_{xz} n)\delta u dS + \int_S (\tau_{xy} l + \sigma_{yy} m + \tau_{yz} n)\delta v dS + \\ &+ \int_S (\tau_{xz} l + \tau_{yz} m + \sigma_{zz} n)\delta w dS + \int_V (B_x \delta u + B_y \delta v + B_z \delta w) dV = \\ &= \int_V \left(\frac{\partial U_o}{\partial \varepsilon_x}\delta\varepsilon_x + \frac{\partial U_o}{\partial \varepsilon_y}\delta\varepsilon_y + \frac{\partial U_o}{\partial \varepsilon_z}\delta\varepsilon_z + \frac{\partial U_o}{\partial \varepsilon_{xy}}\delta\varepsilon_{xy} + \frac{\partial U_o}{\partial \varepsilon_{xz}}\delta\varepsilon_{xz} + \frac{\partial U_o}{\partial \varepsilon_{yz}}\delta\varepsilon_{yz} \right) dV\end{aligned} \tag{8.13}$$

In the above equation we can rearrange the terms of the surface integrals:

$$\begin{aligned}&\int_S \left[\left(\sigma_{xx}\delta u + \tau_{xy}\delta v + \tau_{xz}\delta w\right) l + \left(\tau_{xy}\delta u + \sigma_{yy}\delta v + \tau_{yz}\delta w\right) m + \left(\tau_{xz}\delta u + \tau_{yz}\delta v + \sigma_{zz}\delta w\right) n \right] dS + \\ &+ \int_V \left(B_x \delta u + B_y \delta v + B_z \delta w\right) dV \\ &= \int_V \left(\frac{\partial U_o}{\partial \varepsilon_x}\delta\varepsilon_x + \frac{\partial U_o}{\partial \varepsilon_y}\delta\varepsilon_y + \frac{\partial U_o}{\partial \varepsilon_z}\delta\varepsilon_z + \frac{\partial U_o}{\partial \varepsilon_{xy}}\delta\varepsilon_{xy} + \frac{\partial U_o}{\partial \varepsilon_{xz}}\delta\varepsilon_{xz} + \frac{\partial U_o}{\partial \varepsilon_{yz}}\delta\varepsilon_{yz} \right) dV\end{aligned} \tag{8.14}$$

Let's use notations:

$$a_x = \sigma_{xx}\delta u + \tau_{xy}\delta v + \tau_{xz}\delta w \tag{8.15}$$

$$a_y = \tau_{xy}\delta u + \sigma_{yy}\delta v + \tau_{yz}\delta w \tag{8.16}$$

$$a_z = \tau_{xz}\delta u + \tau_{yz}\delta v + \sigma_{zz}\delta w \tag{8.17}$$

Then, Eq. (8.14) can be rewritten:

$$\begin{aligned}&\int_S \left(a_x l + a_y m + a_z n\right) dS + \int_V \left(B_x \delta u + B_y \delta v + B_z \delta w\right) dV = \\ &= \int_V \left(\frac{\partial U_o}{\partial \varepsilon_x}\delta\varepsilon_x + \frac{\partial U_o}{\partial \varepsilon_y}\delta\varepsilon_y + \frac{\partial U_o}{\partial \varepsilon_z}\delta\varepsilon_z + \frac{\partial U_o}{\partial \varepsilon_{xy}}\delta\varepsilon_{xy} + \frac{\partial U_o}{\partial \varepsilon_{xz}}\delta\varepsilon_{xz} + \frac{\partial U_o}{\partial \varepsilon_{yz}}\delta\varepsilon_{yz} \right) dV\end{aligned} \tag{8.18}$$

With the aid of the Gauss divergence theorem (see Appendix) we can convert the surface integral in Eq. (8.18) into a volume integral:

$$\int_S \left(a_x l + a_y m + a_z n\right) dS = \int_V \left(\frac{\partial a_x}{\partial x} + \frac{\partial a_y}{\partial y} + \frac{\partial a_z}{\partial z} \right) dV \tag{8.19}$$

Then, Eq. (8.8) yields:

$$\begin{aligned}&\int_V \left(\frac{\partial a_x}{\partial x} + \frac{\partial a_y}{\partial y} + \frac{\partial a_z}{\partial z} \right) dV + \int_V \left(B_x \delta u + B_y \delta v + B_z \delta w\right) dV = \\ &= \int_V \left(\frac{\partial U_o}{\partial \varepsilon_x}\delta\varepsilon_x + \frac{\partial U_o}{\partial \varepsilon_y}\delta\varepsilon_y + \frac{\partial U_o}{\partial \varepsilon_z}\delta\varepsilon_z + \frac{\partial U_o}{\partial \varepsilon_{xy}}\delta\varepsilon_{xy} + \frac{\partial U_o}{\partial \varepsilon_{xz}}\delta\varepsilon_{xz} + \frac{\partial U_o}{\partial \varepsilon_{yz}}\delta\varepsilon_{yz} \right) dV\end{aligned} \tag{8.20}$$

Taking into account Eqs. (8.15)–(8.17), the following equation can be obtained:

$$\int_V \left[\frac{\partial}{\partial x}\left(\sigma_{xx}\delta u + \tau_{xy}\delta v + \tau_{xz}\delta w\right) + \frac{\partial}{\partial y}\left(\tau_{xy}\delta u + \sigma_{yy}\delta v + \tau_{yz}\delta w\right) + \right.$$
$$\left. + \frac{\partial}{\partial z}\left(\tau_{xz}\delta u + \tau_{yz}\delta v + \sigma_{zz}\delta w\right) + B_x\delta u + B_y\delta v + B_z\delta w \right] dV = \tag{8.21}$$
$$= \int_V \left(\frac{\partial U_o}{\partial \varepsilon_x}\delta\varepsilon_x + \frac{\partial U_o}{\partial \varepsilon_y}\delta\varepsilon_y + \frac{\partial U_o}{\partial \varepsilon_z}\delta\varepsilon_z + \frac{\partial U_o}{\partial \varepsilon_{xy}}\delta\varepsilon_{xy} + \frac{\partial U_o}{\partial \varepsilon_{xz}}\delta\varepsilon_{xz} + \frac{\partial U_o}{\partial \varepsilon_{yz}}\delta\varepsilon_{yz} \right) dV$$

Making the derivatives in the left side of Eq. (8.21) and rearranging the terms, the following equation can be derived:

$$\int_V \left[\left(\frac{\partial \sigma_{xx}}{\partial x} + \frac{\partial \sigma_{yx}}{\partial y} + \frac{\partial \sigma_{zx}}{\partial z} + B_x \right)\delta u + \left(\frac{\partial \sigma_{xy}}{\partial x} + \frac{\partial \sigma_{yy}}{\partial y} + \frac{\partial \sigma_{zy}}{\partial z} + B_y \right)\delta v + \right.$$
$$+ \left(\frac{\partial \sigma_{xz}}{\partial x} + \frac{\partial \sigma_{yz}}{\partial y} + \frac{\partial \sigma_{zz}}{\partial z} + B_z \right)\delta w + \sigma_{xx}\frac{\partial \delta u}{\partial x} + \sigma_{yy}\frac{\partial \delta v}{\partial y} + \sigma_{zz}\frac{\partial \delta w}{\partial z} +$$
$$\left. + \tau_{xy}\left(\frac{\partial \delta u}{\partial y} + \frac{\partial \delta v}{\partial x} \right) + \tau_{xz}\left(\frac{\partial \delta u}{\partial z} + \frac{\partial \delta w}{\partial x} \right) + \tau_{yz}\left(\frac{\partial \delta v}{\partial z} + \frac{\partial \delta w}{\partial y} \right) \right] dV = \tag{8.22}$$
$$= \int_V \left(\frac{\partial U_o}{\partial \varepsilon_x}\delta\varepsilon_x + \frac{\partial U_o}{\partial \varepsilon_y}\delta\varepsilon_y + \frac{\partial U_o}{\partial \varepsilon_z}\delta\varepsilon_z + \frac{\partial U_o}{\partial \varepsilon_{xy}}\delta\varepsilon_{xy} + \frac{\partial U_o}{\partial \varepsilon_{xz}}\delta\varepsilon_{xz} + \frac{\partial U_o}{\partial \varepsilon_{yz}}\delta\varepsilon_{yz} \right) dV$$

With the aid of the equilibrium equations from the stress theory (Chapter 1)

$$\frac{\partial \sigma_{xx}}{\partial x} + \frac{\partial \tau_{yx}}{\partial y} + \frac{\partial \tau_{zx}}{\partial z} + B_x = 0 \tag{8.23}$$

$$\frac{\partial \tau_{xy}}{\partial x} + \frac{\partial \sigma_{yy}}{\partial y} + \frac{\partial \tau_{zy}}{\partial z} + B_y = 0 \tag{8.24}$$

$$\frac{\partial \tau_{xz}}{\partial x} + \frac{\partial \tau_{yz}}{\partial y} + \frac{\partial \sigma_{zz}}{\partial z} + B_z = 0 \tag{8.25}$$

and the strain-displacement relations:

$$\varepsilon_{xx} = \frac{\partial u}{\partial x} \tag{8.26}$$

$$\varepsilon_{yy} = \frac{\partial v}{\partial y} \tag{8.27}$$

$$\varepsilon_{zz} = \frac{\partial w}{\partial z} \tag{8.28}$$

$$\varepsilon_{xy} = \frac{1}{2}\left(\frac{\partial u}{\partial y} + \frac{\partial v}{\partial x}\right) \tag{8.29}$$

$$\varepsilon_{yz} = \frac{1}{2}\left(\frac{\partial v}{\partial z} + \frac{\partial w}{\partial y}\right) \tag{8.30}$$

$$\varepsilon_{xz} = \frac{1}{2}\left(\frac{\partial u}{\partial z} + \frac{\partial w}{\partial x}\right) \tag{8.31}$$

Eq. (8.22), which expresses the first law of thermodynamics, can be written as follows:

$$\begin{aligned}&\int_V \left(\sigma_x\delta\varepsilon_x + \sigma_y\delta\varepsilon_y + \sigma_z\delta\varepsilon_z + 2\tau_{xy}\delta\varepsilon_{xy} + 2\tau_{yz}\delta\varepsilon_{yz} + 2\tau_{xz}\delta\varepsilon_{xz}\right)dV= \\ &= \int_V \left(\frac{\partial U_o}{\partial\varepsilon_x}\delta\varepsilon_x + \frac{\partial U_o}{\partial\varepsilon_y}\delta\varepsilon_y + \frac{\partial U_o}{\partial\varepsilon_z}\delta\varepsilon_z + \frac{\partial U_o}{\partial\varepsilon_{xy}}\delta\varepsilon_{xy} + \frac{\partial U_o}{\partial\varepsilon_{xz}}\delta\varepsilon_{xz} + \frac{\partial U_o}{\partial\varepsilon_{yz}}\delta\varepsilon_{yz}\right)dV\end{aligned} \tag{8.32}$$

Eq. (8.32) yields the following formulae correlating the stress components with the strain energy density:

$$\sigma_x = \frac{\partial U_o}{\partial\varepsilon_x} \tag{8.33}$$

$$\sigma_y = \frac{\partial U_o}{\partial\varepsilon_y} \tag{8.34}$$

$$\sigma_z = \frac{\partial U_o}{\partial\varepsilon_z} \tag{8.35}$$

$$\tau_{xy} = \frac{1}{2}\frac{\partial U_o}{\partial\varepsilon_{xy}} \tag{8.36}$$

$$\tau_{xz} = \frac{1}{2}\frac{\partial U_o}{\partial\varepsilon_{xz}} \tag{8.37}$$

$$\tau_{yz} = \frac{1}{2}\frac{\partial U_o}{\partial\varepsilon_{yz}} \tag{8.38}$$

In linear elastic materials the stress and strain is correlated by Hooke's law:

$$\sigma_i = E\varepsilon_i \quad i = x, y, z \tag{8.39}$$

$$\tau_{ij} = G\varepsilon_{ij} \quad ij = xy, yz, xz \tag{8.40}$$

Therefore the formulae

$$\sigma_i = \frac{\partial U_o}{\partial \varepsilon_i} \tag{8.41}$$

$$\tau_{ij} = \frac{1}{2}\frac{\partial U_o}{\partial \varepsilon_{ij}} \tag{8.42}$$

can be transformed to the following equations, respectively:

$$E\varepsilon_i = E\frac{\partial U_o}{\partial \sigma_i} \tag{8.43}$$

$$G\varepsilon_{ij} = \frac{1}{2}G\frac{\partial U_o}{\partial \tau_{ij}} \tag{8.44}$$

or

$$\varepsilon_i = \frac{\partial U_o}{\partial \sigma_i} \tag{8.45}$$

$$\varepsilon_{ij} = \frac{1}{2}\frac{\partial U_o}{\partial \tau_{ij}} \tag{8.46}$$

The above equations can be written as follows:

$$\varepsilon_x = \frac{\partial U_o}{\partial \sigma_x} \tag{8.47}$$

$$\varepsilon_y = \frac{\partial U_o}{\partial \sigma_y} \tag{8.48}$$

$$\varepsilon_z = \frac{\partial U_o}{\partial \sigma_z} \tag{8.49}$$

$$\varepsilon_{xy} = \frac{1}{2}\frac{\partial U_o}{\partial \tau_{xy}} \tag{8.50}$$

$$\varepsilon_{xz} = \frac{1}{2}\frac{\partial U_o}{\partial \tau_{xz}} \tag{8.51}$$

$$\varepsilon_{yz} = \frac{1}{2}\frac{\partial U_o}{\partial \tau_{yz}} \tag{8.52}$$

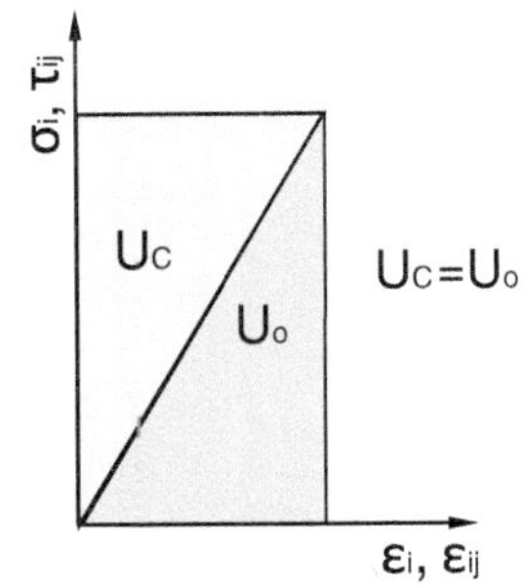

FIG. 8.2

In linear elastic materials, the strain energy density is equal to the complementary strain energy density.

Because the derivation of Eqs. (8.47)–(8.51) is based on Hooke's law, this implies that the strain energy density U_o is equal to the complementary strain energy density U_c (Fig. 8.2).

8.2 THE PRINCIPLE OF VIRTUAL WORK

If virtual displacements are applied on an elastic body subjected to external forces under equilibrium, the work done by the external forces with the virtual displacements is equal to the work done by stresses with strains, that is the variation of the strain energy. This principle is known as the principle of virtual work.

If δu, δv and δw are the virtual displacements of the point P in Fig. 8.1, the work done by the external forces with the virtual displacements can be obtained by the following equation:

$$W = \int_S \left(\sigma_{Px}\delta u + \sigma_{Py}\delta v + \sigma_{Pz}\delta w\right) dS \tag{8.53}$$

With the aid of Eqs. (8.10)–(8.12), the above equation yields:

$$\begin{aligned} W = \int_S \Big[& \left(\sigma_{xx}\cdot l + \tau_{xy}\cdot m + \tau_{xz}\cdot n\right)\delta u + \left(\tau_{xy}\cdot l + \sigma_{yy}\cdot m + \tau_{yz}\cdot n\right)\delta v \\ & + \left(\tau_{xz}\cdot l + \tau_{yz}\cdot m + \sigma_{zz}\cdot n\right)\delta w \Big] dS \end{aligned} \tag{8.54}$$

After some rearrangements of the terms, the above equation can be transformed to the following:

$$W = \int_S \left[a_x l + a_y m + a_z n\right] dS \tag{8.55}$$

In the above equation, the notations a_x, a_y and a_z are given in Eqs. (8.15)–(8.17). With the aid of the Gauss divergence theorem, Eq. (8.55) yields:

$$W = \int_V \left[\frac{\partial a_x}{\partial x} + \frac{\partial a_y}{\partial y} + \frac{\partial a_z}{\partial z}\right] dV \tag{8.56}$$

or

$$W = \int_V \left[\frac{\partial}{\partial x}\left(\sigma_{xx}\delta u + \tau_{xy}\delta v + \tau_{xz}\delta w\right) + \frac{\partial}{\partial y}\left(\tau_{xy}\delta u + \sigma_{yy}\delta v + \tau_{yz}\delta w\right) + \frac{\partial}{\partial z}\left(\tau_{xz}\delta u + \tau_{yz}\delta v + \sigma_{zz}\delta w\right)\right] dV \tag{8.57}$$

Taking the derivatives and rearranging the terms of the above equation, the following result can be obtained:

$$\begin{aligned} W = \int_V \Bigg[& \left(\frac{\partial \sigma_{xx}}{\partial x} + \frac{\partial \tau_{xy}}{\partial y} + \frac{\partial \tau_{xz}}{\partial z}\right)\delta u + \left(\frac{\partial \tau_{xy}}{\partial x} + \frac{\partial \sigma_{yy}}{\partial y} + \frac{\partial \tau_{yz}}{\partial z}\right)\delta v + \\ & + \left(\frac{\partial \tau_{xz}}{\partial x} + \frac{\partial \tau_{yz}}{\partial y} + \frac{\partial \sigma_{zz}}{\partial z}\right)\delta w + \sigma_{xx}\frac{\partial \delta u}{\partial x} + \sigma_{yy}\frac{\partial \delta v}{\partial y} + \sigma_{zz}\frac{\partial \delta w}{\partial z} + \\ & + \tau_{xy}\left(\frac{\partial \delta u}{\partial y} + \frac{\partial \delta v}{\partial x}\right) + \tau_{xz}\left(\frac{\partial \delta u}{\partial z} + \frac{\partial \delta w}{\partial x}\right) + \tau_{yz}\left(\frac{\partial \delta v}{\partial z} + \frac{\partial \delta w}{\partial y}\right)\Bigg] dV \end{aligned} \tag{8.58}$$

With the aid of the equilibrium equations (Eqs. 8.23–8.25) and the strain-displacement relations equations (Eqs. 8.26–8.31), the above equation yields:

$$\begin{aligned} W = \int_V [& -B_x\delta u - B_y\delta v - B_z\delta w + \sigma_{xx}\delta\varepsilon_{xx} + \sigma_{yy}\delta\varepsilon_{yy} + \sigma_{zz}\delta\varepsilon_{zz} + \\ & + 2\tau_{xy}\delta\varepsilon_{xy} + 2\tau_{xz}\delta\varepsilon_{xz} + 2\tau_{yz}\delta\varepsilon_{yz}] dV \end{aligned} \tag{8.59}$$

With the aid of Eq. (8.53), the above equation yields the quantitative expression of the principle of virtual work:

$$\begin{aligned} & \int_S \left(\sigma_{Px}\delta u + \sigma_{Py}\delta v + \sigma_{Pz}\delta w\right) dS + \int_V \left[B_x\delta u + B_y\delta v + B_z\delta w\right] dV = \\ & \int_V \left[\sigma_{xx}\delta\varepsilon_{xx} + \sigma_{yy}\delta\varepsilon_{yy} + \sigma_{zz}\delta\varepsilon_{zz} + 2\tau_{xy}\delta\varepsilon_{xy} + 2\tau_{xz}\delta\varepsilon_{xz} + 2\tau_{yz}\delta\varepsilon_{yz}\right] dV \end{aligned} \tag{8.60}$$

Because only the equilibrium equations (Eqs. 8.23–8.25) were used for the derivation of the above equation, the principle of virtual work is valid to both linear and nonlinear elastic materials. It should be noted that the principle of virtual work is equivalent to the equilibrium equations that are written in an alternative expression. In several cases, the principle of virtual work has more convenient applicability than the equilibrium equations.

8.3 THE PRINCIPLE OF MINIMUM POTENTIAL ENERGY

With the aid of Eqs. (8.33)–(8.38), the principle of virtual work given in Eq. (8.60) can take the following form:

$$\int_S \left(\sigma_{Px}\delta u + \sigma_{Py}\delta v + \sigma_{Pz}\delta w\right) dS + \int_V \left[B_x \delta u + B_y \delta v + B_z \delta w\right] dV =$$

$$\int_V \left[\frac{\partial U_o}{\partial \varepsilon_{xx}}\delta\varepsilon_{xx} + \frac{\partial U_o}{\partial \varepsilon_{yy}}\delta\varepsilon_{yy} + \frac{\partial U_o}{\partial \varepsilon_{zz}}\delta\varepsilon_{zz} + \frac{\partial U_o}{\partial \varepsilon_{xy}}\delta\varepsilon_{xy} + \frac{\partial U_o}{\partial \varepsilon_{xz}}\delta\varepsilon_{xz} + \frac{\partial U_o}{\partial \varepsilon_{yz}}\delta\varepsilon_{yz}\right] dV \tag{8.61}$$

Considering Eq. (8.8), the above equation yields:

$$\int_S \left(\sigma_{Px}\delta u + \sigma_{Py}\delta v + \sigma_{Pz}\delta w\right) dS + \int_V \left[B_x \delta u + B_y \delta v + B_z \delta w\right] dV = \int_V \delta U_o dV \tag{8.62}$$

Therefore, according to Eq. (8.62), the quantity

$$\delta\Pi = \underbrace{-\int_S \left(\sigma_{Px}\delta u + \sigma_{Py}\delta v + \sigma_{Pz}\delta w\right) dS}_{\text{work done by the external forces}} - \underbrace{\int_V \left[B_x \delta u + B_y \delta v + B_z \delta w\right] dV}_{\text{work done by the body forces}} + \underbrace{\int_V \delta U_o dV}_{\text{the change of the elastic energy}} \tag{8.63}$$

is

$$\delta\Pi = 0 \tag{8.64}$$

The quantity Π expresses the balance of the work done by the external forces, the work done by the body forces, and the change of the elastic energy. The quantity Π is called *potential energy* of the system. In Eq. (8.63), the work done by the external and the body forces has a negative sign (–) because it is consumed to increase the elastic energy $U = \int_V \delta U_o dV$.

Because the equilibrium equations of stresses were used for the derivation of Eq. (8.64), this implies that among all possible states of a structural system, those that satisfy the equilibrium equations yield stationary potential energy with respect to small possible displacements. On the other hand, the use of the Eqs. (8.33)–(8.38) implies that the principle of minimum potential energy is valid for linear elastic systems (conservative systems) only.

An alternative expression of Eq. (8.64) that has a convenient applicability to structure under point forces is:

$$\frac{\partial \Pi}{\partial d_i} = 0 \qquad i = 1,2,\ldots, n \tag{8.65}$$

where $d_i \quad i = 1,2,\ldots, n$ are displacements on the nodes i.

For linear elastic structures the potential Π in Eq. (8.65) can be written as $\Pi = -W + U$ where W is the work of the external forces (and body forces) and U is the elastic energy of the system.

8.4 CASTIGLIANO'S THEOREM

As it has been stated previously, the strain energy density of a linear elastic system is equal to the complementary one. Therefore:

$$\delta U = \delta U_C = \int_V \left[\varepsilon_x \delta\sigma_{xx} + \varepsilon_y \delta\sigma_{yy} + \varepsilon_z \delta\sigma_{zz} + 2\varepsilon_{xy}\delta\tau_{xy} + 2\varepsilon_{xz}\delta\tau_{xz} + 2\varepsilon_{yz}\delta\tau_{yz}\right] dV \qquad (8.66)$$

With the aid of the strain-displacement relations (Eqs. 26–31), the above equation yields:

$$\begin{aligned}\delta U = \int_V \Bigg[& \frac{\partial u}{\partial x}\delta\sigma_{xx} + \frac{\partial v}{\partial y}\delta\sigma_{yy} + \frac{\partial w}{\partial z}\delta\sigma_{zz} + \left(\frac{\partial u}{\partial y} + \frac{\partial v}{\partial x}\right)\delta\tau_{xy} + \left(\frac{\partial u}{\partial z} + \frac{\partial w}{\partial x}\right)\delta\tau_{xz} \\ & + \left(\frac{\partial v}{\partial z} + \frac{\partial w}{\partial y}\right)\delta\tau_{yz}\Bigg] dV\end{aligned} \qquad (8.67)$$

After some simple algebraic operations, Eq. (8.67) can be written in the following form:

$$\begin{aligned}\delta U = \int_V \Bigg[& \left(\frac{\partial(u\delta\sigma_{xx})}{\partial x} - u\frac{\partial(\delta\sigma_{xx})}{\partial x}\right) + \left(\frac{\partial(v\delta\sigma_{yy})}{\partial y} - v\frac{\partial(\delta\sigma_{yy})}{\partial y}\right) + \left(\frac{\partial(w\delta\sigma_{zz})}{\partial z} - w\frac{\partial(\delta\sigma_{zz})}{\partial z}\right) + \\ & + \left(\frac{\partial(u\delta\tau_{xy})}{\partial y} - u\frac{\partial(\delta\tau_{xy})}{\partial y}\right) + \left(\frac{\partial(v\delta\tau_{xy})}{\partial x} - v\frac{\partial(\delta\tau_{xy})}{\partial x}\right) + \left(\frac{\partial(v\delta\tau_{yz})}{\partial z} - v\frac{\partial(\delta\tau_{yz})}{\partial z}\right) + \\ & + \left(\frac{\partial(w\delta\tau_{yz})}{\partial y} - w\frac{\partial(\delta\tau_{yz})}{\partial y}\right) + \left(\frac{\partial(w\delta\tau_{xz})}{\partial x} - w\frac{\partial(\delta\tau_{xz})}{\partial x}\right) + \left(\frac{\partial(u\delta\tau_{xz})}{\partial z} - u\frac{\partial(\delta\tau_{xz})}{\partial z}\right)\Bigg] dV\end{aligned} \qquad (8.68)$$

Let's use the following notations:

$$\beta_x = u\delta\sigma_{xx} \quad \beta_y = u\delta\tau_{xy} \quad \beta_z = u\delta\tau_{xz} \qquad (8.69)$$

$$\gamma_x = v\delta\tau_{xy} \quad \gamma_y = v\delta\sigma_{yy} \quad \gamma_z = v\delta\tau_{yz} \qquad (8.70)$$

$$\zeta_x = w\delta\tau_{xz} \quad \zeta_y = w\delta\tau_{yz} \quad \zeta_z = w\delta\sigma_{zz} \qquad (8.71)$$

Then, Eq. (8.68) can be written as:

$$\delta U = \int_V \left[\left(\frac{\partial(\beta_x)}{\partial x} + \frac{\partial(\beta_y)}{\partial y} + \frac{\partial(\beta_z)}{\partial z} \right) + \left(\frac{\partial(\gamma_x)}{\partial x} + \frac{\partial(\gamma_y)}{\partial y} + \frac{\partial(\gamma_z)}{\partial z} \right) + \left(\frac{\partial(\zeta_x)}{\partial x} + \frac{\partial(\zeta_y)}{\partial y} + \frac{\partial(\zeta_z)}{\partial z} \right) \right] dV -$$
$$\int_V u \left(\frac{\partial(\delta\sigma_{xx})}{\partial x} + \frac{\partial(\delta\tau_{xy})}{\partial y} + \frac{\partial(\delta\tau_{zx})}{\partial z} \right) + v \left(\frac{\partial(\delta\tau_{xy})}{\partial x} + \frac{\partial(\delta\sigma_{yy})}{\partial y} + \frac{\partial(\delta\tau_{yz})}{\partial z} \right) \tag{8.72}$$
$$+ w \left(\frac{\partial(\delta\tau_{xz})}{\partial x} + \frac{\partial(\delta\tau_{yz})}{\partial y} + \frac{\partial(\delta\sigma_{zz})}{\partial z} \right) \Bigg] dV$$

With the aid of the equilibrium equations (Eqs. 23–25), the above equation yields:

$$\delta U = \int_V \left[\left(\frac{\partial(\beta_x)}{\partial x} + \frac{\partial(\beta_y)}{\partial y} + \frac{\partial(\beta_z)}{\partial z} \right) + \left(\frac{\partial(\gamma_x)}{\partial x} + \frac{\partial(\gamma_y)}{\partial y} + \frac{\partial(\gamma_z)}{\partial z} \right) + \left(\frac{\partial(\zeta_x)}{\partial x} + \frac{\partial(\zeta_y)}{\partial y} + \frac{\partial(\zeta_z)}{\partial z} \right) \right] dV -$$
$$\int_V u(-\delta B_x) + v(-\delta B_y) + w(-\delta B_z) \big] dV \tag{8.73}$$

Taking into account the Gauss divergence theorem, Eq. (8.73) can be transformed to the following equation:

$$\delta U = \int_V \left[\left(\beta_x l + \beta_y m + \beta_z n \right) + \left(\gamma_x l + \gamma_y m + \gamma_z n \right) \right.$$
$$\left. + \left(\zeta_x l + \zeta_y m + \zeta_z n \right) \right] dS + \int_V u(\delta B_x) + v(\delta B_y) + w(\delta B_z) \big] dV \tag{8.74}$$

Considering the notations in Eqs. (8.69)–(8.71), the following equation can be obtained:

$$\delta U = \int_S \left[u \left(\delta\sigma_{xx} l + \delta\tau_{xy} m + \delta\tau_{xz} n \right) + v \left(\delta\tau_{xy} l + \delta\sigma_{yy} m + \delta\tau_{yz} n \right) \right.$$
$$\left. + w \left(\delta\tau_{xz} l + \delta\tau_{yz} m + \delta\sigma_{zz} n \right) \right] dS + \tag{8.75}$$
$$\int_V u(\delta B_x) + v(\delta B_y) + w(\delta B_z) \big] dV$$

Assuming the absence of body forces and considering Eqs. (8.10)–(8.12), Eq. (8.75) yields:

$$\delta U = \int_S u \delta\sigma_{Px} dS + \int_S v \delta\sigma_{Py} dS + \int_S w \delta\sigma_{Pz} dS \tag{8.76}$$

In case that a solid is subjected to point forces (Fig. 8.3), the above equation can be transformed to the following relationship:

$$\delta U = \sum_{i=1}^{n} u_i \cdot \delta P_{xi} + \sum_{i=1}^{n} v_i \cdot \delta P_{yi} + \sum_{i=1}^{n} w_i \cdot \delta P_{zi} \tag{8.77}$$

Eq. (8.77) implies that:

$$u_i = \frac{\partial U}{\partial P_{xi}} \tag{8.78}$$

$$v_i = \frac{\partial U}{\partial P_{yi}} \tag{8.79}$$

$$w_i = \frac{\partial U}{\partial P_{zi}} \tag{8.80}$$

Generally, displacements and forces can also be slopes θ_i and the corresponding moments M_i (Fig. 8.3):

$$\theta_i = \frac{\partial U}{\partial M_i} \tag{8.81}$$

It should be stressed that the used assumption $\delta U = \delta U_C$ is valid to linear elastic materials only. Therefore Castigliano's theorem can be applied for analyzing structures

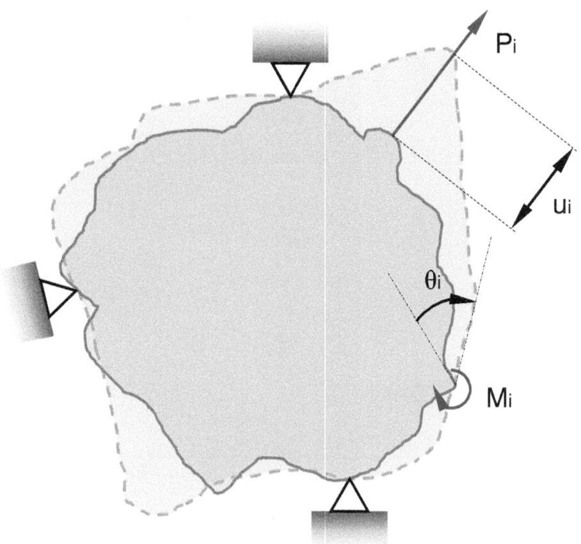

FIG. 8.3

Elastic medium subjects to point forces (and moments).

with linear elastic behavior. The applicability of Castigliano's theorem is more convenient than the principles of virtual work and minimum potential energy. Castigliano's theorem and the principle of minimum potential energy can be used for analyzing both statically determinate and indeterminate structures, and the results satisfy the equilibrium equations of the structure. In the following subchapters, the expression for the elastic energy for axial tension, bending, torsion, and shear will be derived aiming to implementing the formulae (Eqs. 78–81) in engineering applications.

8.4.1 AXIAL LOADING

The energy of a bar under an axial load is (see Fig. 8.4):

$$U = \int_V \frac{1}{2}\sigma\varepsilon dV$$

or

$$U = \int_0^L \frac{1}{2}\sigma\left(\frac{\sigma}{E}\right)Adx$$

or

$$U = \int_0^L \frac{\left(\frac{N}{A}\right)^2 A}{2E}dx$$

Finally, the elastic energy under axial loading can be obtained by the following formula:

$$U = \int_0^L \frac{N^2}{2EA}dx \tag{8.82}$$

For a bar or beam of a uniform cross section, Eq. (8.82) yields:

$$U = \frac{N^2L}{2EA} \tag{8.83}$$

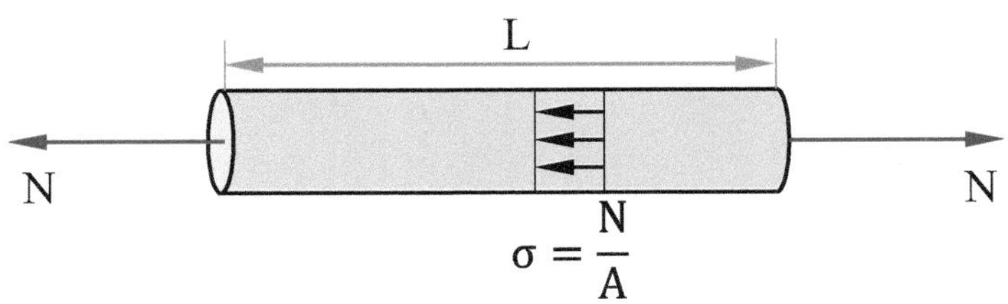

FIG. 8.4

Component under axial tension.

and Castigliano's theorem for axial loading can be obtained by the following relationship:

$$u_N = \frac{\partial U}{\partial N} = \frac{NL}{EA} \tag{8.84}$$

8.4.2 BENDING

The elastic energy stored in a beam subjected to bending (Fig. 8.5) can be calculated by integrating the elastic energy of a material element over the volume of the beam:

$$U_b = \int_V \frac{1}{2}\sigma\varepsilon dV \tag{8.85}$$

Considering Hooke's law, the above equation yields:

$$U_b = \int_A \int_L \frac{1}{2}\sigma\left(\frac{\sigma}{E}\right)dx \cdot dA \tag{8.86}$$

Because the bending-induced normal stresses (Fig. 8.5) are given by the following formula

$$\sigma = \frac{M}{I}y \tag{8.87}$$

the following equation can be obtained:

$$U_b = \int_A \left(\int_L \frac{M^2}{2EI^2}dx \right) \cdot y^2 dA \tag{8.88}$$

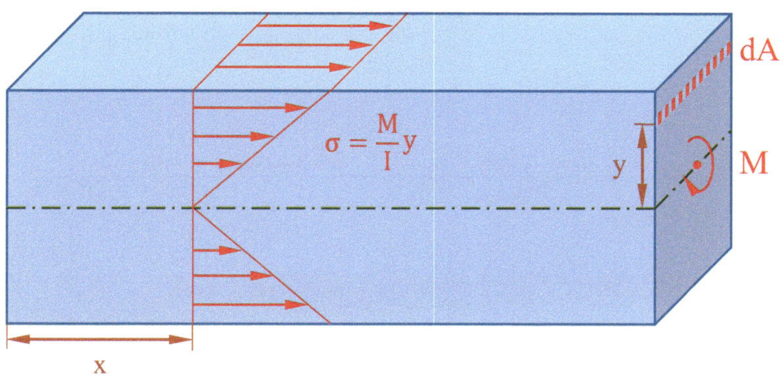

FIG. 8.5

Beam under bending.

Considering the definition of the moment of inertia

$$I = \int_A y^2 dA \tag{8.89}$$

the elastic energy stored in a beam subjected to bending can be obtained by the following relationship:

$$U_b = \int_L \frac{M^2}{2EI} dx \tag{8.90}$$

Then, Castigliano's theorem for calculating the displacement in the direction of a force F yields:

$$u_F = \frac{\partial U}{\partial F} = \frac{1}{EI} \int M(x) \frac{\partial M(x)}{\partial F} dx \tag{8.91}$$

8.4.3 TORSION

The torsion-induced elastic energy (Fig. 8.6) can be calculated by the following formula:

$$U_T = \int \frac{1}{2} M_T d\theta \tag{8.92}$$

where M_T is the torsional moment. Considering the yellow and the green triangles in Fig. 8.6, the following geometric relationships

$$ds = \gamma \ dx \tag{8.93}$$

$$ds = R \ d\theta \tag{8.94}$$

yield:

$$\gamma = R \frac{d\theta}{dx} \tag{8.95}$$

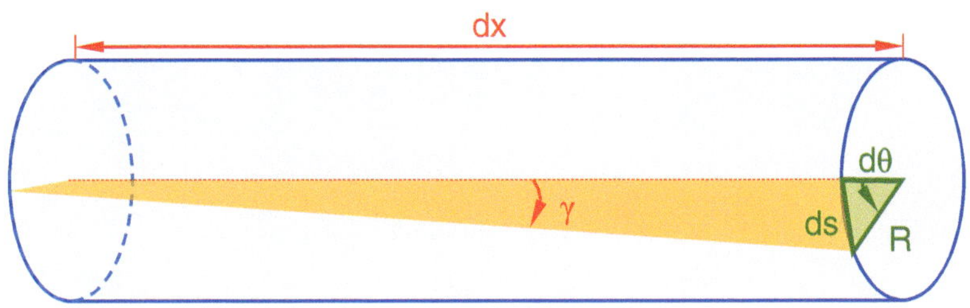

FIG. 8.6

Beam under torsion.

We recall Hooke's law for shear stress:

$$\tau = G\gamma \tag{8.96}$$

and the formula for torsion-induced shear stresses

$$\tau = \frac{M_T}{I_P}R \tag{8.97}$$

Combining Eqs. (8.95)-(8.97), the following result can be obtained:

$$d\theta = \frac{M_T}{GI_P}dx \tag{8.98}$$

where I_p is the polar moment of inertia.
With the aid of Eq. (8.98), Eq. (8.92) yields the torsion-induced elastic energy:

$$U_T = \int_0^L \frac{M_T^2}{2GI_P}dx \tag{8.99}$$

If we apply Castigliano's theorem for calculating the contribution of torsion on the displacement in the direction of a force F, the following result can be obtained:

$$u_F = \frac{\partial U_T}{\partial F} = \frac{1}{GI_P}\int_0^L M_T\frac{\partial M_T}{\partial F}dx \tag{8.100}$$

8.4.4 SHEAR

The elastic energy stored in a beam subjected to bending-induced shear stresses (Fig. 8.7) can be calculated by integrating the elastic energy of a material element over the volume of the beam:

$$U_S = \int_V \frac{1}{2}\tau\,\gamma dV \tag{8.101}$$

where γ is the shear strain (Fig. 8.8) that can be provided by Hooke's law

$$\gamma = \frac{\tau}{G} \tag{8.102}$$

Then, Eq. (8.101) yields

$$U_S = \int_V \frac{\tau^2}{2G}dV \tag{8.103}$$

Let's recall the formula for the bending-induced (Fig. 8.7) shear stress:

$$\tau(y) = \frac{Q}{I_z b(y)}S_Z(y) \tag{8.104}$$

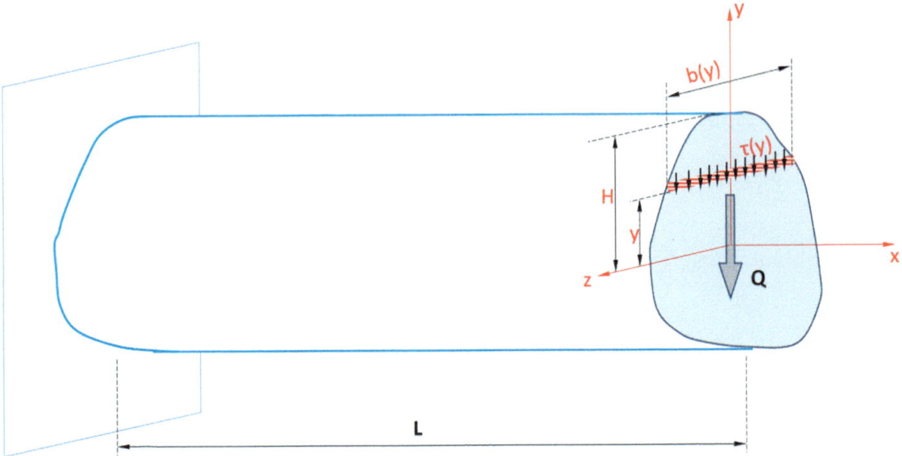

FIG. 8.7

Beam under bending-induced shear stresses.

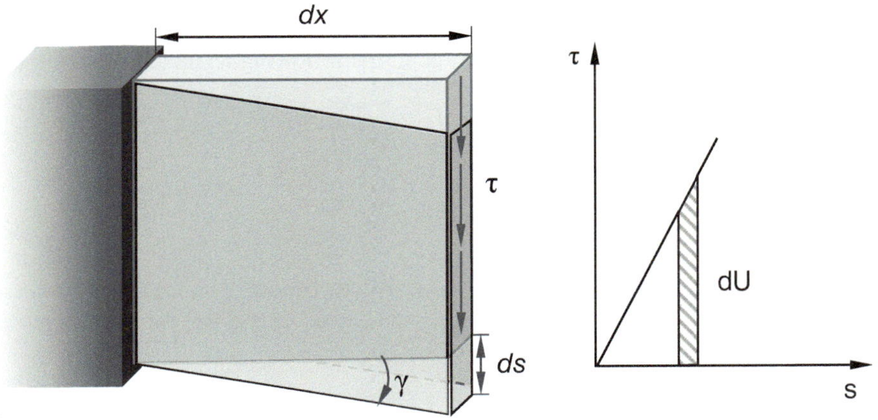

FIG. 8.8

Hooke's law for a material element under shear.

where

$$S_z(y) = \int_y^H y\, dA \tag{8.105}$$

is the static moment of inertia of the area [y,H] (see Fig. 8.7).

Combining Eqs. (8.103) and (8.105), the following formula can be obtained:

$$U_S = \iint_{AL} \frac{Q^2}{2GI_z^2}\left[\frac{S_z^2(y)}{b^2(y)} dA\right] dx \tag{8.106}$$

If we use the notation k for the following quantity that is a shape function of the beam's cross section

$$k = \frac{A}{I_z^2} \int_A \frac{S_z^2(y)}{b^2(y)} dA \tag{8.107}$$

Eq. (8.106) yields:

$$U_S = \int_0^L k \frac{Q^2}{2GA} dx \tag{8.108}$$

If we apply Castigliano's theorem for calculating the contribution of bending-induced shear stresses on the displacement in the direction of a force F, the following result can be obtained:

$$u_F = \frac{\partial U_S}{\partial F} = \frac{1}{GA} \int_0^L k\, Q \frac{\partial Q}{\partial F} dx \tag{8.109}$$

It should be noted that the contribution of shear stresses to the displacement is small and can be neglected in engineering applications.

EXERCISE 1

Prove that the principle of virtual work is equivalent to the equilibrium equations for the truss shown in the figure below:

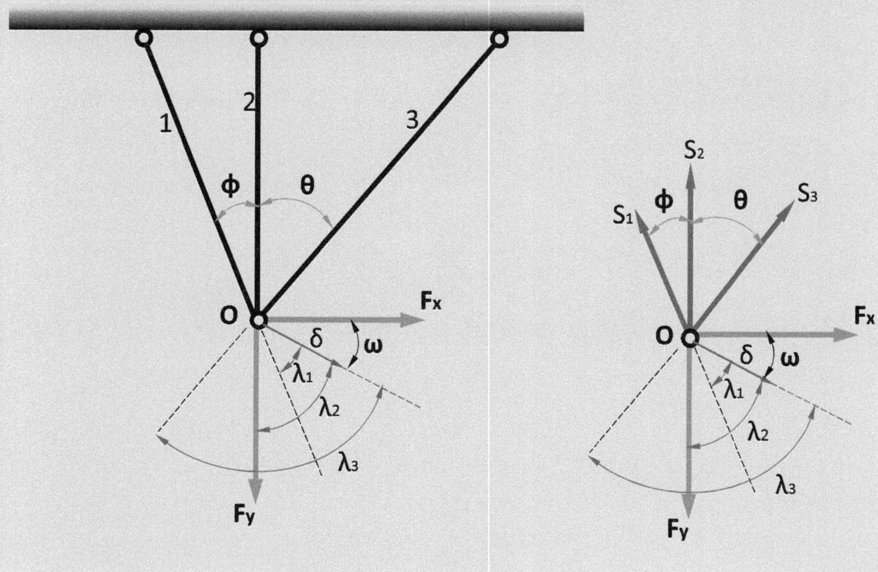

(Continued)

EXERCISE 1—CONT'D

Solution

For a virtual displacement δ in an arbitrary direction with inclination angle ω, the work of the external forces *Fx* and *Fy* is:

$$W_{ef} = F_x(\delta \cos \omega) + F_y(\delta \sin \omega)$$

The work of the internal forces S_1, S_2, and S_3 can be obtained by the following equation:

$$W_{if} = S_1(\delta \cos \lambda_1) + S_2(\delta \cos \lambda_2) + S_3(\delta \cos \lambda_3)$$

According to the principle of virtual work, the above equations should be equal:

$$W_{if} = W_{ef}$$

or

$$S_1(\cos \lambda_1) + S_2(\cos \lambda_2) + S_3(\cos \lambda_3) = F_x(\cos \omega) + F_y(\sin \omega)$$

With the aid of the following trigonometric identities

$$\cos \lambda_1 = \cos\left(\frac{\pi}{2} - \omega - \phi\right) = \sin(\omega + \phi) = \sin\omega \cos \phi + \cos \omega \sin \phi$$

$$\cos \lambda_2 = \cos\left(\frac{\pi}{2} - \omega\right) = \sin \omega$$

$$\cos \lambda_3 = \cos\left(\frac{\pi}{2} - \omega + \theta\right) = -\sin(\theta - \omega) = -(\sin\theta \cos \omega - \cos \theta \sin \omega)$$

the principle of the virtual works yields:

$$S_1(\sin\omega \cos \phi + \cos \omega \sin \phi) + S_2 \sin \omega - S_3(\sin\theta \cos \omega - \cos \theta \sin \omega) = F_x(\cos \omega) + F_y(\sin \omega)$$

Rearranging the terms of the above equation, the following relationship can be obtained:

$$(S_1 \sin \phi - S_3 \sin\theta)\cos \omega + (S_1 \cos \phi + S_2 + S_3 \cos \theta)\sin \omega = F_x \cos \omega + F_y \sin \omega$$

or

$$F_x = S_1 \sin \phi - S_3 \sin\theta$$

$$F_y = S_1 \cos \phi + S_2 + S_3 \cos \theta$$

The above relationships are the equilibrium equations of the node O.

EXERCISE 2

A node is supported by two parallel bars of different material and cross-section area. The node is subjected to a vertical force F. Calculate the forces S_1 and S_2 of the bars.

Solution

The force equilibrium equation of the node is:

$$S_1 + S_2 = F$$

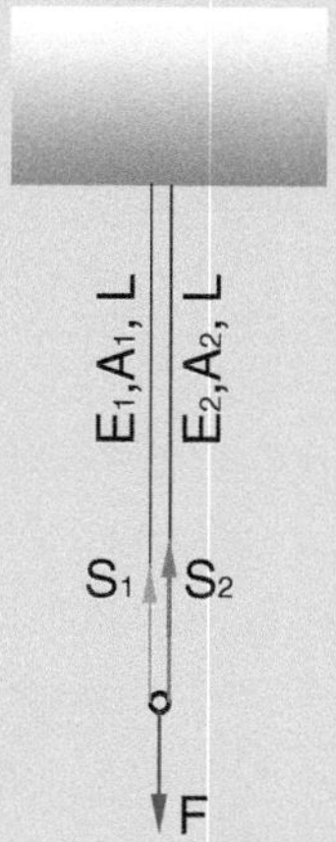

This equation is not sufficient to calculate both forces S_1 and S_2. The solution of this problem can be achieved with the aid of the principle of minimum potential energy. To this end, the elastic energy of the bars and the work of the external forces should be calculated first:

Elastic energy of bar 1: $U_1 = \frac{1}{2}\sigma_1\varepsilon_1 V_1$

Elastic energy of bar 1: $U_2 = \frac{1}{2}\sigma_2\varepsilon_2 V_2$

Work of the external force: $W = F \cdot \Delta L$

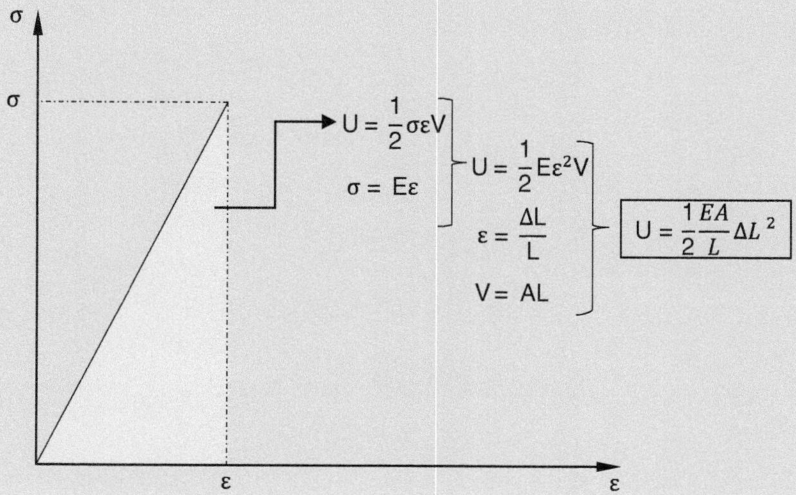

(Continued)

EXERCISE 2—CONT'D

Then, the total potential energy of the system is:

$$\Pi = U - W$$

or

$$\Pi = \frac{1}{2}\frac{E_1A_1}{L}\Delta L^2 + \frac{1}{2}\frac{E_2A_2}{L}\Delta L^2 - F\Delta L$$

We observe that the potential energy Π is a quadratic function of ΔL. We can obtain then the value of ΔL that minimizes the potential energy of the system:

$$\frac{\partial \Pi}{\partial \Delta L} = 0$$

or

$$\frac{1}{2}\frac{E_1A_1}{L}2\Delta L + \frac{1}{2}\frac{E_2A_2}{L}2\Delta L - F = 0$$

The above equation yields:

$$\Delta L = \frac{F}{\dfrac{E_1A_1}{L} + \dfrac{E_2A_2}{L}}$$

With the aid of the above results, the forces S_1 and S_2 can now easily be obtained by Hooke's law of each bar:

$$\sigma_1 = E_1\varepsilon_1$$

$$\sigma_2 = E_2\varepsilon_2$$

or

$$\frac{S_1}{A_1} = E_1\frac{\Delta L}{L}$$

$$\frac{S_2}{A_2} = E_2\frac{\Delta L}{L}$$

Finally, the above equations yield:

$$S_1 = \frac{E_1A_1}{E_1A_1 + E_2A_2}F$$

$$S_2 = \frac{E_2A_2}{E_1A_1 + E_2A_2}F$$

We can see that the above results satisfy the force equilibrium equation:

$$S_1 + S_2 = F$$

EXERCISE 3

Assuming that the horizontal beam is stiff, calculate the axial forces of the bars a and b with the principle of minimum potential energy (MPE).

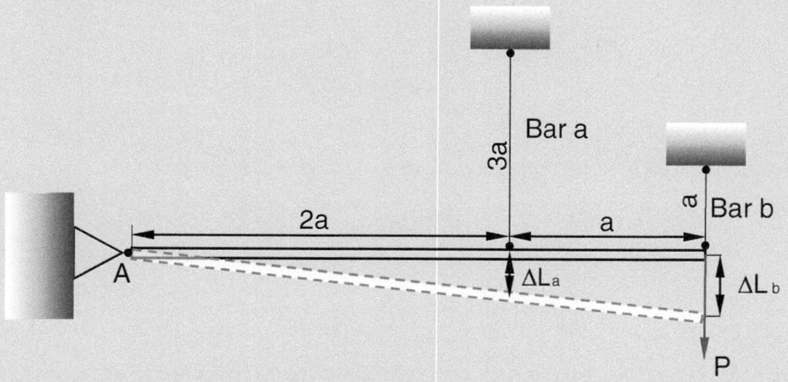

Solution

Because the horizontal beam is stiff, we can ignore its elastic energy. Then, the potential energy of the system is:

$$\Pi = U - W = \frac{1}{2}\frac{EA}{3a}\Delta L_a^2 + \frac{1}{2}\frac{EA}{a}\Delta L_b^2 - P\Delta L_b$$

In the above equation, we can use the following identity from trigonometry:

$$\frac{\Delta L_a}{2a} = \frac{\Delta L_b}{3a}$$

Then:

$$\Pi = \frac{1}{2}\frac{EA}{3a}\frac{4}{9}\Delta L_b^2 + \frac{1}{2}\frac{EA}{a}\Delta L_b^2 - P\Delta L_b$$

We calculate the value ΔL_b for which the potential energy is minimum:

$$\frac{\partial \Pi}{\partial \Delta L_b} = \frac{4}{27}\frac{EA}{a}\Delta L_b + \frac{EA}{a}\Delta L_b - P = 0$$

or

$$\Delta L_b = \frac{27}{31}\frac{a}{EA}P$$

Considering the identity from trigonometry, the following elongation of the bar "α" can be obtained:

$$\Delta L_a = \frac{2}{3}\frac{27}{31}\frac{a}{EA}P = \frac{54}{93}\frac{a}{EA}P$$

(Continued)

EXERCISE 3—CONT'D

With the aid of Hooke's law, the forces of the bars "a" and "b" can be calculated by the following equations:

$$S_a = \frac{EA}{3a}\Delta L_a = \frac{EA}{3a}\left(\frac{54}{93}\frac{a}{EA}P\right) = \frac{6}{31}P$$

$$S_b = \frac{EA}{a}\Delta L_b = \frac{EA}{a}\left(\frac{27}{31}\frac{a}{EA}P\right) = \frac{27}{31}P$$

Obviously, the above results satisfy the equilibrium equation of moments:

$$\sum M_A = S_a \cdot 2a + S_b \cdot 3a - P \cdot 3a = \frac{6}{31}P \cdot 2 + \frac{27}{31}P \cdot 3 - 3P = 0$$

EXERCISE 4

With the aid of Castigliano's theorem, determine the maximum vertical deflection at the end of the beam in the following figure.

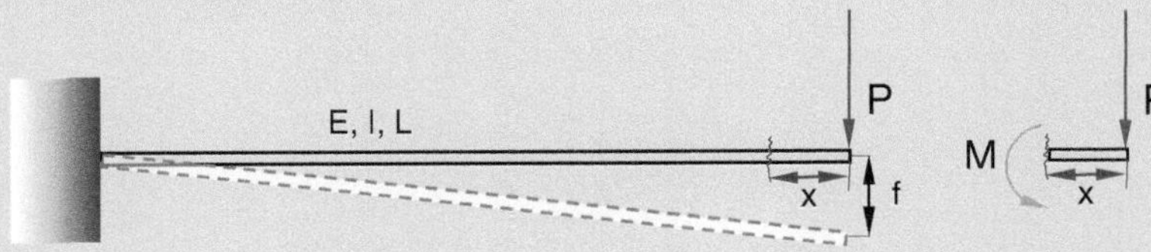

Solution

Because bending is the predominant loading case for a beam, the contribution of shear forces into the elastic energy is negligible. Therefore only the bending moments can be considered to contribute to Castigliano's theorem for the beam:

$$U = \int_L \frac{M^2}{2EI}dx$$

The deflection can be determined using Castigliano's theorem:

$$f = \frac{\partial U}{\partial P} = \frac{1}{EI}\int_0^L M\frac{\partial M}{\partial P}dx$$

From the moment equilibrium in the arbitrary point x

$$M - Px = 0$$

we can calculate the bending moment distribution.

$$M = Px$$

The derivative of the bending moment with respect to the force P is

$$\frac{\partial M}{\partial P} = x$$

Then, Castigliano's theorem yields:

$$f = \frac{1}{EI}\int_0^L (Px)(x)dx$$

or

$$f = \frac{PL^3}{3EI}$$

EXERCISE 5

Calculate the rotation φ for the beam under pure bending.

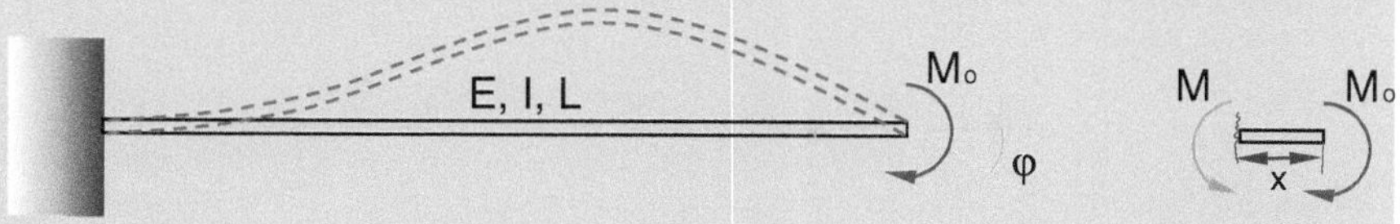

Solution

The elastic energy for bending is:

$$U = \int_L \frac{M^2}{2EI} dx$$

We apply Castigliano's theorem:

$$\varphi = \frac{\partial U}{\partial M_o} = \frac{1}{EI} \int M \frac{\partial M}{\partial M_o} dx$$

From the moment equilibrium in the arbitrary point x

$$M = M_o$$

we get

$$\frac{\partial M}{\partial M_o} = 1$$

Then, Castigliano's theorem yields:

$$\varphi = \frac{\partial U}{\partial M_o} = \frac{1}{EI} \int M \frac{\partial M}{\partial M_o} dx = \frac{1}{EI} \int_0^L M_o \cdot 1 \cdot dx = \frac{M_o L}{EI}$$

EXERCISE 6

Determine the vertical deflection of point A for the structure in the following figure.

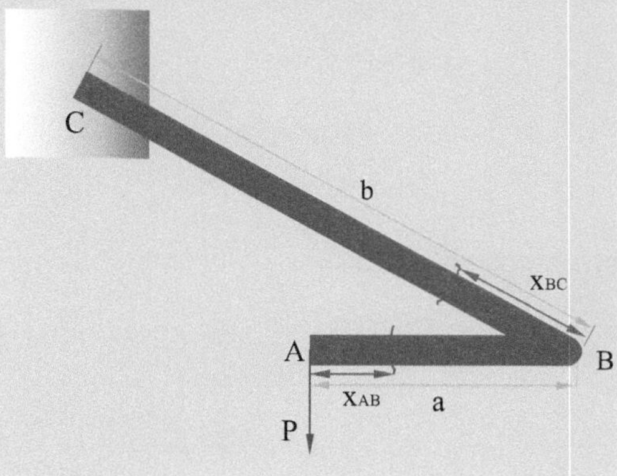

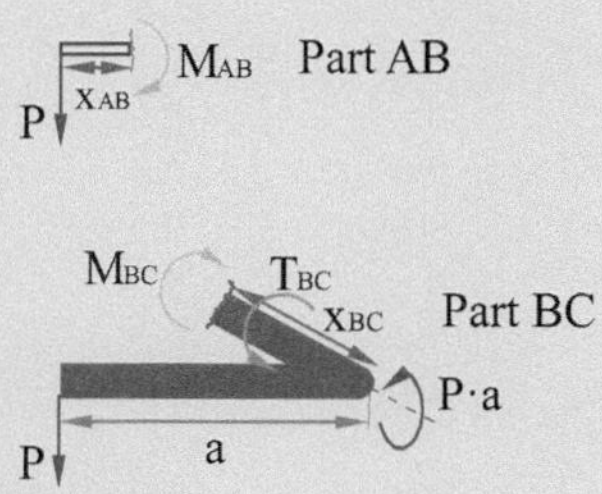

(Continued)

EXERCISE 6—CONT'D

Solution

The vertical deflection f of point A can be determined using Castigliano's theorem:

$$f = \frac{\partial U}{\partial P}$$

We observe that the structure has two parts, the beam AB subjected to bending and the beam BC subjected to bending and torsion. To obtain the elastic energy of the beams, we should determine the bending moment for the beam AB and the bending and torsional moments for the beam BC.

Part *AB* (Bending):

$$M_{AB} = Px$$

$$\frac{\partial M_{AB}}{\partial P} = x$$

Part *BC* (Bending):

$$M_{BC} = -Px$$

$$\frac{\partial M_{BC}}{\partial P} = -x$$

Part *BC* (Torsion):

$$T_{BC} = -Pa$$

$$\frac{\partial T_{BC}}{\partial P} = -a$$

We recall the formulae for the elastic energy for bending and torsion:

$$U_b = \int_L \frac{M^2}{2EI} dx$$

$$U_T = \int_0^L \frac{M_T^2}{2GI_P} dx$$

Then, the elastic energy of each beam can be calculated by the following equations:

$$U_{M_{AB}} = \int_0^a \frac{M_{AB}^2}{2EI} dx \quad \text{Bending of beam } AB$$

$$U_{M_{BC}} = \int_0^b \frac{M_{BC}^2}{2EI} dx \quad \text{Bending of beam } BC$$

$$U_{T_{BC}} = \int_0^b \frac{T_{BC}^2}{2GI_p} dx \quad \text{Torsion of beam } BC$$

The total elastic energy of the system can be obtained by the superposition of the above equations:

$$U = \int_0^a \frac{M_{AB}^2}{2EI} dx + \int_0^b \frac{M_{BC}^2}{2EI} dx + \int_0^b \frac{T_{BC}^2}{2GI_p} dx$$

Considering the above equation, Castigliano's theorem yields:

$$f = \frac{1}{EI} \int_0^a M_{AB} \frac{\partial M_{AB}}{\partial P} dx + \frac{1}{EI} \int_0^b M_{BC} \frac{\partial M_{BC}}{\partial P} dx + \frac{1}{GI_p} \int_0^b T_{BC} \frac{\partial T_{BC}}{\partial P} dx$$

or

$$f = \frac{1}{EI} \int_0^a (Px)x \; dx + \frac{1}{EI} \int_0^b (Px)x \; dx + \frac{1}{GI_p} \int_0^b (Pa)a \; dx$$

or

$$f = \frac{Pa^3}{3EI} + \frac{Pb^3}{3EI} + \frac{Pa^2b}{GI_p}$$

EXERCISE 7

Calculate the stiffness k of the spring in the following figure. The known data are the diameter of the coil D, the diameter of the wire d, the number of coils n, and the inclination angle of the coils β.

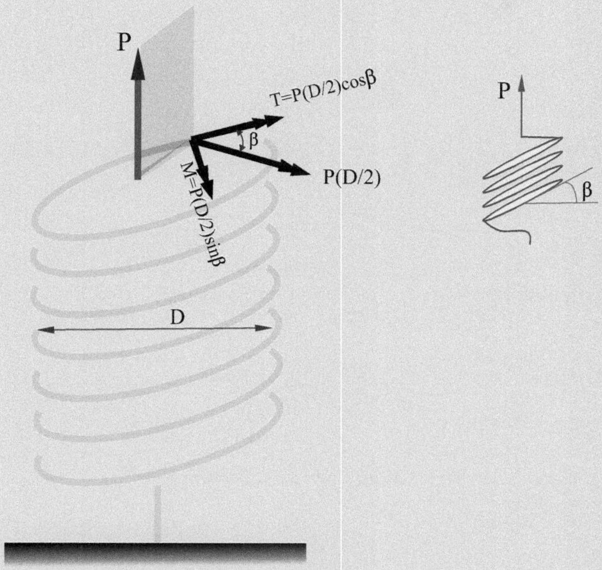

Solution

We start defining the total length of the coil L, the torsional moment T, and the bending moment M:

$$L = n(\pi D)$$

$$T = \left(P\frac{D}{2}\right)\cos\beta$$

$$M = \left(P\frac{D}{2}\right)\sin\beta$$

Then, the elastic energy of the system can be defined:

$$U = \int_0^L \frac{M^2}{2EI}dx + \int_0^L \frac{T^2}{2GI_p}dx$$

(Continued)

EXERCISE 7—CONT'D

or

$$U = \frac{M^2}{2EI} n\pi D + \frac{T^2}{2GI_p} n\pi D$$

or

$$U = \left(\frac{(PD)^2}{4EI} \sin^2 \beta + \frac{(PD)^2}{4GI_p} \cos^2 \beta \right) \frac{n\pi D}{2}$$

or

$$U = \frac{n\pi D^3}{8} P^2 \left(\frac{\sin^2 \beta}{EI} + \frac{\cos^2 \beta}{GI_p} \right)$$

Applying Castigliano's theorem, the following formula can be obtained for the vertical deflection of the upper end of the spring:

$$f = \frac{\partial U}{\partial P} = \frac{n\pi D^3}{8} 2P \left(\frac{\sin^2 \beta}{EI} + \frac{\cos^2 \beta}{GI_p} \right)$$

For a small angle β, the following simplifications are reasonable:

$$\sin^2 \beta \approx 0$$

$$\cos^2 \beta \approx 1$$

Then,

$$f = P \frac{n\pi D^3}{4} \frac{1}{GI_p}$$

Considering the formula for the polar moment of inertia of circular cross section

$$I_p = \frac{\pi d^4}{32}$$

the formula for the deflection yields:

$$f = P \frac{n\pi D^3}{4} \frac{32}{G\pi d^4}$$

Then, the elastic coefficient of the spring can be obtained by the following formula:

$$k = \frac{P}{f} = \frac{Gd^4}{8nD^3}$$

EXERCISE 8

Solve Exercise 3 with Castigliano's theorem, that is, determine the axial forces on the bars a and b, where the member AC is a stiff beam.

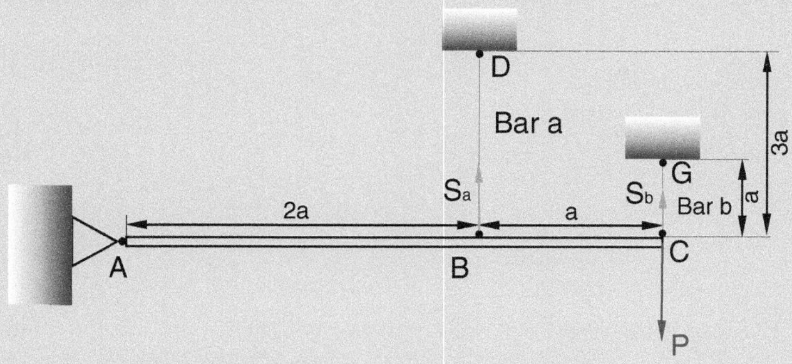

Solution

Equilibrium of the moments at point A yields:

$$\sum M_A = S_a\ 2a + S_b\ 3a - P\ 3a = 0$$

or

$$S_a = \frac{3}{2}P - \frac{3}{2}S_b$$

We apply Castigliano's theorem at point G, where we know the displacement is null:

$$u_G = \frac{\partial U}{\partial S_b} = \underbrace{\frac{a}{EA}S_b\frac{\partial S_b}{\partial S_b}}_{\substack{\text{derivative of elastic}\\ \text{energy of bar b with}\\ \text{respect to the force } S_b}} + \underbrace{\frac{3a}{EA}S_a\frac{\partial S_a}{\partial S_b}}_{\substack{\text{derivative of elastic}\\ \text{energy of bar a with}\\ \text{respect to the force } S_b}} = 0$$

or

$$\frac{a}{EA}S_b + \frac{3a}{EA}\left(\frac{3}{2}P - \frac{3}{2}S_b\right)\left(-\frac{3}{2}\right) = 0$$

The above equation yields:

$$S_b = \frac{27}{31}P$$

Considering the above result in the equilibrium equation of moments, the value of S_a can be obtained:

$$S_a = \frac{6}{31}P$$

EXERCISE 9

The beam AC in the previous example is now flexible. With the aid of Castigliano's theorem, determine the axial forces of the bars a and b, without the assumption of the stiff beam.

Solution

The equilibrium equation of the moments

$$\sum M_A = S_a \ 2a + S_b \ 3a - P \ 3a = 0$$

yield:

$$S_a = \frac{3}{2}P - \frac{3}{2}S_b$$

We can now apply Castigliano's theorem considering the bending-induced elastic energy of the beam and the axial tension-induced elastic energy of the bars. Then, the following formula should replace the formula of Castigliano's theorem in Exercise 8.

$$u_G = \frac{\partial U}{\partial S_b} = \frac{a}{EA} S_b \frac{\partial S_b}{\partial S_b} + \frac{3a}{EA} S_a \frac{\partial S_a}{\partial S_b} + \frac{1}{EI}\int_0^a M_{BC}\frac{\partial M_{BC}}{\partial S_b}dx + \frac{1}{EI}\int_a^{3a} M_{BA}\frac{\partial M_{BA}}{\partial S_b}dx = 0$$

We calculate the bending moments M_{BC} and M_{BA} (see figure below).

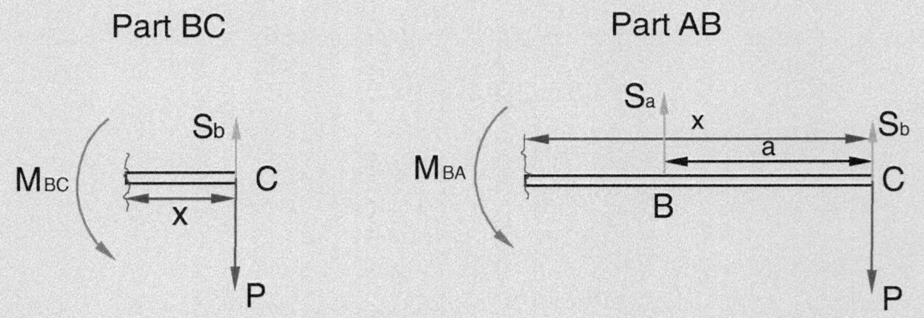

$$M_{BC} = (P - S_b)x$$

$$\frac{\partial M_{BC}}{\partial S_b} = -x$$

$$M_{BA} = (P - S_b)x - \left(\frac{3}{2}P - \frac{3}{2}S_b\right)(x - a)$$

$$\frac{\partial M_{BA}}{\partial S_b} = \frac{1}{2}x - \frac{3}{2}a$$

We continue the solution following the steps of Exercise 8.

EXERCISE 10

Determine an expression for the vertical displacement distribution along the cantilever beam of the following figure.

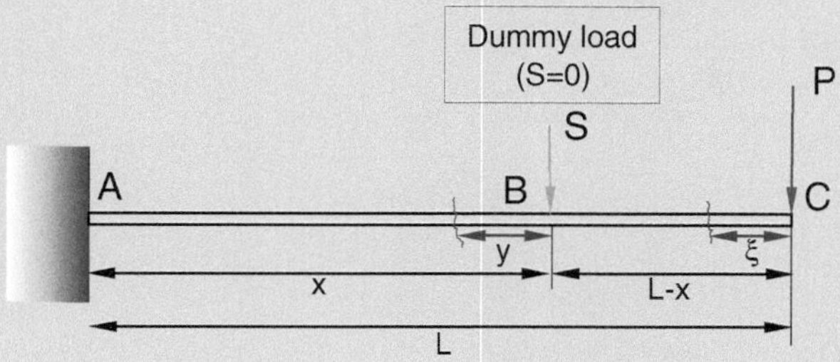

Solution

We must calculate the vertical displacement of any arbitrary point B located in a variable distance x from the supported end A. Because there is not any real force on B, we assume a dummy force S to apply Castigliano's theorem. Point B divides the entire beam into two segments (AB and BC) where we should derive the bending moment formulae separately.

Bending moment in BC:

$$M_{BC} = P\xi$$

$$\frac{\partial M_{BC}}{\partial S} = 0$$

Bending moment in AB:

$$M_{AB} = P[(L - x) + y] + S \cdot y$$

$$\frac{\partial M_{AB}}{\partial S} = y$$

The vertical deflection $W(x)$ of the point B can now be calculated with Castigliano's theorem. Because bending is the predominant loading case for a beam, the contribution of shear forces into the elastic energy is negligible. Therefore only the bending moments can be considered to contribute to Castigliano's theorem for the beam.

$$W(x) = \frac{\partial U}{\partial S} = \frac{1}{EI}\int_0^{L-x} M_{BC}\frac{\partial M_{BC}}{\partial S}d\xi + \frac{1}{EI}\int_0^{x} M_{BA}\frac{\partial M_{BA}}{\partial S}dy$$

or

$$W(x) = \frac{1}{EI}\int_0^{x} \{P[(L - x) + y] + S \cdot y\} \cdot y dy$$

Because $S = 0$, the above equation yields:

$$W(x) = \frac{P}{EI}\frac{x^2(3L - x)}{6}$$

We can confirm this result, for $x = L$:

$$W(L) = \frac{P}{EI}\frac{L^2(3L - L)}{6}$$

or

$$W_C = \frac{PL^3}{3EI}$$

EXERCISE 11

In the truss shown in figure below calculate the following:

A. Support reactions
B. Forces of truss members
C. Horizontal displacement of node C
D. Slope (rotation) of member CB

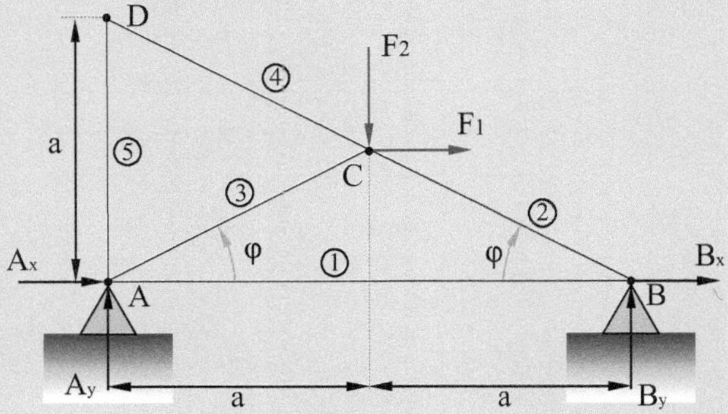

Solution

Because there are three equilibrium equations available for this plane structure $\left(\sum F_x = 0, \sum F_y = 0, \sum M = 0\right)$, but four support reactions are unknown, the structure is statically indeterminate. Therefore, apart from the three equilibrium equations, we can use Castigliano's theorem.

Force equilibrium equations of the structure:

$$\sum F_x = 0 \Rightarrow A_x + B_x = -F_1$$

$$\sum F_y = 0 \Rightarrow A_y + B_y = F_2$$

$$\sum M_A = 0 \Rightarrow 2aB_y - F_2 a - F_1 \frac{a}{2} = 0$$

The system of the above algebraic equations yields:

$$B_y = \frac{F_1}{4} + \frac{F_2}{2}$$

$$A_y = \frac{F_2}{2} - \frac{F_1}{4}$$

$$A_x = -F_1 - B_x$$

The force B_x is still unknown.

We are going to use the displacement equation, obtained with Castigliano's theorem, in a point where we know the displacements are zero (supports of the structure). We can select to apply the condition that the horizontal displacement of the node B is null, that is, $u_B = 0$. For the implementation of Castigliano's theorem, we should calculate the axial force of all the truss members and then derive the equation of the elastic energy of the truss. For calculation of the axial force of the truss members, we can apply the equilibrium equations of each node separately.

Equilibrium equations of the node B:

$$\left.\begin{aligned} \sum F_y &= 0: S_2 \sin\varphi + B_y = 0 \\ \sum F_x &= 0: S_1 + S_2 \cos\varphi = B_x \end{aligned}\right\}$$

EXERCISE 11—CONT'D

The above equations yield:

$$S_1 = B_x + \frac{F_1 + 2F_2}{4 \sin \varphi} \cos \varphi$$

$$S_2 = -\frac{F_1 + 2F_2}{4 \sin \varphi}$$

Equilibrium equations of the node *D*:

$$\sum F_x = 0: \ S_4 \cos \varphi = 0$$

$$\sum F_y = 0: \ S_4 \sin \varphi + S_5 = 0$$

From these equations, the following results can be obtained:

$$S_4 = 0$$

$$S_5 = 0$$

Equilibrium equations of the node *C*:

$$\sum F_x = 0: \quad S_2 \cos \varphi + F_1 = S_3 \cos \varphi + \cancel{S_4} \cos \varphi$$
$$\sum F_y = 0: \quad \cancel{S_4} \sin \varphi - F_2 = S_3 \sin \varphi + S_2 \sin \varphi$$

or

$$S_3 = \frac{1}{2}\left(\frac{F_1}{\cos \varphi} - \frac{F_2}{\sin \varphi}\right)$$

Castigliano's theorem:
The elastic energy of the truss is given by the following equation:

$$U = \frac{S_1^2 L_1}{2EA} + \frac{S_2^2 L_2}{2EA} + \frac{S_3^2 L_3}{2EA}$$

And applying Castigliano's theorem on point *B*, we set the horizontal displacement null:

$$u_B = \frac{\partial U}{\partial B_x} = 0$$

or

$$S_1 \frac{\partial S_1}{\partial B_x} L_1 + S_2 \cancel{\frac{\partial S_2}{\partial B_x}} L_2 + S_3 \cancel{\frac{\partial S_3}{\partial B_x}} L_3 = 0$$

Considering the result for the force S_1, the following result for B_x can be obtained:

$$B_x = -\frac{\cot \varphi}{4}(F_1 + 2F_2)$$

(Continued)

EXERCISE 11—CONT'D

Results for support reactions

With the aid of the previous results, we summarize below the findings for the support reactions:

$$A_x = -F_1 - B_x$$

or

$$A_x = \frac{3\cot\varphi - 4}{4}F$$

$$A_y = \frac{F_2}{2} - \frac{F_1}{4}$$

$$B_x = -\frac{\cot\varphi}{4}(F_1 + 2F_2)$$

$$B_y = \frac{F_1}{4} + \frac{F_2}{2}$$

Results for the forces of the truss members

Considering the above results for the support reactions and the results from the equilibrium equations of the nodes, the following findings for the axial forces of the bars can be obtained:

$$S_1 = B_x - \frac{\cot\varphi}{4}(F_1 + 2F_2) = 0$$

$$S_2 = -\frac{F_1 + 2F_2}{4\sin\varphi}$$

$$S_3 = \frac{1}{2}\left(\frac{F_1}{\cos\varphi} - \frac{F_2}{\sin\varphi}\right)$$

$$S_4 = 0$$

$$S_5 = 0$$

Horizontal displacement of the node *C*

The horizontal displacement u_C of the node C can be calculated with the aid of Castigliano's theorem:

$$u_C = \frac{\partial U}{\partial F}$$

or

$$u_C = \frac{1}{EA}\left(S_2\frac{\partial S_2}{\partial F}L_2 + S_3\frac{\partial S_3}{\partial F}L_3\right)$$

Considering the results for the forces S_2 and S_3, the following formula can be obtained:

$$u_C = \frac{1}{EA}\left[\left(-\frac{3}{4\sin\varphi}F\right)\left(-\frac{3}{4\sin\varphi}\right)L_2 + \left(\frac{1}{2}\left(\frac{1}{\cos\varphi} - \frac{1}{\sin\varphi}\right)F\right)\left(\frac{1}{2}\left(\frac{1}{\cos\varphi} - \frac{1}{\sin\varphi}\right)\right)L_3\right]$$

or

$$u_C = \frac{F}{EA}\left[\frac{9L_2}{16\sin^2\varphi} + \frac{L_3}{4}\left(\frac{1}{\cos\varphi} - \frac{1}{\sin\varphi}\right)^2\right]$$

EXERCISE 11—CONT'D

Rotation of the member *CB*

To calculate the rotation of the member *CB*, we apply a dummy moment *M* (a moment equal to zero) on member *CB*. Because the members of the truss are not beams, only point forces on the joints can be applied. Therefore we add two opposite forces on points *B* and *C* that yield a resultant equal to the dummy moment *M* (see figure below). Note that because the forces of the members 1, 4, and 5 are null, the problem can be simplified by ignoring them.

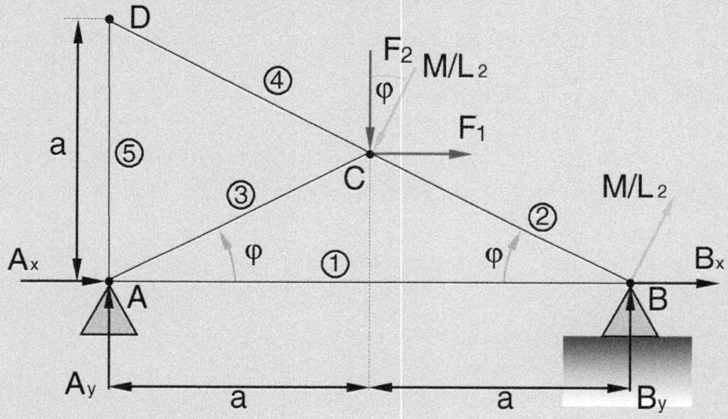

We should now consider the external forces F_1, F_2, and M/L_2 to calculate the support reactions of the truss and the forces of the bars. Then, the rotation of the bar *CB* can be calculated with the aid of Castigliano's theorem $\varphi_{BC} = \left.\frac{\partial U}{\partial M}\right|_{M=0}$.

Equilibrium equations of the forces in x-direction of the nodes *B* and *C*:

$$\textit{Node } B\text{: } -S_2 \cos\varphi + B_x + \frac{M}{L_2}\sin\varphi = 0$$

$$\textit{Node } C\text{: } S_3 \cos\varphi + \frac{M}{L_2}\sin\varphi = S_2\cos\varphi + F_1$$

The system of the above equations yield:

$$S_2 = \frac{B_x}{\cos\varphi} + \frac{M}{L_2}\tan\varphi$$

$$S_3 = \frac{B_x}{\cos\varphi} + \frac{F_1}{\cos\varphi}$$

Because the horizontal displacement of the node *B* is null, Castigliano's theorem can be applied for calculating B_x:

$$u_B = \frac{\partial U}{\partial B_x} = 0$$

or

$$\frac{1}{EA}\left(S_2\frac{\partial S_2}{\partial B_x}L_2 + S_3\frac{\partial S_3}{\partial B_x}L_3\right) = 0$$

Considering the findings for S_2 and S_3, the above equation yields:

$$B_x = -\frac{M\sin\varphi + F_1 L_3}{L_2 + L_3}$$

(Continued)

EXERCISE 11—CONT'D

Considering the above finding for B_x, the results for S_2, S_3 are:

$$S_2 = -\frac{M \sin\varphi + F_1 L_3}{(L_2 + L_3)\cos\varphi} + \frac{M}{L_2}\tan\varphi$$

$$S_3 = -\frac{M \sin\varphi + F_1 L_3}{(L_2 + L_3)\cos\varphi} + \frac{F_1}{\cos\varphi}$$

With the aid of the above results, the rotation of the bar BC can be obtained by Castigliano's theorem:

$$\varphi_{BC} = \left.\frac{\partial U}{\partial M}\right|_{M=0} = \frac{1}{EA}\left.\left(S_2\frac{\partial S_2}{\partial M}L_2 + S_3\frac{\partial S_3}{\partial M}L_3\right)\right|_{M=0}$$

EXERCISE 12

In the curved beam shown in the figure below, determine:

A. Deflection of point B in vertical direction
B. Deflection of point B in horizontal direction
C. Slope of the cross section (rotation) at point B

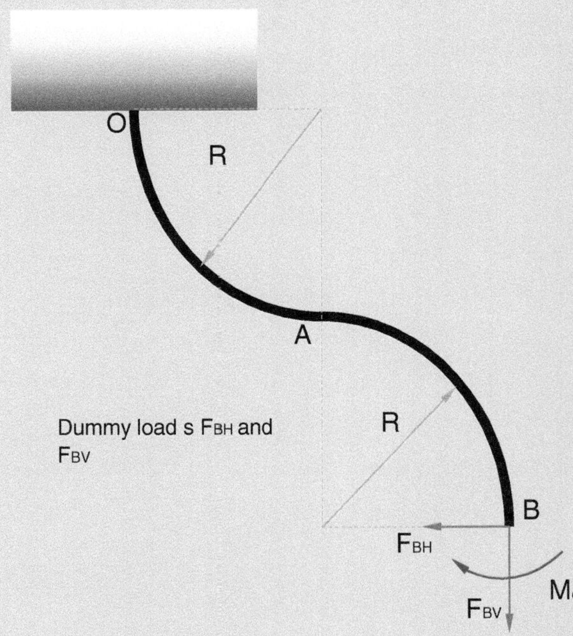

Solution

The solution can be obtained with the aid of Castigliano's theorem. To this end, the elastic energy of the beam should be determined. Because bending is the predominant loading case for the beam, the contribution of shear and axial forces into the elastic energy is negligible. Therefore only the bending moments can be considered to contribute to Castigliano's theorem for the beam OAB. Point A divides the beam into two segments AO and AB. The bending moments for these segments should be calculated separately.

EXERCISE 12—CONT'D

Bending moment distribution M_{AB}

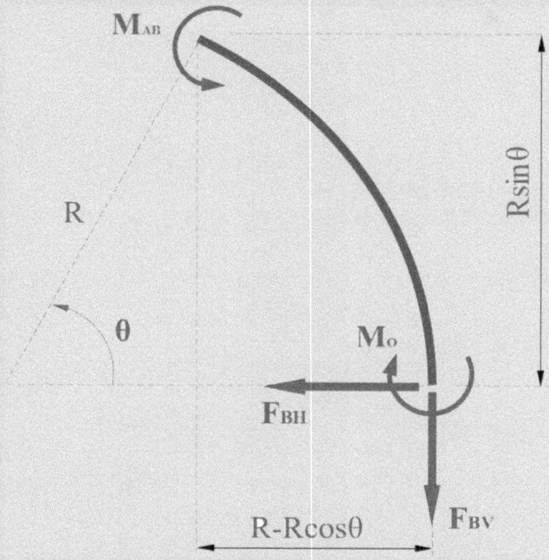

$$M_{AB} = M_o + F_{BV}(R - R\cos\theta) + F_{BH}R\sin\theta$$

$$\frac{\partial M_{AB}}{\partial F_{BV}} = R - R\cos\theta = R(1-\cos\theta)$$

$$\frac{\partial M_{AB}}{\partial F_{BH}} = R\sin\theta$$

Bending moment distribution M_{AO}

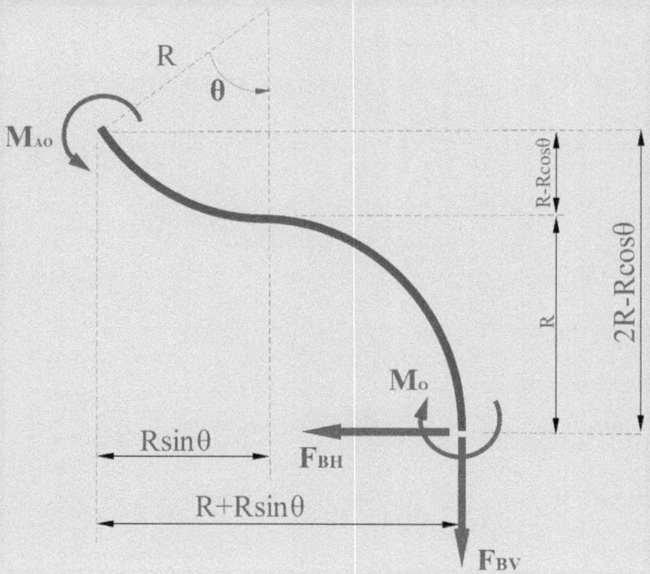

(Continued)

EXERCISE 12—CONT'D

$$M_{AO} = M_o + F_{BV}(R + R\sin\theta) + F_{BH}(2R - R\cos\theta)$$

$$\frac{\partial M_{AO}}{\partial F_{BV}} = R(1 + \sin\theta)$$

$$\frac{\partial M_{AO}}{\partial F_{BH}} = R(2 - \cos\theta)$$

Vertical displacement of point *B*:

We apply Castigliano's theorem to obtain the vertical displacement in point *B*:

$$u_B = \left.\frac{\partial U}{\partial F_{BV}}\right|_{F_{BV}=0} = \left.\frac{1}{EI}\left(\int_0^{\pi/2} M_{AB}\frac{\partial M_{AB}}{\partial F_{BV}}R\,d\theta + \int_0^{\pi/2} M_{AO}\frac{\partial M_{AO}}{\partial F_{BV}}R\,d\theta\right)\right|_{F_{BV}=0}$$

or

$$u_B = \frac{1}{EI}\left(\int_0^{\pi/2}\left(M_o + \cancel{F_{BV}}(R - R\cos\theta) + \cancel{F_{BH}}R\sin\theta\right)R(1-\cos\theta)R\,d\theta + \right.$$
$$\left. + \int_0^{\pi/2}\left(M_o + \cancel{F_{BV}}(R + R\sin\theta) + \cancel{F_{BH}}(R - R\cos\theta)\right)R(1+\sin\theta)R\,d\theta\right)$$

The above equation yields:

$$u_B = \frac{1}{EI}\left(\int_0^{\pi/2} M_oR^2(1 - \cos\theta)d\theta + \int_0^{\pi/2} M_oR^2(1 + \sin\theta)d\theta\right) = \frac{M_oR^2\pi}{EI}$$

Horizontal displacement of point *B*:

We apply Castigliano's theorem to obtain the horizontal displacement in point *B*:

$$f_B = \left.\frac{\partial U}{\partial F_{BH}}\right|_{F_{BH}=0} = \left.\frac{1}{EI}\left(\int_0^{\pi/2} M_{AB}\frac{\partial M_{AB}}{\partial F_{BH}}R\,d\theta + \int_0^{\pi/2} M_{AO}\frac{\partial M_{AO}}{\partial F_{BH}}R\,d\theta\right)\right|_{F_{BH}=0}$$

or

$$f_B = \frac{1}{EI}\left(\int_0^{\pi/2}\left(M_o + \cancel{F_{BV}}(R - R\cos\theta) + \cancel{F_{BH}}R\sin\theta\right)R\sin\theta R\,d\theta + \right.$$
$$\left. + \int_0^{\pi/2}\left(M_o + \cancel{F_{BV}}(R + R\sin\theta) + \cancel{F_{BH}}(R - R\cos\theta)\right)R(2-\cos\theta)R\,d\theta\right)$$

The above equation yields:

$$f_B = \frac{1}{EI}\left(\int_0^{\pi/2} M_oR^2\sin\theta\,d\theta + \int_0^{\pi/2} M_oR^2(2 - \cos\theta)d\theta\right) = \frac{M_oR^2\pi}{EI}$$

Slope of the cross section at point *B*:

Now, we calculate the slope of the cross section at point *B*. Because there are no additional forces in the structures, we can determine the moments

$$M_{AB} = M_{AO} = M_o$$

$$\frac{\partial M_{AB}}{\partial M_o} = \frac{\partial M_{AO}}{\partial M_o} = 1$$

EXERCISE 12—CONT'D

and apply Castigliano's theorem:

$$\varphi_B = \frac{\partial U}{\partial M_o} = \frac{1}{EI}\left(\int_0^{\pi/2} M_{AB}\frac{\partial M_{AB}}{\partial M_o}R\ d\theta + \int_0^{\pi/2} M_{AO}\frac{\partial M_{AO}}{\partial M_o}R\ d\theta\right)$$

or

$$\varphi_B = \frac{1}{EI}\left(\int_0^{\pi/2} M_o R\ d\theta + \int_0^{\pi/2} M_o R\ d\theta\right) = \frac{M_o R\pi}{EI}$$

APPENDIX: GAUSS DIVERGENCE THEOREM

Considering two vectors $\vec{a}$ and $\vec{N}$ such as in Fig. A.8.1, the Gauss divergence theorem states:

$$\int_S \left(\vec{a}\cdot\vec{N}\right)dS = \int_V (\mathrm{div}\vec{a})\ dV$$

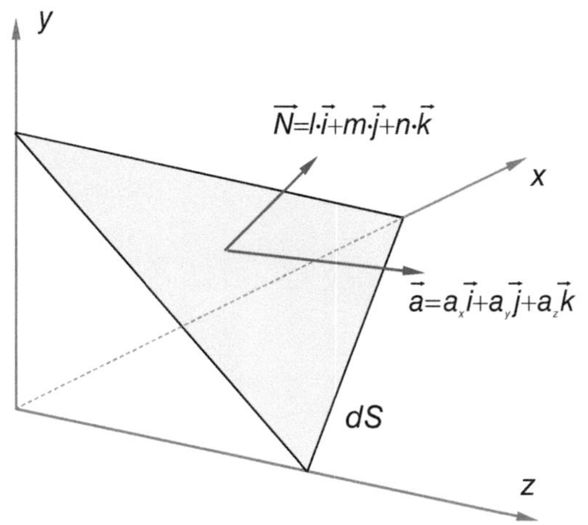

Fig. A.8.1:

The above expression can be written analytically as follows:

$$\int_S \left(a_x l + a_y m + a_z n\right)dS = \int_V \left(\frac{\partial a_x}{\partial x} + \frac{\partial a_y}{\partial y} + \frac{\partial a_z}{\partial z}\right) dV$$

REFERENCES

[1] Pavlou D. Essentials of the Finite Element Method. Academic Press; 2015.
[2] Arthur P, Boresi Richard Schmidt J. , Advanced Mechanics of Materials. Wiley; 2002.
[3] Timoshenko SP, Goodier JN. Theory of elasticity. McGraw-Hill; 1970.
[4] Fenner RT. Engineering elasticity. Ellis Horwood ltd; 1986.
[5] Bittencourt Marco. Computational solid mechanics. CRC press; 2015.
[6] Ugural Ansel, Fenster Saul. Advanced Mechanics of Materials and Applied Elasticity. Pearson; 2019.
[7] Solecki Roman, Conant RJay. Advanced Mechanics of Materials. Oxford University Press; 2003.
[8] Cook Robert. Advanced Mechanics of Materials. Pearson; 1998.
[9] Bickford William. Advanced Mechanics of Materials. Longman Pub Group; 1998.
[10] Anand Lallit, Govindjee Sanjay. Continuum Mechanics of Solids. OUP Oxford; 2020.

CHAPTER

Flat Plates

9

Chapter Outline

Structural components with a thickness much smaller than the length and width where their middle surface lies in a flat plane are described as plates. Plates are widely used in civil and mechanical engineering structures (buildings, bridges, offshore platforms, etc.) and are capable to sustain bending. Design codes provide certain specifications for the maximum deflection and flexural stress for plates. The aim of this chapter is the displacement and stress analysis of flat plates.

9.1 CYLINDRICAL PURE BENDING OF FLAT RECTANGULAR PLATES (SINGLE CURVATURE)

The bending analysis of flat plates can be performed with the same concept and under the same assumptions as beams:

- The cross sections of the plate remain plane after bending.
- The mid-pane of the plate remains free of flexural stresses during bending.
- Normal stresses in the direction normal to the plate surface can be neglected (pure bending).

The deformed shape of a bended plate may have single or double curvature. Bended plate with a single curvature is called cylindrical bending. In the cylindrical

Advanced Mechanics of Solids and Structures. DOI: http://doi.org/10.1016/B978-0-443-23778-2.00009-3

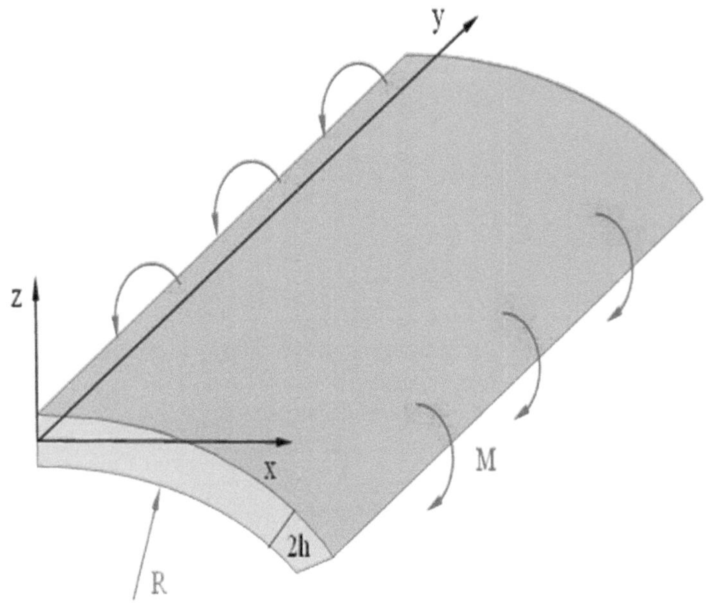

FIG. 9.1

Rectangular plate under cylindrical bending.

bending of the rectangular plate in Fig. 9.1, the bending moments are uniformly distributed to two opposite edges. In bending analysis of plates, the bending moment per unit length M is used for the calculations.

As it was explained in bending of beams, the following formula can be used:

$$\varepsilon_{xx}(z) = \frac{z}{R} \tag{9.1}$$

Because one-dimensional bending is assumed, $\varepsilon_{yy} = 0$. Therefore Hooke's law

$$\varepsilon_{yy} = \frac{1}{E}(\sigma_{yy} - \nu\sigma_{xx}) \tag{9.2}$$

yields

$$\sigma_{yy} = \nu\sigma_{xx} \tag{9.3}$$

With the aid of the above finding, the following formula for strain ε_{xx}

$$\varepsilon_{xx} = \frac{1}{E}(\sigma_{xx} - \nu\sigma_{yy}) \tag{9.4}$$

can be transformed to:

$$\varepsilon_{xx} = \frac{\sigma_{xx}}{E}(1 - \nu^2) \tag{9.5}$$

Combining Eqs. (9.1) and (9.5)

$$\frac{\sigma_{xx}}{E}(1 - \nu^2) = \frac{z}{R} \tag{9.6}$$

the following result can be obtained:

$$\sigma_{xx} = \frac{E}{R(1 - \nu^2)}z \tag{9.7}$$

The above equation can be utilized in the bending moment equilibrium:

$$M = \int_{-h}^{h} \sigma_{xx} z \, dz \tag{9.8}$$

or

$$M = \frac{E}{R(1 - \nu^2)} \int_{-h}^{h} z^2 \, dz \tag{9.9}$$

The above equation yields

$$M = \frac{2Eh^3}{3R(1 - \nu^2)} \tag{9.10}$$

Using the definition of curvature versus the vertical deflection w

$$\frac{1}{R} = -\frac{d^2w}{dx^2} \tag{9.11}$$

Eq. (9.10) can be written in the following form:

$$M = -\frac{2Eh^3}{3(1 - \nu^2)} \frac{d^2w}{dx^2} \tag{9.12}$$

Let's use the notation D for the following quantity that is called flexural rigidity

$$D = \frac{2Eh^3}{3(1 - \nu^2)} \tag{9.13}$$

Then, Eq. (9.12) can be written as

$$M = -D\frac{d^2w}{dx^2} \tag{9.14}$$

PURE BENDING OF PLATES IN TWO DIRECTIONS (DOUBLE CURVATURE)

Let's assume a rectangular plate under bending moment distribution in the four edges (Fig. 9.2). The deformed shape has now two curvatures, and the following equations can be written:

$$\varepsilon_{xx} = \frac{z}{R_x} = -z\frac{\partial^2 w}{\partial x^2} \tag{9.15}$$

$$\varepsilon_{yy} = \frac{z}{R_y} = -z\frac{\partial^2 w}{\partial y^2} \tag{9.16}$$

With the aid of Eqs. (9.15) and (9.16), Hooke's law for stresses σ_{xx} and σ_{yy}

$$\sigma_{xx} = \frac{E}{1-\nu^2}(\varepsilon_{xx} - \nu\varepsilon_{yy}) \tag{9.17}$$

$$\sigma_{yy} = \frac{E}{1-\nu^2}(\varepsilon_{yy} - \nu\varepsilon_{xx}) \tag{9.18}$$

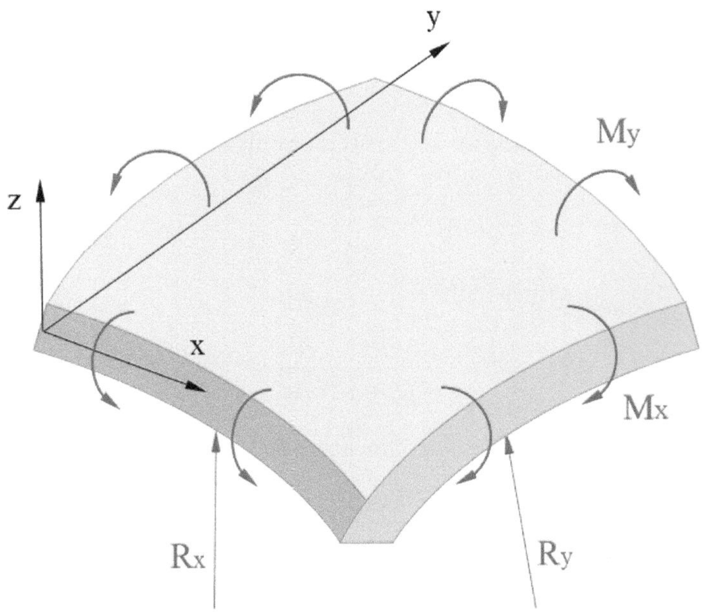

FIG. 9.2

Pure bending of a rectangular plate in two dimensions (double curvature).

yield

$$\sigma_{xx} = \frac{E}{1-\nu^2} z \left(-\frac{\partial^2 w}{\partial x^2} - \nu \frac{\partial^2 w}{\partial y^2} \right) \tag{9.19}$$

$$\sigma_{yy} = \frac{E}{1-\nu^2} z \left(-\frac{\partial^2 w}{\partial y^2} - \nu \frac{\partial^2 w}{\partial x^2} \right) \tag{9.20}$$

and the moments per unit length M_x and M_y can be obtained by the following equations:

$$M_x = \int_{-h}^{h} \sigma_{xx} z \, dz = -\frac{E}{1-\nu^2} \frac{2h^3}{3} \left(\frac{\partial^2 w}{\partial x^2} + \nu \frac{\partial^2 w}{\partial y^2} \right) \tag{9.21}$$

$$M_y = \int_{-h}^{h} \sigma_{yy} z \, dz = -\frac{E}{1-\nu^2} \frac{2h^3}{3} \left(\frac{\partial^2 w}{\partial y^2} + \nu \frac{\partial^2 w}{\partial x^2} \right) \tag{9.22}$$

Using the definition of the flexural rigidity given in Eq. (9.13), the above equations can be written in the following form:

$$M_x = D \left(\frac{\partial^2 w}{\partial x^2} + \nu \frac{\partial^2 w}{\partial y^2} \right) \tag{9.23}$$

$$M_y = -D \left(\frac{\partial^2 w}{\partial y^2} + \nu \frac{\partial^2 w}{\partial x^2} \right) \tag{9.24}$$

9.2 DEFLECTION AND BENDING STRESSES OF PLATES UNDER GENERAL LOADING

Fig. 9.3 demonstrates a rectangular plate element with dimensions δx and δy and thickness $2h$. The plate element is subjected to a general pressure $p(x, y)$. This model will be used for the determination of the vertical deflection distribution $w(x, y)$ and the bending stresses.

As Fig. 9.3 shows, the internal forces can be classified into the following three types:

- Pure bending moments per unit length M_x, M_y
- Twisting moments per unit lengthM_{xy}, M_{yx}
- Shear forces per unit length S_x, S_y

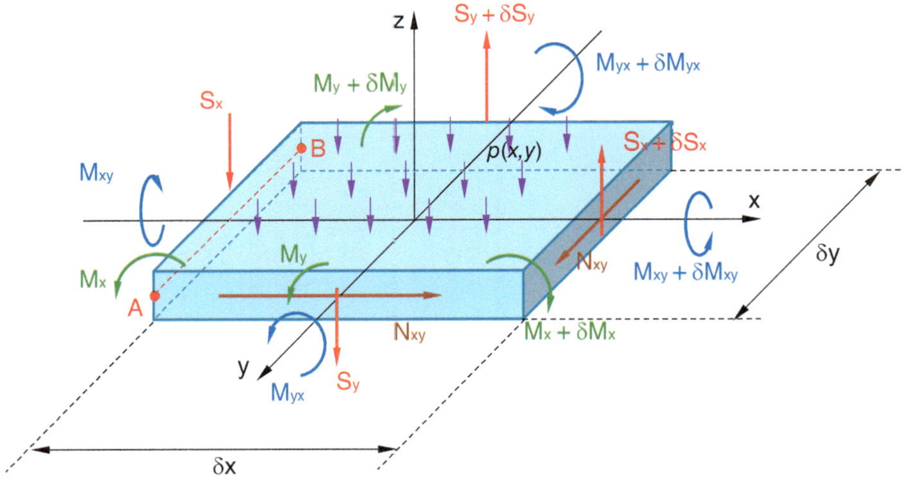

FIG. 9.3

A rectangular plate element under general pressure distribution $p(x, y)$.

For the derivation of the differential equations for calculating the plate deflection distribution and stresses, the equilibrium equations of the forces of the elementary plate element (Fig. 9.3) will be used:

Equilibrium of forces in z-direction:

Taking the equilibrium of the forces in z-direction, the following equation will be obtained:

$$(S_x + \delta S_x)\delta y - S_x\ \delta y + \left(S_y + \delta S_y\right)\delta x - S_y\ \delta x + p(x, y)\delta x\delta y = 0 \qquad (9.25)$$

or

$$\delta S_x \delta y + \delta S_y \delta x - p(x, y)\delta x\delta y = 0 \qquad (9.26)$$

Dividing the above equation by the term $\delta x\delta y$ and using the formulae

$$\delta S_x = \frac{\partial S_x}{\partial x}\delta x \qquad (9.27)$$

$$\delta S_y = \frac{\partial S_y}{\partial y}\delta y \qquad (9.28)$$

the following equation can be obtained:

$$\frac{\partial S_x}{\partial x} + \frac{\partial S_y}{\partial y} + p(x, y) = 0 \qquad (9.29)$$

Equilibrium of moments of the edge AB

$$(M_x + \delta M_x)\delta y - M_x\ \delta y + \left(M_{yx} + \delta M_{yx}\right)\delta x - M_{yx}\ \delta x -$$
$$(S_x + \delta S_x)\delta y \cdot \delta x - \left(S_y + \delta S_y\right)\delta x\frac{\delta x}{2} + S_y\ \delta x\frac{\delta x}{2} + p(x, y) \tag{9.30}$$
$$\delta x \delta y\frac{\delta x}{2} = 0$$

Ignoring the products of more than two increments, that is, $\delta(\bullet)\delta(\bullet)\delta(\bullet) \approx 0$, the following equation can be obtained:

$$\frac{\partial M_x}{\partial x} + \frac{\partial M_{yx}}{\partial y} - S_x = 0 \tag{9.31}$$

Similarly, the equilibrium of the moments of an edge perpendicular to AB yields:

$$\frac{\partial M_{xy}}{\partial x} - \frac{\partial M_y}{\partial y} + S_y = 0 \tag{9.32}$$

Considering the equations

$$M_{xy} = \int_{-h}^{h} -\sigma_{xy} z\ dz \tag{9.33}$$

$$M_{yx} = \int_{-h}^{h} \sigma_{yx} z\ dz \tag{9.34}$$

$$\sigma_{yx} = \sigma_{xy} \tag{9.35}$$

the following result can be obtained:

$$M_{xy} = -M_{yx} \tag{9.36}$$

With the aid of Eq. (9.36), Eqs. (9.31) and (9.32) can be transformed to:

$$\frac{\partial M_x}{\partial x} - \frac{\partial M_{xy}}{\partial y} - S_x = 0 \tag{9.37}$$

$$\frac{\partial M_{xy}}{\partial x} - \frac{\partial M_y}{\partial y} + S_y = 0 \tag{9.38}$$

or

$$\frac{\partial}{\partial x}\left[\frac{\partial M_x}{\partial x} - \frac{\partial M_{xy}}{\partial y} - S_x\right] = 0 \tag{9.39}$$

$$\frac{\partial}{\partial y}\left[\frac{\partial M_{xy}}{\partial x} - \frac{\partial M_y}{\partial y} + S_y\right] = 0 \tag{9.40}$$

Deducting Eq. (9.40) from Eq. (9.39), the following differential equation can be obtained:

$$\frac{\partial^2 M_x}{\partial x^2} - 2\frac{\partial^2 M_{xy}}{\partial x \partial y} + \frac{\partial^2 M_y}{\partial y^2} = \underbrace{\frac{\partial S_x}{\partial x} + \frac{\partial S_y}{\partial y}}_{-p(x,y)} \tag{9.41}$$

With the aid of Eq. (9.29), the above equation yields:

$$\frac{\partial^2 M_x}{\partial x^2} - 2\frac{\partial^2 M_{xy}}{\partial x \partial y} + \frac{\partial^2 M_y}{\partial y^2} = -p(x, y) \tag{9.42}$$

Correlation of M_{xy} to displacements:
Combining the shear strain versus displacement formula

$$\varepsilon_{xy} = \frac{\partial u}{\partial y} + \frac{\partial v}{\partial x} \tag{9.43}$$

with Hooke's law

$$\sigma_{xy} = G\varepsilon_{xy} = \frac{E}{2(1+\nu)}\varepsilon_{xy} \tag{9.44}$$

the following result can be obtained:

$$\sigma_{xy} = \frac{E}{2(1+\nu)}\left(\frac{\partial u}{\partial y} + \frac{\partial v}{\partial x}\right) \tag{9.45}$$

Combining the definition of strain

$$\varepsilon_{xx} = \frac{\partial u}{\partial x} \tag{9.46}$$

with Eq. (9.15), the following formula for the displacement u in the x-direction resulted:

$$u = -z\frac{\partial w}{\partial x} \tag{9.47}$$

Similarly, combination of the definition of strain

$$\varepsilon_{yy} = \frac{\partial v}{\partial y} \tag{9.48}$$

with Eq. (9.16) yields:

$$v = -z\frac{\partial w}{\partial y} \tag{9.49}$$

With the aid of Eqs. (9.47) and (9.49), Eq. (9.45) can correlate the shear stress σ_{xy} to the vertical displacement w:

$$\sigma_{xy} = -\frac{Ez}{1+\nu}\frac{\partial^2 w}{\partial x \partial y} \tag{9.50}$$

Then, the moment M_{xy} is

$$M_{xy} = \int_{-h}^{h} -\sigma_{xy} z \, dz = -\int_{-h}^{h} -\frac{Ez}{1+\nu}\frac{\partial^2 w}{\partial x \partial y} z \, dz \tag{9.51}$$

or

$$M_{xy} = \frac{E}{1+\nu}\frac{\partial^2 w}{\partial x \partial y}\int_{-h}^{h} z^2 \, dz \tag{9.52}$$

The last equation yields

$$M_{xy} = \frac{2Eh^3}{3(1+\nu)}\frac{\partial^2 w}{\partial x \partial y} \tag{9.53}$$

or

$$M_{xy} = D(1-\nu)\frac{\partial^2 w}{\partial x \partial y} \tag{9.54}$$

Taking the following derivatives to Eqs. (9.23), (9.24), and (9.54),

$$\frac{\partial^2 M_x}{\partial x^2} = D\left(\frac{\partial^4 w}{\partial x^4} + \nu\frac{\partial^4 w}{\partial x^2 \partial y^2}\right) \tag{9.55}$$

$$\frac{\partial^2 M_y}{\partial y^2} = D\left(\frac{\partial^4 w}{\partial y^4} + \nu\frac{\partial^4 w}{\partial x^2 \partial y^2}\right) \tag{9.56}$$

$$\frac{\partial^2 M_{xy}}{\partial x \partial y} = D(1-\nu)\frac{\partial^4 w}{\partial x^2 \partial y^2} \tag{9.57}$$

and replacing the above results to Eq. (9.42), the following biharmonic equation can be obtained:

$$\frac{\partial^4 w}{\partial x^4} + 2\frac{\partial^4 w}{\partial x^2 \partial y^2} + \frac{\partial^4 w}{\partial y^4} = \frac{p(x, y)}{D} \tag{9.58}$$

Eq. (9.42) provides the vertical displacement distribution $w = w(x, y)$. Then, the bending stresses of a plate can be obtained by Eqs. (9.19), (9.20), and (9.50) and the bending strains by Eqs. (9.15) and (9.16). Finally, the bending moment distributions are provided by Eqs. (9.23), (9.24), and (9.54), and the shear forces can be obtained by Eqs. (9.31) and (9.32).

9.3 BOUNDARY CONDITIONS IN PLATE PROBLEMS

The solution of Eq. (9.58) requires the boundary conditions of the plate. For a plate of arbitrary shape, the orientation of the free edge at a point P is specified by the vectors $\vec{n}$, $\vec{s}$ (Fig. 9.4). The expressions of the bending moments in terms of the variables n, s can be obtained by adapting Eqs. (9.24) and (9.54):

$$M_n = -D\left(\frac{\partial^2 w}{\partial n^2} + \nu\frac{\partial^2 w}{\partial s^2}\right) \tag{9.59}$$

$$M_{ns} = D(1-\nu)\frac{\partial^2 w}{\partial n \partial s} \tag{9.60}$$

Similarly, for the shear forces in the free edge the following formula obtained by Eq. (9.31) can be used:

$$S_n = \frac{\partial M_n}{\partial n} + \frac{\partial M_{ns}}{\partial s} \tag{9.61}$$

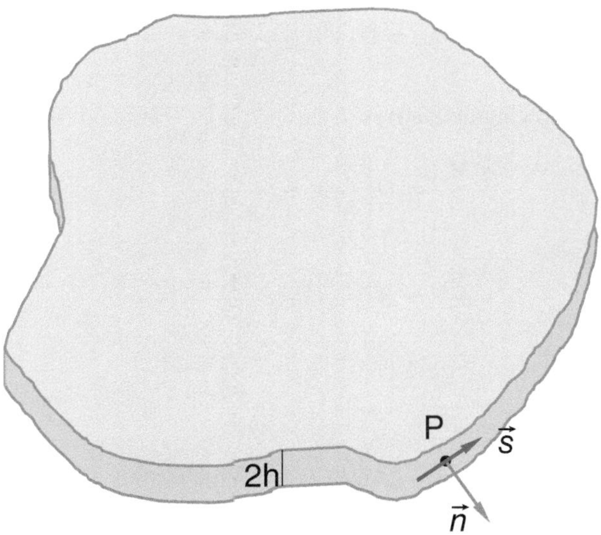

FIG. 9.4

Plate of arbitrary shape.

The above equation can be expressed in terms of w with the aid of Eqs. (9.59) and (9.60):

$$S_n = -D\frac{\partial}{\partial n}\left(\frac{\partial^2 w}{\partial n^2} + \frac{\partial^2 w}{\partial s^2}\right) \tag{9.62}$$

Depending to the type of plate support, the following three boundary conditions exist:

- Fixed supported edge

In fixed supported edge, both deflection and slope of the plate are null:

$$w = 0 \tag{9.63}$$

$$\frac{\partial w}{\partial n} = 0 \tag{9.64}$$

- Simply supported edge

In simply supported edge the deflection w and the bending moment M_n is null:

$$w = 0 \tag{9.65}$$

$$M_n = -D\left(\frac{\partial^2 w}{\partial n^2} + \nu\frac{\partial^2 w}{\partial s^2}\right) = 0 \tag{9.66}$$

- Free edge

In a free edge, no bending or twisting moments nor shear forces are applied. Therefore:

$$M_n = -D\left(\frac{\partial^2 w}{\partial n^2} + \nu\frac{\partial^2 w}{\partial s^2}\right) = 0 \tag{9.67}$$

$$M_{ns} = D(1 - \nu)\frac{\partial^2 w}{\partial n \partial s} = 0 \tag{9.68}$$

$$S_n = -D\frac{\partial}{\partial n}\left(\frac{\partial^2 w}{\partial n^2} + \frac{\partial^2 w}{\partial s^2}\right) = 0 \tag{9.69}$$

9.4 AXISYMMETRIC PLATES

Bending of circular plates can be easily analyzed if we use the polar coordinate system (Fig. 9.5). The load q can be a function of the variables (r, θ), or can be axisymmetric too, that is, $q = q(r)$.

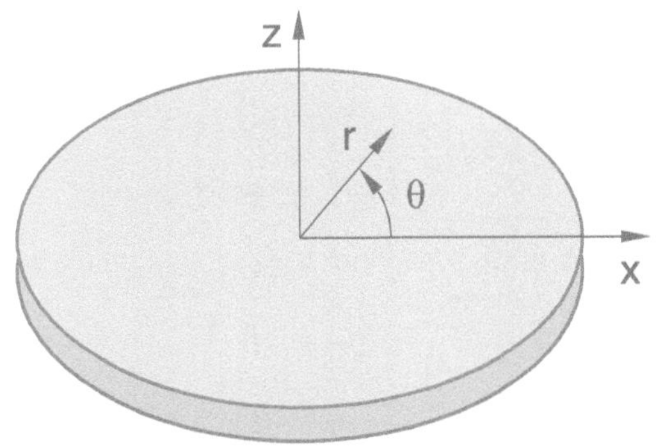

FIG. 9.5

Circular plate in a polar coordinate system.

The biharmonic Eq. (9.58) and the formulae for bending moments and shear forces in the polar coordinate system have the following form:

$$\left(\frac{\partial^2}{\partial r^2}+\frac{1}{r}\frac{\partial}{\partial r}+\frac{1}{r^2}\frac{\partial^2}{\partial \theta^2}\right)\left(\frac{\partial^2 w}{\partial r^2}+\frac{1}{r}\frac{\partial w}{\partial r}+\frac{1}{r^2}\frac{\partial^2 w}{\partial \theta^2}\right)=\frac{p(r,\theta)}{D} \tag{9.70}$$

$$M_r=-D\left[\frac{\partial^2 w}{\partial r^2}+\nu\left(\frac{1}{r}\frac{\partial w}{\partial r}+\frac{1}{r^2}\frac{\partial^2 w}{\partial \theta^2}\right)\right] \tag{9.71}$$

$$M_\theta=-D\left[\frac{1}{r}\frac{\partial w}{\partial r}+\frac{1}{r^2}\frac{\partial^2 w}{\partial \theta^2}+\nu\frac{\partial^2 w}{\partial r^2}\right] \tag{9.72}$$

$$M_{r\theta}=D(1-\nu)\left(\frac{1}{r}\frac{\partial^2 w}{\partial r\partial \theta}-\frac{1}{r}\frac{\partial w}{\partial \theta}\right) \tag{9.73}$$

$$S_r=-D\frac{\partial}{\partial r}(\nabla^2 w) \tag{9.74}$$

$$S_\theta=-D\frac{1}{r}\frac{\partial}{\partial \theta}(\nabla^2 w) \tag{9.75}$$

If the load is axisymmetric, there is no variation with angular coordinate, that is, $q=q(r)$. Therefore all the derivatives with respect to the angular coordinate are $\partial^n(\bullet)/\partial\theta^n=0$, and the above equations can be simplified:

$$\left(\frac{d^2}{dr^2}+\frac{1}{r}\frac{d}{dr}\right)\left(\frac{d^2 w}{dr^2}+\frac{1}{r}\frac{dw}{dr}\right)=\frac{p(r)}{D} \tag{9.76}$$

$$M_r = -D\left[\frac{d^2w}{dr^2} + \frac{\nu}{r}\frac{dw}{dr}\right] \tag{9.77}$$

$$M_\theta = -D\left[\frac{1}{r}\frac{dw}{dr} + \nu\frac{d^2w}{dr^2}\right] \tag{9.78}$$

$$M_{r\theta} = 0 \tag{9.79}$$

$$S_r = -D\frac{d}{dr}\left[\frac{1}{r}\frac{d}{dr}\left(r\frac{dw}{dr}\right)\right] \tag{9.80}$$

$$S_\theta = 0 \tag{9.81}$$

The corresponding bending stresses can be provided by the following equations:

$$\sigma_{rr} = \frac{M_r}{I}z \tag{9.82}$$

$$\sigma_{\theta\theta} = \frac{M_\theta}{I}z \tag{9.83}$$

$$\sigma_{zr} = \frac{S_r}{2I}(h^2 - z^2) \tag{9.84}$$

where

$$I = \frac{2h^3}{3} \tag{9.85}$$

The flexural rigidity D is given by Eq. (9.13). If the thickness h of the plate is constant, Eq. (9.76) can be written in the following form

$$\frac{1}{r}\frac{d}{dr}\left\{r\frac{d}{dr}\left[\frac{1}{r}\frac{d}{dr}\left(r\frac{dw}{dr}\right)\right]\right\} = \frac{p(r)}{D} \tag{9.86}$$

and can be easily solved by four successive integrations.

If the thickness is variable, that is, $h = h(r)$, the governing equation for axisymmetric plates under axisymmetric loading can be written as:

$$\frac{d^2}{dr^2}\left\{D\left(\frac{d^2w}{dr^2} + \frac{\nu}{r}\frac{dw}{dr}\right) + \frac{1}{r}\frac{d}{dr}\left[D\left(\frac{1}{r}\frac{dw}{dr} + \nu\frac{d^2w}{dr^2}\right)\right]\right\} = p(r) \tag{9.87}$$

where

$$D = D(r) = \frac{2Eh^3(r)}{3(1 - \nu^2)} \tag{9.88}$$

EXERCISE 1

Derive the formulae for calculating the deflection, bending moments, shear force distribution, and the bending stresses at $r = 0$ and $r = R$ of a uniformly loaded circular plate that is clamped at the perimeter (Fig. 9.6).

Solution

Eq. (9.86) can be written in the following form:

$$\frac{d}{dr}\left\{r\frac{d}{dr}\left[\frac{1}{r}\frac{d}{dr}\left(r\frac{dw}{dr}\right)\right]\right\} = \frac{p}{D}r$$

Integration of the above equation yields

$$r\frac{d}{dr}\left[\frac{1}{r}\frac{d}{dr}\left(r\frac{dw}{dr}\right)\right] = \frac{pr^2}{2D} + C_1$$

or

$$\frac{d}{dr}\left[\frac{1}{r}\frac{d}{dr}\left(r\frac{dw}{dr}\right)\right] = \frac{pr}{2D} + \frac{C_1}{r}$$

Integration of the last equation results:

$$\frac{1}{r}\frac{d}{dr}\left(r\frac{dw}{dr}\right) = \frac{pr^2}{4D} + C_1 \ln r + C_2$$

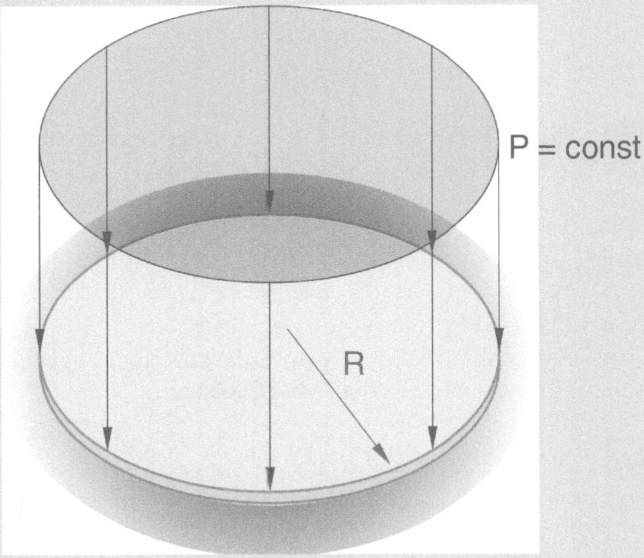

FIG. 9.6

Clamped circular plate under uniform pressure distribution.

EXERCISE 1—CONT'D

or

$$\frac{d}{dr}\left(r\frac{dw}{dr}\right) = \frac{pr^3}{4D} + C_1 r \ln r + C_2 r$$

Taking one more integration, we get:

$$r\frac{dw}{dr} = \frac{pr^4}{16D} + C_1\left(\frac{r^2}{2}\ln r - \frac{r^2}{4}\right) + C_2\frac{r^2}{2} + C_3$$

or

$$\frac{dw}{dr} = \frac{pr^3}{16D} + C_1\left(\frac{r}{2}\ln r - \frac{r}{4}\right) + C_2\frac{r}{2} + \frac{C_3}{r}$$

By integrating the last equation, the following result can be obtained:

$$w = \frac{pr^4}{64D} + C_1\left(\frac{r^2}{4}\ln r - \frac{r^2}{4}\right) + C_2\frac{r^2}{4} + C_3\ln r + C_4$$

The constants of integrations C_1, C_2, C_3, *and* C_4 can be calculated with the aid of the boundary conditions.

The resulted formula for w contains the term $\ln r$ that is infinite for $r = 0$. Therefore this term should be eliminated by setting

$$C_1 = C_3 = 0$$

and the general solution takes the following simplified form:

$$w = \frac{pr^4}{64D} + C_2\frac{r^2}{4} + C_4$$

The constants C_2 *and* C_4 can be calculated by the following boundary conditions:

$$w(R) = 0$$

$$\left.\frac{dw}{dr}\right|_{r=R} = 0$$

Considering the simplified solution, the above conditions yield:

$$\frac{pR^4}{64D} + C_2\frac{R^2}{4} + C_4 = 0$$

$$\frac{pR^3}{16D} + C_2\frac{R}{2} = 0$$

or

$$C_2 = -\frac{pR^2}{8D}$$

$$C_4 = \frac{pR^4}{64D}$$

(Continued)

EXERCISE 1—CONT'D

Then, the deflection profile of the plate is:

$$w = \frac{p}{32D}(R^2 - r^2)^2$$

With the aid of the last equation, the bending moments and shear force can be obtained by the following equations:

$$M_r = -D\left[\frac{d^2w}{dr^2} + \frac{\nu}{r}\frac{dw}{dr}\right] = \frac{p}{16}[(1+\nu)R^2 - (3+\nu)r^2]$$

$$M_\theta = -D\left[\frac{1}{r}\frac{dw}{dr} + \nu\frac{d^2w}{dr^2}\right] = \frac{p}{16}[(1+\nu)R^2 - (1+3\nu)r^2]$$

$$S_r = -D\frac{d}{dr}\left[\frac{1}{r}\frac{d}{dr}\left(r\frac{dw}{dr}\right)\right] = -\frac{pr}{2}$$

The stresses at $r = 0$ and $r = R$ at $z = \pm h$ can be determined by Eqs. (9.82) and (9.83). Then, the following results can be obtained:

$$\sigma_{rr}(0) = \pm\frac{3p(1+\nu)R^2}{32h^2}$$

$$\sigma_{rr}(R) = \mp\frac{3pR^2}{16h^2}$$

$$\sigma_{\theta\theta}(0) = \frac{3p(1+\nu)R^2}{32h^2}$$

$$\sigma_{\theta\theta}(R) = \mp\frac{\nu pR^2}{16h^2}$$

9.5 PLATE ON ELASTIC FOUNDATION UNDER AXISYMMETRIC LOADING

Plates are usually used in foundations where they rest on an elastic substrate. Apart from large structural plates on elastic foundation, thin plates or layers of metal, ceramic, or synthetic materials are bonded in the surface of machine structural parts or electronic devices to improve their mechanical, thermal, electronic, or tribological properties. In all these applications the substrate of the thin plate can be simulated as a Winkler-type foundation that reacts with a pressure proportional to the deflection of the plate at each point. For an infinite plate under axisymmetric loading (Fig. 9.7), the loading term in Eq. (9.76) is

$$p(r) = q(r) - k\,w(r) \tag{9.89}$$

where k is the modulus of the elastic foundation and expresses its stiffness.

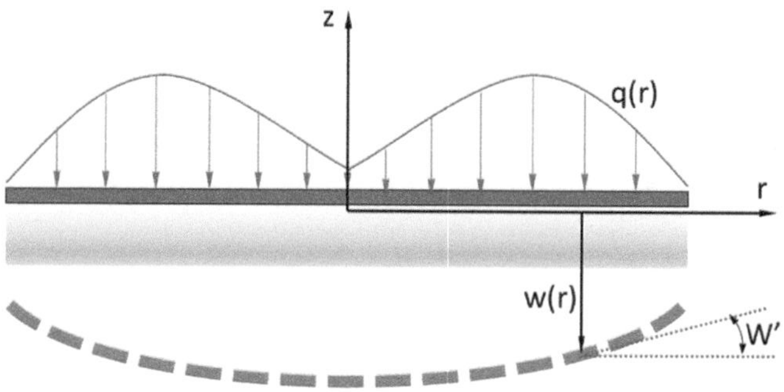

FIG. 9.7

Plate on elastic foundation under axisymmetric loading.

With the aid of Eq. (9.89), Eq. (9.76) takes the following form:

$$\left(\frac{d^2}{dr^2} + \frac{1}{r}\frac{d}{dr}\right)\left(\frac{d^2w}{dr^2} + \frac{1}{r}\frac{dw}{dr}\right) = \frac{p(r) - kw}{D} \tag{9.90}$$

or

$$\left(\frac{d^2}{dr^2} + \frac{1}{r}\frac{d}{dr}\right)\left(\frac{d^2w}{dr^2} + \frac{1}{r}\frac{dw}{dr}\right) + \lambda w = \frac{p(r)}{D} \tag{9.91}$$

where

$$\lambda = \frac{k}{D} \tag{9.92}$$

If we use the operator

$$B_n = \frac{d^2}{dr^2} + \frac{1}{r}\frac{d}{dr} - \frac{n^2}{r^2} \tag{9.93}$$

Eq. (9.91) can be written in the following form:

$$B_0 B_0 w(r) + \lambda w = \frac{p(r)}{D} \tag{9.94}$$

The term of the elastic foundation makes the solution of Eq. (9.91) quite complex. To this end, a technique based on Hankel integral transforms will be employed.

9.5.1 HANKEL INTEGRAL TRANSFORM FOR THE SOLUTION OF AN AXISYMMETRIC PLATE ON ELASTIC FOUNDATION

The definition of Hankel transform and its inverse form employed in the proposed solution is expressed by the following equations:

$$\overline{f}_n(\xi) = H_n\{f(r);\ \xi\} = \int_0^\infty rf(r)J_n(\xi r)dr \tag{9.95}$$

and

$$f(r) = H_n^{-1}\left\{\overline{f}_n(\xi);\ r\right\} = \int_0^\infty \xi\,\overline{f}_n(\xi)J_n(\xi r)d\xi \tag{9.96}$$

where $J_n(x)$ is the Bessel function of n-th order of second kind, whereas $H_n\{f(r);\ \xi\}$ and $H_n^{-1}\left\{\overline{f}_n(\xi);\ r\right\}$ are the Hankel transform and inverse Hankel transform operator, respectively.

Taking the operator H_0 to Eq. (9.91) we get:

$$H_0\{B_0B_0w(r);\ \xi\} + \lambda H_0\{w(r\};\ \xi\} = \frac{1}{D}H_0\{p(r);\ \xi\} \tag{9.97}$$

Let's now recall the following property of Hankel transform:

$$H_0\{B_0f;\ \xi\} = -\xi^2H_0\{f;\ \xi\} \tag{9.98}$$

Then, Eq. (9.97) can be written as:

$$-\xi^2H_0\{B_0w(r);\ \xi\} + \lambda H_0\{w(r\};\ \xi\} = \frac{1}{D}H_0\{p(r);\ \xi\} \tag{9.99}$$

or

$$-\xi^2(-\xi^2)H_0\{w(r);\ \xi\} + \lambda H_0\{w(r\};\ \xi\} = \frac{1}{D}H_0\{p(r);\ \xi\} \tag{9.100}$$

The above equation yields:

$$\xi^4H_0\{w(r);\ \xi\} + \lambda H_0\{w(r);\ \xi\} = \frac{1}{D}H_0\{p(r);\ \xi\} \tag{9.101}$$

or

$$H_0\{w(r);\ \xi\} = \frac{\overline{p}(\xi)}{\xi^4 + \lambda} \tag{9.102}$$

where

$$\overline{p}(\xi) = \frac{1}{D}H_0\{p(r);\ \xi\} \tag{9.103}$$

Considering Eq. (9.96), the analytical solution of the differential equation (Eq. 9.91) can be written:

$$w(r) = H_0^{-1}\left\{\frac{\overline{p}(\xi)}{\xi^4 + \lambda}; r\right\} \tag{9.104}$$

where $H_0^{-1}\{\}$ is the inverse Hankel's transform of order zero, given by Eq. (9.96).

Eq. (9.104) represents the deflection distribution of an infinite circular plate on elastic foundation under axisymmetric loading $p(r)$ lying on the infinite area $0 < r < \infty$.

Solution for an infinite circular plate on elastic foundation under uniform loading q_o lying on the finite area

$$r_i < r < r_o$$

For the case of a uniform loading q_o lying on the finite area $r_i < r < r_o$ (Fig. 9.8), the function $p(r)$ can be written:

$$p(r) = \begin{cases} q_o & for \;\; r_i \le r \le r_o \\ 0 & for \;\; r < r_i \;\; and \;\; r > r_o \end{cases} \tag{9.105}$$

Fig. 9.8 Infinite circular plate on elastic foundation under uniform loading qo lying on the finite area $r_i < r < r_o$.

Using the Heaviside functions, Eq. (9.105) can be expressed as follows:

$$p(r) = q_0\big(H(r - r_i) - H(r - r_0)\big) \tag{9.106}$$

where $H\{\bullet\}$ denotes the Heaviside function. To determine the Hankel transform for this loading case, the following property can be used:

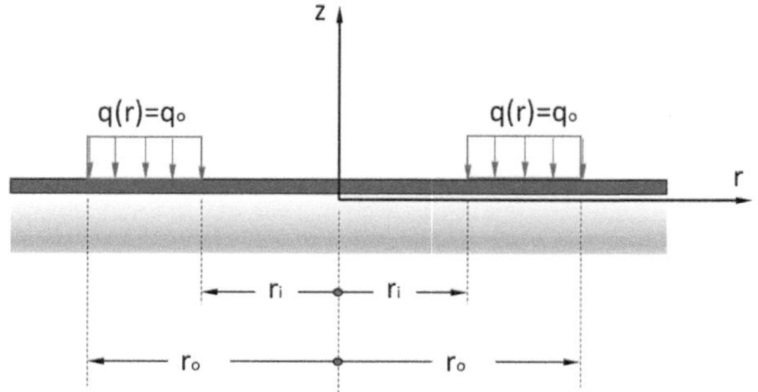

FIG. 9.8

Infinite circular plate on elastic foundation under uniform loading qo lying on the finite area $r_i < r < r_o$.

$$H_0\{H(r-a);\xi\} = J_1(\xi a) \tag{9.107}$$

Then,

$$\overline{p}(\xi) = \frac{q_o}{D}\left[J_1(\xi r_i) - J_1(\xi r_o)\right] \tag{9.108}$$

With the aid of the above equation, the deflection distribution due to uniform loading q_o acting on the area $r_i \leq r \leq r_o$ can be obtained by Eq. (9.104):

$$w(r) = \frac{q_0}{D}H_0^{-1}\left\{\frac{J_1(\xi r_i) - J_1(\xi r_o)}{\xi^4 + \lambda}; r\right\} \tag{9.109}$$

EXERCISE 2

Determine the displacement distribution of an infinite plate on elastic foundation with $k = 10{,}000\ kN/m^3$ under uniform loading $q_o = 5\ kN/m$ acting on the area $0 \leq r \leq 2.0\ m$. The plate is made by steel with elasticity modulus E = 210 GPa, and Poisson's ratio $\nu = 0.3$ and has thickness $2h = 2.0\ cm$.

Solution

The flexural rigidity of the plate is

$$D = \frac{2Eh^3}{3(1-\nu^2)} = \frac{2 \cdot 210 \cdot 10^9 \cdot (10^{-2})^3}{3(1-0.3^2)} = 153846\ Pa \cdot m^3$$

The coefficient λ is

$$\lambda = \frac{k}{D} = \frac{10000 \cdot 10^3}{153846} = 65$$

The displacement distribution can be obtained by the following equation:

$$w(r) = \frac{q_0}{D}H_0^{-1}\left\{\frac{J_1(\xi r_i) - J_1(\xi r_o)}{\xi^4 + \lambda}; r\right\} = \frac{5 \cdot 10^3}{153846}H_0^{-1}\left\{\frac{J_1(0) - J_1(2\xi)}{\xi^4 + 65}; r\right\}$$

or

$$w = \frac{5 \cdot 10^3}{153846}\int_0^\infty \xi\frac{J_1(0) - J_1(2\xi)}{\xi^4 + 65}J_0(\xi r)d\xi$$

The integration in the above equation has been carried out numerically with the software Mathematica. The result has been demonstrated in Fig. 9.9.

9.6 CYLINDRICAL SHELLS UNDER AXISYMMETRIC LOADING

Analysis of thin shells under arbitrary loading is complex for analytical solutions. For analyzing such problems, numerical techniques are usually employed. However,

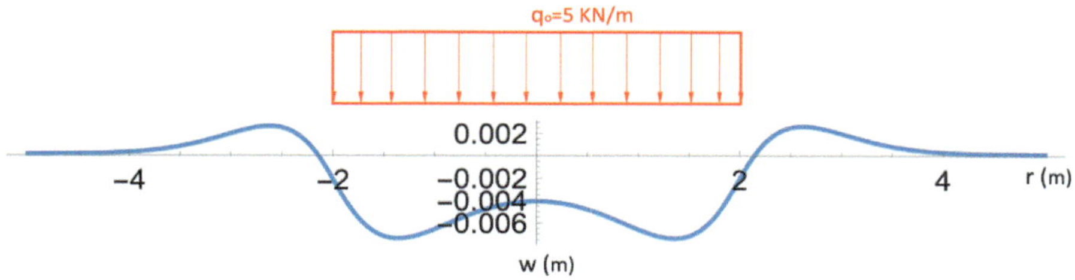

FIG. 9.9

Displacement distribution in a plate on elastic foundation under uniform loading in the area $0 \leq r \leq 2.0\ m$.

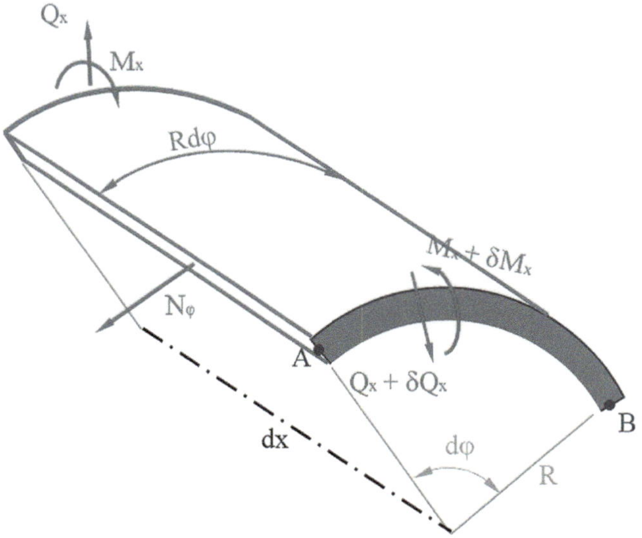

FIG. 9.10

Element of a cylindrical shell under surface pressure P.

many engineering applications (e.g., silos, pipes) can be modeled as cylindrical shells under axisymmetric loading. Fig. 9.10 demonstrates an element of a cylindrical shell under surface pressure P and the internal forces per unit length on its boundary.

Equilibrium equations for this shell element yields:

$$\left(Q_x + \frac{\partial Q_x}{\partial x}dx\right)R\ d\varphi + \left(N_\varphi dx\right)d\varphi - Q_x R\ d\varphi - P(R\ d\varphi)dx = 0 \qquad (9.110)$$

or

$$\frac{dQ_x}{dx} + \frac{1}{R}N_\varphi = P \qquad (9.111)$$

Taking the equilibrium of bending moments in AB of the shell element of Fig. 9.10, the following equation can be obtained:

$$\left(M_x + \frac{\partial M_x}{\partial x}dx\right)R\,d\varphi - M_x R d\varphi + P(Rd\varphi)dx\frac{dx}{2} - Q_x R d\varphi dx = 0 \qquad (9.112)$$

Neglecting the term $(dx)^2 \approx 0$, we get

$$\frac{dM_z}{dx} = Q_x \qquad (9.113)$$

Combining Eqs. (9.111) and (9.113), the following result can be obtained:

$$\frac{d^2M_x}{dx^2} + \frac{1}{R}N_\varphi = P \qquad (9.114)$$

The hoop force per unit length N_φ is the resultant of the hoop stress σ_φ distribution on the shell's wall of thickness t:

$$N_\varphi = \sigma_\varphi t \qquad (9.115)$$

Considering Hooke's law

$$\sigma_\varphi = E\varepsilon_\varphi \qquad (9.116)$$

and the definition of the hoop strain

$$\varepsilon_\varphi = \frac{w}{R} \qquad (9.117)$$

Eq. (9.115) yields

$$N_\varphi = E\frac{w}{R}t \qquad (9.118)$$

Then, Eq. (9.114) can be written in the following form:

$$\frac{d^2M_x}{dx^2} + \frac{Et}{R^2}w = P \qquad (9.119)$$

Because

$$M_x = D\frac{d^2w}{dx^2} \qquad (9.120)$$

Eq. (9.119) yields

$$D\frac{d^4w}{dx^4} + \frac{Et}{R^2}w = P \qquad (9.121)$$

If we use the following notation,

$$k = \frac{Et}{R^2} \tag{9.122}$$

Eq. (9.121) takes the form

$$D\frac{d^4w}{dx^4} + kw = P \tag{9.123}$$

Therefore the cylindrical shells under internal pressure can be analyzed with the theory of the beam of elastic foundation.

REFERENCES

[1] Pavlou DG. Green's function for a pre-stressed thin plate on an elastic foundation under axisymmetric loading. Engineering Analysis with Boundary Elements 2005;29(5).

[2] Pavlou D, et al. An analytical solution of the annular plate on elastic foundation. Structural Engineering and Mechanics 2005;Volume 20(Number 2):209–23. https://doi.org/10.12989/sem.2005.20.2.209

[3] Pavlou D. Elastodynamic analysis of a thin layer bonded on a visco-elastic medium under combined in-plane and pulse lateral loads. Mechanics research communications 2011;Volume 38(Issue 8):546–52.

[4] Pavlou DG. Boundary element analysis of a plate on an elastic foundation loaded by a twisted annular rigid stamp. Proceedings of the Institution of Mechanical Engineers, Part J. Journal of Engineering Tribology 2004;218(6):549–56. https://journals.sagepub.com/doi/10.1243/1350650042794761

[5] Fenner RT. Engineering Elasticity. Ellis Horwood Ltd; 1986.

[6] Boresi AP, Schmidt RJ. Advanced Mechanics of Materials. Wiley; 2003.

[7] Timoshenko SP, Woinowsky-Krieger S. Theory of plates and shells. New York: McGraw-Hill; 1959.

[8] Hetenyi M. Beams on Elastic Foundation. Ann Arbor: The University of Michigan Press; 1961.

[9] Sneddon IN. The use of Integral Transforms. New York: McGraw-Hill; 1972.

[10] Sneddon IN. Fourier Transforms. New York: Dover Publications; 1995.

[11] Pavlou D. Dynamic response of a plate on elastic foundation under moving vertical and in-plane loads. International Review of Mechanical Engineering (IREME) 2015; Volume 9(6):548–52.

[12] Palermo Jr. L. Plate bending analysis using the classical or Reissner-Mindlin models. Engineering Analysis with Boundary Elements 2003;27:603–9.

[13] Utku M, Citipitioglu E, Inceleme I. Circular plates on elastic foundations modelled with annular plates. Computers and Structures 2000;78:365–74.

[14] Chandrashekhara K, Antony J. Elastic analysis of an annular slab-soil interaction problem using hybrid method. Comput and Geotech 1997;20:161–76.

[15] Wolfram S. MATHEMATICA. USA: Wolfram Media; 2003.

[16] Timoshenko SP, Goodier JN. Theory of Elasticity. McGraw-Hill; 1951.

CHAPTER

Airy Stress Function and Stress Concentration 10

Chapter Outline

Airy stress function is a powerful tool for analytical stress analysis. Unlike other methods, it provides exact solutions for stress field derivation. Airy stress function is based on the equilibrium equations in terms of stresses (differential equations of motion) and the compatibility equations of strains.

Let's recall the equilibrium equations in terms of stresses in plane stress conditions:

$$\frac{\partial \sigma_{xx}}{\partial x} + \frac{\partial \tau_{xy}}{\partial y} = 0 \tag{10.1}$$

$$\frac{\partial \tau_{xy}}{\partial x} + \frac{\partial \sigma_{yy}}{\partial y} = 0 \tag{10.2}$$

We select a function $\Phi(x, y)$ that fulfils the following condition:

$$\sigma_{xx} = \frac{\partial^2 \Phi}{\partial y^2} \tag{10.3}$$

With the aid of the above equation, Eq. (10.1) yields:

$$\frac{\partial^3 \Phi}{\partial x \partial y^2} + \frac{\partial \tau_{xy}}{\partial y} = 0 \tag{10.4}$$

Advanced Mechanics of Solids and Structures. DOI: http://doi.org/10.1016/B978-0-443-23778-2.00010-X

or

$$\tau_{xy} = -\frac{\partial^2 \Phi}{\partial x \partial y} \tag{10.5}$$

Combining Eqs. (10.2) and (10.5), the following formula can be obtained:

$$\frac{\partial^3 \Phi}{\partial x^2 \partial y} + \frac{\partial \sigma_{yy}}{\partial y} = 0 \tag{10.6}$$

or

$$\sigma_{yy} = \frac{\partial^2 \Phi}{\partial x^2} \tag{10.7}$$

Eqs. (10.3), (10.5), and (10.7) show that the stresses field σ_{xx}, σ_{yy}, and τ_{xy} can be determined with the aid of only one function, the function $\Phi(x, y)$. This function was derived by the equilibrium equations (Eqs. 10.1) and 10.2). The function $\Phi(x, y)$ is called Airy stress function.

To determine Airy stress function, the following compatibility equation of strains can utilize:

$$\frac{\partial^2 \varepsilon_{xx}}{\partial y^2} + \frac{\partial^2 \varepsilon_{yy}}{\partial x^2} = \frac{\partial^2 \gamma_{xy}}{\partial x \ \partial y} \tag{10.8}$$

With the aid of the following constitutive equation,

$$\begin{Bmatrix} \varepsilon_x \\ \varepsilon_y \\ \gamma_{xy} \end{Bmatrix} = \begin{bmatrix} \frac{1}{E} & -\frac{\nu}{E} & 0 \\ -\frac{\nu}{E} & \frac{1}{E} & 0 \\ 0 & 0 & \frac{2(1+\nu)}{E} \end{bmatrix} \begin{Bmatrix} \sigma_x \\ \sigma_y \\ \tau_{xy} \end{Bmatrix} \tag{10.9}$$

Eq. (10.8) yields:

$$\frac{\partial^2}{\partial y^2}\left[\frac{1}{E}(\sigma_{xx} - \nu\sigma_{yy})\right] + \frac{\partial^2}{\partial x^2}\left[\frac{1}{E}(\sigma_{yy} - \nu\sigma_{xx})\right] = \frac{\partial^2}{\partial x \partial y}\left[\frac{2(1+\nu)}{E}\tau_{xy}\right] \tag{10.10}$$

Utilizing Eqs. (10.3), (10.5), and (10.7) in the above equation, the following formula can be obtained:

$$\frac{\partial^4 \Phi}{\partial y^4} - \nu\frac{\partial^4 \Phi}{\partial x^2 \partial y^2} + \frac{\partial^4 \Phi}{\partial x^4} - \nu\frac{\partial^4 \Phi}{\partial x^2 \partial y^2} = -2(1+\nu)\frac{\partial^4 \Phi}{\partial x^2 \partial y^2} \tag{10.11}$$

or

$$\frac{\partial^4\Phi}{\partial x^4} + 2\frac{\partial^4\Phi}{\partial x^2\partial y^2} + \frac{\partial^4\Phi}{\partial y^4} = 0 \tag{10.12}$$

This is a biharmonic equation in terms of Airy stress function $\Phi(x, y)$. Eq. (10.12) can be written alternatively as:

$$\left(\frac{\partial^2}{\partial y^2} + \frac{\partial^2}{\partial x^2}\right)\left(\frac{\partial^2\Phi}{\partial y^2} + \frac{\partial^2\Phi}{\partial x^2}\right) = 0 \tag{10.13}$$

or

$$\nabla^2\nabla^2\Phi = 0 \tag{10.14}$$

or

$$\nabla^4\Phi = 0 \tag{10.15}$$

The biharmonic equation satisfies the equilibrium equation of stress and the compatibility equations of strains. To find the exact solution of the biharmonic equation, the stress and displacements should satisfy the boundary conditions as well.

10.1 BIHARMONIC EQUATION IN POLAR COORDINATES

For axisymmetric problems, it is convenient to use the biharmonic equation in terms of polar coordinates r and θ. If we transform the Cartesian coordinates x and y to polar coordinates r and θ in Eq. (10.13), the following result can be obtained:

$$\left(\frac{\partial^2}{\partial r^2} + \frac{1}{r}\frac{\partial}{\partial r} + \frac{1}{r^2}\frac{\partial^2}{\partial\theta^2}\right)\left(\frac{\partial^2\Phi}{\partial r^2} + \frac{1}{r}\frac{\partial\Phi}{\partial r} + \frac{1}{r^2}\frac{\partial^2\Phi}{\partial\theta^2}\right) = 0 \tag{10.16}$$

If the solution is a function of only r (in axisymmetric structures under axisymmetric loading), the above equation yields:

$$\left(\frac{\partial^2}{\partial r^2} + \frac{1}{r}\frac{\partial}{\partial r}\right)\left(\frac{\partial^2\Phi}{\partial r^2} + \frac{1}{r}\frac{\partial\Phi}{\partial r}\right) = 0 \tag{10.17}$$

or

$$\frac{\partial^4\Phi}{\partial r^4} + \frac{2}{r}\frac{\partial^3\Phi}{\partial r^3} - \frac{1}{r^2}\frac{\partial^2\Phi}{\partial r^2} + \frac{1}{r^3}\frac{\partial\Phi}{\partial r} = 0 \tag{10.18}$$

The general solution of the above equation is

$$\Phi = A\log(r) + Br^2\log(r) + Cr^2 + D \tag{10.19}$$

where A, B, C, and D are constants to be determined by the boundary conditions.

The radial and circumferential stress components can be obtained by transforming Eqs. (10.3), (10.5), and (10.7) in polar coordinates:

$$\sigma_r = \frac{1}{r}\frac{\partial \Phi}{\partial r} + \frac{1}{r^2}\frac{\partial^2 \Phi}{\partial \theta^2} \tag{10.20}$$

$$\sigma_\theta = \frac{\partial^2 \Phi}{\partial r^2} \tag{10.21}$$

$$\tau_{r\theta} = -\frac{\partial}{\partial r}\left(\frac{1}{r}\frac{\partial \Phi}{\partial \theta}\right) \tag{10.22}$$

In axisymmetric structures under axisymmetric loading, the stresses are a function of only r. For this case, the above equation yields:

$$\sigma_r = \frac{1}{r}\frac{\partial \Phi}{\partial r} \tag{10.23}$$

$$\sigma_\theta = \frac{\partial^2 \Phi}{\partial r^2} \tag{10.24}$$

$$\tau_{r\theta} = 0 \tag{10.25}$$

With the aid of Eq. (10.19) the following results can be obtained:

$$\sigma_r = 2C + \frac{A}{r^2} + B[1 + 2\log(r)] \tag{10.26}$$

$$\sigma_\theta = 2C - \frac{A}{r^2} + B[3 + \log(r)] \tag{10.27}$$

$$\tau_{r\theta} = 0 \tag{10.28}$$

10.2 GENERAL SOLUTION OF THE BIHARMONIC EQUATION IN POLAR COORDINATES

We try the general solution $\Phi = R(r)\cos\theta$ in the biharmonic Eq. (10.16):

$$\nabla^4\Phi = \underbrace{\left(\frac{\partial^2}{\partial r^2} + \frac{1}{r}\frac{\partial}{\partial r} + \frac{1}{r^2}\frac{\partial^2}{\partial\theta^2}\right)}_{\nabla^2}\underbrace{\left(\frac{\partial^2\Phi}{\partial r^2} + \frac{1}{r}\frac{\partial\Phi}{\partial r} + \frac{1}{r^2}\frac{\partial^2\Phi}{\partial\theta^2}\right)}_{\nabla^2\Phi} = 0$$

Let's determine the terms of the above equation separately:

$$\nabla^2\Phi = \frac{1}{r^2}R(r)(-\cos\theta) + \frac{1}{r}\frac{dR}{dr}\cos\theta + \frac{d^2R}{dr^2}\cos\theta =$$
$$= \left(-\frac{1}{r^2}R + \frac{1}{r}\frac{dR}{dr} + \frac{d^2R}{dr^2}\right)\cos\theta$$

$$\nabla^4\Phi = \nabla^2(\nabla^2\Phi) = 0$$

or

$$\nabla^4\Phi = \frac{1}{r^2}\frac{\partial^2}{\partial\theta^2}\left[\left(-\frac{1}{r^2}R + \frac{1}{r}\frac{dR}{dr} + \frac{d^2R}{dr^2}\right)\cos\theta\right] +$$
$$+ \frac{1}{r}\frac{\partial}{\partial r}\left[\left(-\frac{1}{r^2}R + \frac{1}{r}\frac{dR}{dr} + \frac{d^2R}{dr^2}\right)\cos\theta\right] +$$
$$+ \frac{\partial^2}{\partial r^2}\left[\left(-\frac{1}{r^2}R + \frac{1}{r}\frac{dR}{dr} + \frac{d^2R}{dr^2}\right)\cos\theta\right] = 0$$

or

$$\nabla^4\Phi = -\frac{1}{r^2}\left(-\frac{1}{r^2}R + \frac{1}{r}\frac{dR}{dr} + \frac{d^2R}{dr^2}\right)\cos\theta +$$
$$+ \frac{1}{r}\left(\frac{2}{r^3}R - \frac{2}{r^2}\frac{dR}{dr} + \frac{1}{r}\frac{d^2R}{dr^2} + \frac{d^3R}{dr^3}\right)\cos\theta +$$
$$+ \left(-\frac{6}{r^4}R + \frac{6}{r^3}\frac{dR}{dr} - \frac{3}{r^2}\frac{d^2R}{dr^2} + \frac{1}{r}\frac{d^3R}{dr^3} + \frac{d^4R}{dr^4}\right)\cos\theta = 0$$

Because the right side of above equation is equal to 0, the term $\cos\theta$ can be eliminated. Then, the following equation can be obtained:

$$\frac{d^4R}{dr^4} + \frac{2}{r}\frac{d^3R}{dr^3} - \frac{3}{r^2}\frac{d^2R}{dr^2} + \frac{3}{r^3}\frac{dR}{dr} - \frac{3}{r^4}R = 0 \tag{10.29}$$

or

$$\frac{d}{dr}\left\{\frac{1}{r}\frac{d}{dr}\left[r\frac{d}{dr}\left(\frac{1}{r}\frac{d}{dr}(rR)\right)\right]\right\} = 0 \tag{10.30}$$

The above equation can be solved by four successive integrations. The first integration yields:

$$\frac{d}{dr}\left[r\frac{d}{dr}\left(\frac{1}{r}\frac{d}{dr}(rR)\right)\right] = A_1 r$$

Taking the second integration and dividing both sides by r, we get:

$$\frac{d}{dr}\left(\frac{1}{r}\frac{d}{dr}(rR)\right) = A_1\frac{r}{2} + \frac{B_1}{r}$$

Taking the third integration and multiplying both sides by r, the following equation can be obtained:

$$\frac{d}{dr}(rR) = A_1\frac{r^3}{4} + B_1 r\ln r + C_1 r\,(*)$$

Taking the fourth integration in above equation, and dividing both sides by r, the following solution can be obtained:

$$R = \underbrace{\frac{A_1}{16}}_{D}r^3 + \underbrace{\frac{B_1}{2}}_{C}r\ln r \underbrace{-\frac{1}{4}(B_1 - 2C_1)r}_{B} + \underbrace{D_1}_{A}\frac{1}{r}$$

Using the notations $A = D_1$, $B = (B_1 - 2C_1)/4$, $C = B_1/2$, and $D = A_1/16$, the above equation can be written in the following form:

$$R = Ar^{-1} + Br + Cr\ln r + Dr^3 \tag{10.31}$$

Then the following formula for Airy stress function can be obtained:

$$\Phi = R(r)\cos\theta = (Ar^{-1} + Br + Cr\ln r + Dr^3)\cos\theta \tag{10.32}$$

With the aid of the above general solution, the formulae for the stress components in polar coordinates can be obtained by Eqs. (10.20)–(10.22):

$$\sigma_r = \left(-\frac{2A}{r^3} + \frac{C}{r} + 2Dr\right)\cos\theta \tag{10.33}$$

$$\sigma_\theta = \left(\frac{2A}{r^3} + \frac{C}{r} + 6Dr\right)\cos\theta \tag{10.34}$$

$$\tau_{r\theta} = \left(-\frac{2A}{r^3} + \frac{C}{r} + 2Dr\right)\sin\theta \tag{10.35}$$

EXERCISE 1

The stress function $\Phi(x, y) = A\left(xy - \frac{xy^3}{B}\right)$ is used to describe the stress field in the beam of Fig. 10.1. In this function, A and B are given parameters. Determine the bending moment, axial force, and shear force distribution.

Solution

We should check first whether the given Airy stress function fulfils the biharmonic equation:

$$\frac{\partial^4\Phi(x, y)}{\partial x^4} = A\frac{\partial^4}{\partial x^4}\left(xy - \frac{xy^3}{B}\right) = 0$$

$$\frac{\partial^4\Phi(x, y)}{\partial y^4} = A\frac{\partial^4}{\partial y^4}\left(xy - \frac{xy^3}{B}\right) = 0$$

$$\frac{\partial^4\Phi(x, y)}{\partial x^2\partial y^2} = A\frac{\partial^4}{\partial x^2\partial y^2}\left(xy - \frac{xy^3}{B}\right) = 0$$

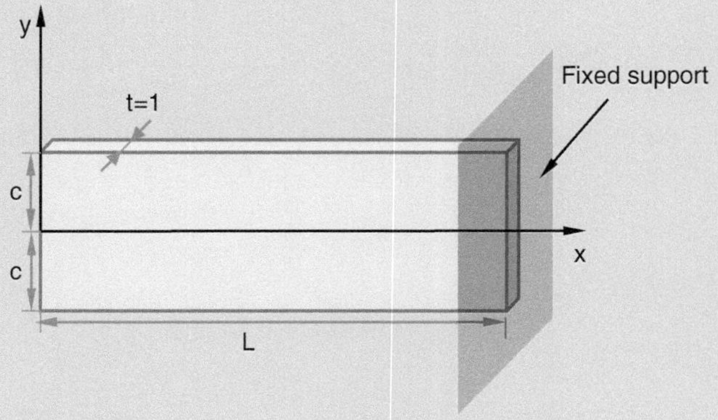

FIG. 10.1

Geometry and coordinate system of the beam.

(Continued)

EXERCISE 1—CONT'D

Therefore the given Airy stress function fulfils the biharmonic equation:

$$\frac{\partial^4\Phi}{\partial x^4} + 2\frac{\partial^4\Phi}{\partial x^2\partial y^2} + \frac{\partial^4\Phi}{\partial y^4} = 0$$

The second step is the calculation of the stresses:

$$\sigma_{xx} = \frac{\partial^2\Phi}{\partial y^2} = -\frac{6Axy}{B}$$

$$\sigma_{yy} = \frac{\partial^2\Phi}{\partial x^2} = 0$$

$$\tau_{xy} = -\frac{\partial^2\Phi}{\partial x\ \partial y} = A\left(1 - \frac{3y^2}{B}\right)$$

10.2.1 AXIAL FORCE DISTRIBUTION

To calculate the axial force F_a, we consider an arbitrary cross section (Fig. 10.2). The axial force is calculated by the following equilibrium:

$$F_a = \int_A \sigma_{xx} dA = \int_{-C}^{C} \sigma_{xx} dy = \int_{-C}^{C} -\frac{6Axy}{B} dy = 0$$

10.2.2 BENDING MOMENT DISTRIBUTION

To calculate the bending moment M, we integrate over a surface the contribution of the axial stresses σ_{xx}, as we can see in Fig. 10.3.

$$M = \int_{-C}^{C} \underbrace{(\sigma_{xx} dA)}_{\text{force}} \underbrace{y}_{\text{distance}} = \int_{-C}^{C} \sigma_{xx} y dy = \int_{-C}^{C} \left(-\frac{6Axy}{B}\right) y dy = -\frac{4Ac^3x}{B}$$

We conclude that the bending moment distribution along the beam is a linear function.

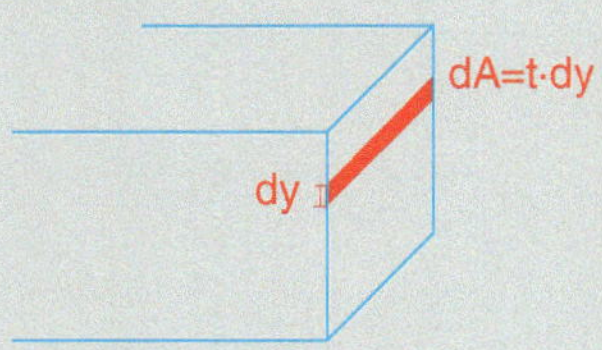

FIG. 10.2

Cross section of the beam.

EXERCISE 1—CONT'D

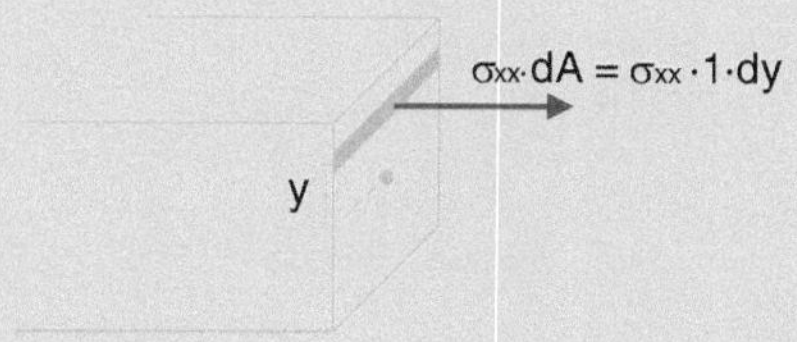

FIG. 10.3

Effect of stresses σ_{xx} on the bending moment.

10.2.3 SHEAR FORCE DISTRIBUTION

$$F_s = \int_A \tau_{xy} dA = \int_{-C}^{C} \tau_{xy} dy = \int_{-C}^{C} A\left(1 - \frac{3y^2}{B}\right) dy = 2Ac\left(1 - \frac{c^2}{B}\right)$$

EXERCISE 2

The following function has six unknown constants

$$\Phi(x, y) = \lambda_1 x^2 + \lambda_2 x^2 y + \lambda_3 y^3 + \lambda_4 x^5 + \lambda_5 y^5 + \lambda_6 x^2 y^3$$

A. Determine the unknown constants $\lambda_i (i = 1,2,\cdots,6)$ in order for the function $\Phi(x, y)$ to be a valid Airy stress function.
B. Determine the formulae for the stress components σ_{xx}, σ_{yy}, and τ_{xy}.
C. Compare the result for σ_{xx} with the results of the elementary beam theory.

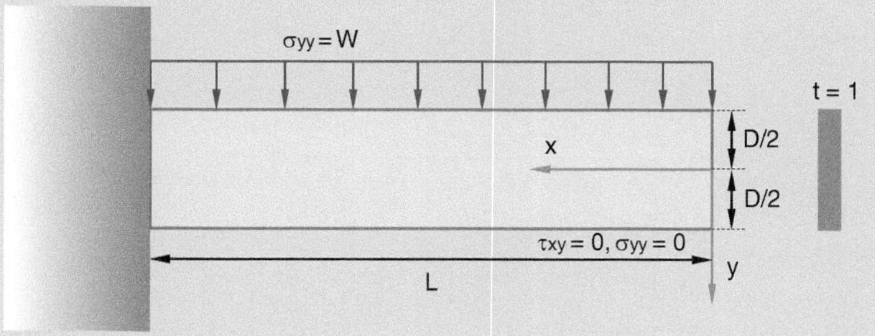

Solution

A. **Step 1: The function $\Phi(x, y)$ should satisfy the biharmonic equation:**

$$\frac{\partial^4 \Phi}{\partial y^4} + \frac{\partial^4 \Phi}{\partial x^4} + 2\frac{\partial^4 \Phi}{\partial x^2 \partial y^2} = 0$$

(Continued)

EXERCISE 2—CONT'D

Inserting the function $\Phi(x, y) = \lambda_1 x^2 + \lambda_2 x^2 y + \lambda_3 y^3 + \lambda_4 x^5 + \lambda_5 y^5 + \lambda_6 x^2 y^3$ into the above equation, the following formula can be obtained:

$$5 \cdot 4 \cdot 3 \cdot 2 \cdot \lambda_4 x + 5 \cdot 4 \cdot 3 \cdot 2 \cdot \lambda_5 y + 2\lambda_6 \cdot 3 \cdot 2y = 0$$

or

$$120\lambda_4 x + 24(\lambda_6 + 5\lambda_5)y = 0$$

The above equation yields:

$$\lambda_4 = 0$$

$$\lambda_6 = -5\lambda_5$$

We calculate the stresses:

$$\sigma_{xx} = \frac{\partial^2 \Phi}{\partial y^2} = 3 \cdot 2\lambda_3 y + 5 \cdot 4\lambda_5 y^3 + 3 \cdot 2\lambda_6 x^2 y$$

or

$$\sigma_{xx} = 6\lambda_3 y + 20\lambda_5 y^3 - 30\lambda_5 x^2 y$$

$$\sigma_{yy} = \frac{\partial^2 \Phi}{\partial x^2} = 2\lambda_1 + 2\lambda_2 y + 2\lambda_6 y^3$$

or

$$\sigma_{yy} = 2\lambda_1 + 2\lambda_2 y - 10\lambda_5 y^3$$

$$\tau_{xy} = -\frac{\partial^2 \Phi}{\partial x \partial y} = -\left(2\lambda_2 x + 2 \cdot 3\lambda_6 x y^2\right)$$

or

$$\tau_{xy} = -2\lambda_2 x + 30\lambda_5 x y^2$$

Step 2: The obtained stress components should satisfy the boundary conditions:

Boundary condition I: $\text{For} y = \frac{D}{2} \rightarrow \tau_{xy} = 0$

Boundary condition II: $\text{For} y = \frac{D}{2} \rightarrow \sigma_{yy} = 0$

Boundary condition III: $\text{For} y = -\frac{D}{2} \rightarrow \sigma_{yy} = W$

Boundary condition IV: $M(x) = \frac{Wx^2}{2} \Rightarrow \int_{-D/2}^{D/2} \left(\underbrace{\sigma_{xx} t dy}_{force} \right) \cdot \underbrace{y}_{distance} = \frac{Wx^2}{2}$

We apply the obtained formulae for stresses into the above boundary conditions:

$$\text{BC-I: } -2\lambda_2 x + 30\lambda_5 x \left(\frac{D}{2}\right)^2 = 0$$

EXERCISE 2—CONT'D

or

$$\lambda_2 = \frac{15}{4}\lambda_5 D^2$$

$$\text{BC–II: } 2\lambda_1 + 2\lambda_2\frac{D}{2} - 10\lambda_5\left(\frac{D}{2}\right)^3 = 0$$

or

$$2\lambda_1 + 2\left(\frac{15}{4}\lambda_5 D^2\right)\frac{D}{2} - 10\lambda_5\left(\frac{D}{2}\right)^3 = 0$$

or

$$\lambda_1 = -\frac{5}{4}\lambda_5 D^3$$

$$\text{BC–III: } 2\lambda_1 + 2\lambda_2\left(-\frac{D}{2}\right) - 10\lambda_5\left(-\frac{D}{2}\right)^3 = W$$

or

$$2\left(-\frac{5}{4}\lambda_5 D^3\right) + 2\left(\frac{15}{4}\lambda_5 D^2\right)\frac{D}{2} - 10\lambda_5\left(\frac{D}{2}\right)^3 = W$$

or

$$\lambda_5 = -\frac{W}{5D^3}$$

Then

$$\lambda_2 = -\frac{3W}{4D}$$

$$\lambda_1 = \frac{W}{4}$$

$$\lambda_6 = -5\lambda_5 = \frac{W}{D^3}$$

$$\text{BC-IV: } M(x) = \frac{Wx^2}{2}$$

or

$$t\int_{-D/2}^{D/2} \sigma_{xx} y \; dy = \frac{Wx^2}{2}$$

(Continued)

EXERCISE 2—CONT'D

or

$$\int_{-D/2}^{D/2} 6\lambda_3 y^2 \, dy + \int_{-D/2}^{D/2} 20\lambda_5 y^4 \, dy - \int_{-D/2}^{D/2} 30\lambda_5 x^2 y^2 \, dy = \frac{Wx^2}{2t}$$

or

$$\underbrace{6\lambda_3 \frac{y^3}{3}\Big|_{-D/2}^{D/2} + 20\lambda_5 \frac{y^5}{5}\Big|_{-D/2}^{D/2}}_{A} - \underbrace{30\lambda_5 x^2 \frac{y^3}{3}\Big|_{-D/2}^{D/2}}_{B} = \frac{Wx^2}{2t}$$

Because the right side of the above equation is a function of the variable x, the left side should be a function of x too. Therefore the term A should be null:

$$6\lambda_3 \frac{y^3}{3}\Big|_{-D/2}^{D/2} + 20\lambda_5 \frac{y^5}{5}\Big|_{-D/2}^{D/2} = 0$$

or

$$\lambda_3 = -\lambda_5 \frac{D^2}{2}$$

Then, the term B should be $B = Wx^2/2$

$$-30\lambda_5 x^2 \frac{y^3}{3}\Big|_{-D/2}^{D/2} = \frac{Wx^2}{2t}$$

or

$$\lambda_5 = -\frac{W}{5tD^3}$$

Then:

$$\lambda_3 = \frac{W}{10tD}$$

B. With the aid of the obtained parameters λi ($i = 1,2,\ldots,6$), we can now determine the stress components:

$$\sigma_{xx} = 6\lambda_3 y + 20\lambda_5 y^3 - 30\lambda_5 x^2 y = 6\frac{W}{10tD}y + 20\left(-\frac{W}{5tD^3}\right)y^3 - 30\left(-\frac{W}{5tD^3}\right)x^2 y$$

$$\sigma_{yy} = 2\lambda_1 + 2\lambda_2 y - 10\lambda_5 y^3 = \frac{W}{2} - \frac{3W}{2D}y + \frac{2W}{tD^3}y^3$$

$$\tau_{xy} = -2\lambda_2 x + 30\lambda_5 xy^2 = \frac{3W}{2D}x - \frac{6W}{tD^2}xy^2$$

EXERCISE 2—CONT'D

C. The classical bending theory is based on the assumption that the cross sections of a beam under bending remain plane. Because of this assumption, the bending stresses σ_{xx} is a linear function of the variable y only:

$$\sigma_{xx} = \frac{M}{I}y$$

Unlike the results of the bending theory, the determination of the bending stress with the aid of Airy stress function is exact. Apart from the linear term of the variable y the exact solution has terms x^2y and y^3:

$$\sigma_{xx} = 6\frac{W}{10tD}y + 20\left(-\frac{W}{5tD^3}\right)y^3 - 30\left(-\frac{W}{5tD^3}\right)x^2y$$

Let's assume the following data for the beam;
L = 2.0 m, t = 10 mm, D = 3*t, W = 100 N/m
We can now calculate the stress σ_{xx} in the location $x = L$ and y = D/2
The solution of the Airy stress function yields:

$$\sigma_{xx} = 6\frac{W}{10tD}y + 20\left(-\frac{W}{5tD^3}\right)y^3 - 30\left(-\frac{W}{5tD^3}\right)x^2y = 1.333313333333333 \times 10^8 \text{ Pa}$$

Now we can use the elementary beam theory. We should first calculate the second moment of inertia and the bending moment on the cross section located at $x = L$:

$$I = \frac{tD^3}{12}$$

$$M_{x=L} = \frac{wL^2}{2} = 5$$

Then the stress with the aid of the classical beam theory is:

$$\sigma_x = \frac{M}{I}y = 1.333333333333333 \times 10^8 \text{ Pa}$$

We see that the elementary beam theory is a good approximation for this type of problem because the error is negligible.

10.3 STRESS CONCENTRATION IN 2D PROBLEMS

When holes exist in plates under in-plane loading, a stress concentration takes place in the vicinity of the holes. The determination of the stress field near circular holes is a complex analytical problem. Airy stress function is a powerful theoretical tool for stress concentration analysis. In this chapter, the use of Airy stress function for circular hole–induced stress concentration will be demonstrated. For the sake of simplicity, the thin, infinite plate under remote uniform axial stresses (Fig. 10.4) will be considered for the solution.

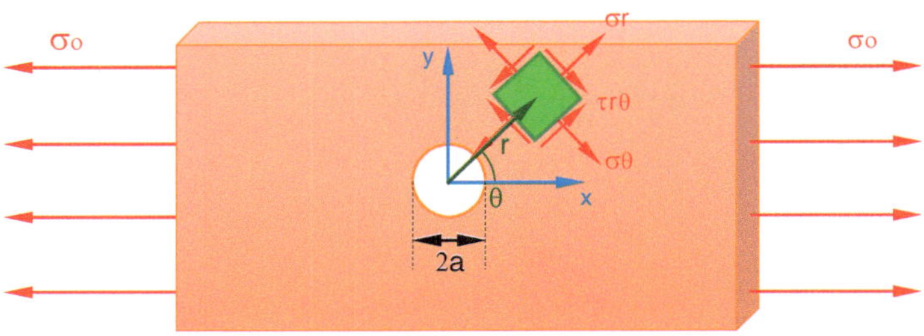

FIG. 10.4

Stress concentration in circular holes.

By definition, the biharmonic Eq. (10.15) satisfies both the equilibrium equation of stresses and the compatibility conditions of strains. Let's now specify the boundary conditions of the problem in Fig. 10.4.

10.3.1 BOUNDARY CONDITIONS

Boundary conditions 1:

The remote stresses in the orthogonal coordinate system x-y are:

$$\sigma_x = \sigma_0$$

$$\sigma_y = 0$$

$$\tau_{xy} = 0$$

Boundary conditions 2:

The stresses (in the polar coordinate system r-θ) at the edge of the hole ($r = a$) are:

$$\sigma_r = 0$$

$$\tau_{r\theta} = 0$$

Because circular holes have axisymmetric shape, we select a polar coordinate system for the solution.

Boundary conditions 1: Remote stresses in the infinite plate at $r \to \infty$

In Fig. 10.5 we see that the polar coordinate system is rotated with respect to the Cartesian one by an angle θ. Then, above boundary conditions 1 can be expressed

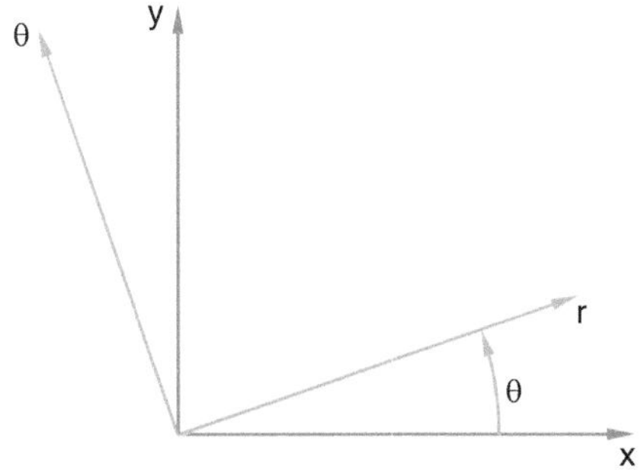

FIG. 10.5

The polar coordinate system r-θ is rotated by an angle θ.

in the rotated (polar) coordinate system with the aid of the following formulae (see Chapter 1):

$$\sigma_{rr} = \frac{\sigma_x + \sigma_y}{2} + \frac{\sigma_x - \sigma_y}{2}\cos 2\theta + \tau_{xy}\sin 2\theta \tag{10.36}$$

$$\tau_{r\theta} = -\frac{1}{2}(\sigma_x - \sigma_y)\sin 2\theta + \tau_{xy}\cos 2\theta \tag{10.37}$$

Because $\sigma_x = \sigma_0,\ \sigma_y = 0,\ \tau_{xy} = 0$, the above equations yield:

$$\sigma_{rr} = \frac{\sigma_0}{2} + \frac{\sigma_0}{2}\cos 2\theta \tag{10.38}$$

$$\tau_{r\theta} = -\frac{1}{2}\sigma_0 \sin 2\theta \tag{10.39}$$

The above equations indicate that the remote stresses have a constant component $\frac{\sigma_0}{2}$ for σ_{rr} and the harmonic components $\frac{\sigma_0}{2}\cos 2\theta$ and $-\frac{1}{2}\sigma_0 \sin 2\theta$ for σ_{rr} and $\tau_{r\theta}$, respectively. Therefore the remote stress state (A) can be analyzed to the stress states (B) and (C) shown in Fig. 10.6. Obviously, the superposition of the stress states (B) and (C) yields the stress state (A).

The problem is now reduced to solving the biharmonic equation in polar coordinates for the stress states (B) and (C) and add the solutions. Unlike stress state (A), the stress states (B) and (C) are axisymmetric. Therefore, for the solution, we should recall the biharmonic Eq. (10.16) for polar coordinates:

$$\left(\frac{\partial^2}{\partial r^2} + \frac{1}{r}\frac{\partial}{\partial r} + \frac{1}{r^2}\frac{\partial^2}{\partial \theta^2}\right)\left(\frac{\partial^2 \Phi}{\partial r^2} + \frac{1}{r}\frac{\partial \Phi}{\partial r} + \frac{1}{r^2}\frac{\partial^2 \Phi}{\partial \theta^2}\right) = 0$$

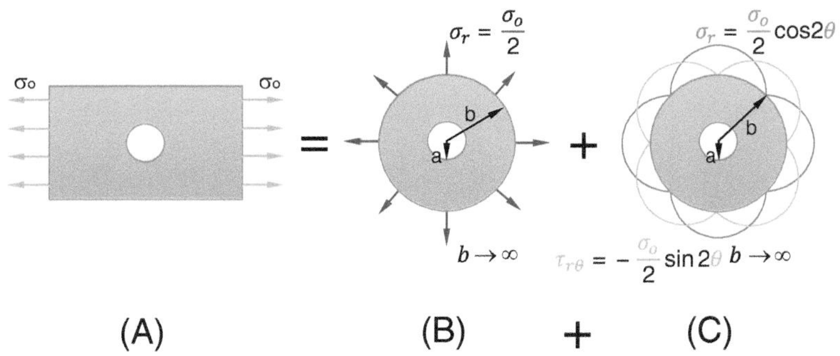

FIG. 10.6

Superposition of stress states (B) and (C) yields the stress state (A).

and the corresponding formulae (10.20)–(10.22) for the stress components in polar coordinates:

$$\sigma_{rr} = \frac{1}{r}\frac{\partial \Phi}{\partial r} + \frac{1}{r^2}\frac{\partial^2 \Phi}{\partial \theta^2} \quad \sigma_{\theta\theta} = \frac{\partial^2 \Phi}{\partial r^2} \quad \tau_{r\theta} = -\frac{\partial}{\partial r}\left(\frac{1}{r}\frac{\partial \Phi}{\partial \theta}\right)$$

It is convenient to solve problems (B) and (C) separately and superpose the solutions afterward.

10.3.2 SOLUTION FOR STRESS STATE (C)

The elastic medium in the stress state (C) is subjected to the following stresses:

$$\sigma_{rr} = \frac{\sigma_0}{2}\cos 2\theta \tag{10.40}$$

$$\tau_{r\theta} = -\frac{1}{2}\sigma_0 \sin 2\theta \tag{10.41}$$

Combining Eq. (10.40) with Eq. (10.20) and Eq. (10.41) with Eq. (10.22), the following equations can be obtained:

$$\sigma_{rr} = \frac{\sigma_o}{2}\cos 2\theta = \frac{1}{r}\frac{\partial \Phi}{\partial r} + \frac{1}{r^2}\frac{\partial^2 \Phi}{\partial \theta^2} \tag{10.42}$$

$$\tau_{r\theta} = -\frac{\sigma_o}{2}\sin 2\theta = -\frac{\partial}{\partial r}\left(\frac{1}{r}\frac{\partial \Phi}{\partial \theta}\right) \tag{10.43}$$

We are looking for a stress function Φ that the second derivative with respect to θ yields $\cos 2\theta$ and the first derivative with respect to θ yields $\sin 2\theta$. Obviously, the function Φ should have the form:

$$\Phi = f(r)\cos 2\theta \tag{10.44}$$

where $f(r)$ is a function of r to be derived.

We introduce the above function Φ into the biharmonic Eq. (10.16):

$$\left(\frac{\partial^2}{\partial r^2} + \frac{1}{r}\frac{\partial}{\partial r} + \frac{1}{r^2}\frac{\partial^2}{\partial \theta^2}\right)\left(\frac{\partial^2 f(r)}{\partial r^2}\cos 2\theta + \frac{1}{r}\frac{\partial f(r)}{\partial r}\cos 2\theta + \frac{1}{r^2}f(r)(-4\cos 2\theta)\right) = 0 \tag{10.45}$$

The above equation yields:

$$\frac{\partial^4 f(r)}{\partial r^4} + \frac{2}{r}\frac{\partial^3 f(r)}{\partial r^3} - \frac{9}{r^2}\frac{\partial^2 f(r)}{\partial r^2} + \frac{9}{r^3}\frac{\partial f(r)}{\partial r} = 0 \tag{10.46}$$

The general solution of Eq. (10.46) is:

$$f(r) = C_1 r^2 + C_2 r^4 + \frac{C_3}{r^2} + C_4 \tag{10.47}$$

where C_1, C_2, C_3, C_4 are constant parameters.

With the aid of Eq. (10.47), the stress function Φ in Eq. (10.44) can be written as:

$$\Phi = f(r)\cos 2\theta = \left(C_1 r^2 + C_2 r^4 + \frac{C_3}{r^2} + C_4\right)\cos 2\theta \tag{10.48}$$

The above equation can be utilized in Eqs. (10.20)–(10.22) for deriving the stress components σ_{rr}, $\sigma_{\theta\theta}$, $\tau_{r\theta}$:

$$\sigma_{rr} = \frac{1}{r}\frac{\partial \Phi}{\partial r} + \frac{1}{r^2}\frac{\partial^2 \Phi}{\partial \theta^2} = -\left(2C_1 + \frac{6C_3}{r^4} + \frac{4C_4}{r^2}\right)\cos 2\theta \tag{10.49}$$

$$\sigma_{\theta\theta} = \frac{\partial^2 \Phi}{\partial r^2} = \left(2C_1 + 12C_2 r^2 + \frac{6C_3}{r^4}\right)\cos 2\theta \tag{10.50}$$

$$\tau_{r\theta} = -\frac{\partial}{\partial r}\left(\frac{1}{r}\frac{\partial \Phi}{\partial \theta}\right) = \left(2C_1 + 6C_2 r^2 - \frac{6C_3}{r^4} - \frac{2C_4}{r^2}\right)\sin 2\theta \tag{10.51}$$

The constants C_1, C_2, C_3, and C_4 can be calculated by the boundary conditions 1 and 2.

Combining Eq. (10.49) with Eq. (10.40) and Eq. (10.51) with Eq. (10.41), and inserting $r \to \infty$ (remote stresses), the following results can be obtained:

$$\sigma_{rr} = \frac{\sigma_o}{2}\cos 2\theta = -\left(2C_1 + \frac{6C_3}{\infty} + \frac{4C_4}{\infty}\right)\cos 2\theta \tag{10.52}$$

$$\tau_{r\theta} = -\frac{\sigma_o}{2}\sin 2\theta = \left(2C_1 + 6C_2\infty^2 - \frac{6C_3}{\infty} - \frac{2C_4}{\infty}\right)\sin 2\theta \tag{10.53}$$

The above equations yield:

$$C_1 = -\frac{\sigma_o}{4} \tag{10.54}$$

$$C_2 = 0 \text{ because the stresses at } r \to \infty \text{ should not be infinite} \tag{10.55}$$

Then, the formulae for stresses given in Eqs. (10.49)–(10.51) can be simplified:

$$\sigma_{rr} = \left(\frac{\sigma_o}{2} - \frac{6C_3}{r^4} - \frac{4C_4}{r^2}\right)\cos 2\theta \tag{10.56}$$

$$\sigma_{\theta\theta} = \left(-\frac{\sigma_o}{2} + \frac{6C_3}{r^4}\right)\cos 2\theta \tag{10.57}$$

$$\tau_{r\theta} = \left(-\frac{\sigma_o}{2} - \frac{6C_3}{r^4} - \frac{2C_4}{r^2}\right)\sin 2\theta \tag{10.58}$$

We can now apply in Eqs. (10.56) and (10.58) the boundary conditions 2 (for $r = a, \sigma_r = 0, \tau_{r\theta} = 0$). Then, the following formulae can be obtained:

$$\sigma_{rr} = 0 = \left(\frac{\sigma_o}{2} - \frac{6C_3}{a^4} - \frac{4C_4}{a^2}\right)\cos 2\theta \tag{10.59}$$

$$\tau_{r\theta} = 0 = \left(-\frac{\sigma_o}{2} - \frac{6C_3}{a^4} - \frac{2C_4}{a^2}\right)\sin 2\theta \tag{10.60}$$

The solution of the algebraic system of Eqs. (10.59) and (10.60) yields:

$$C_3 = -\frac{a^4}{4}\sigma_o \tag{10.61}$$

$$C_4 = \frac{a^2}{2}\sigma_o \tag{10.62}$$

With the aid of the above results, Eqs. (10.56)–(10.58) can provide the stress field of the stress state (C):

$$\sigma_{rr} = \left(\frac{\sigma_o}{2} - \frac{6}{r^4}\left(-\frac{a^4}{4}\sigma_o\right) - \frac{4}{r^2}\left(\frac{a^2}{2}\sigma_o\right)\right)\cos 2\theta$$

$$\sigma_{\theta\theta} = \left(-\frac{\sigma_o}{2} + \frac{6}{r^4}\left(-\frac{a^4}{4}\sigma_o\right)\right)\cos 2\theta$$

$$\tau_{r\theta} = \left(-\frac{\sigma_o}{2} - \frac{6}{r^4}\left(-\frac{a^4}{4}\sigma_o\right) - \frac{2}{r^2}\left(\frac{a^2}{2}\sigma_o\right)\right)\sin 2\theta$$

or

$$\sigma_{rr} = \frac{\sigma_o}{2}\left(1 + 3\left(\frac{a}{r}\right)^4 - 4\left(\frac{a}{r}\right)^2\right)\cos 2\theta \tag{10.63}$$

$$\sigma_{\theta\theta} = \frac{\sigma_o}{2}\left(-1 - 3\left(\frac{a}{r}\right)^4\right)\cos 2\theta \tag{10.64}$$

$$\tau_{r\theta} = \frac{\sigma_o}{2}\left(-1 + 3\left(\frac{a}{r}\right)^4 - 2\left(\frac{a}{r}\right)^2\right)\sin 2\theta \tag{10.65}$$

The above equations are the solution for the stress state (C) and should be combined with the solution for the stress state (B).

10.3.3 SOLUTION FOR STRESS STATE (B)

Because the stresses in the case (B) are independent from θ (due to symmetric solid under symmetric loading), the derivatives $\partial^{(n)}/\partial\theta^{(n)}$ in the biharmonic equation can be eliminated. Then the biharmonic equation in polar coordinates can be simplified

$$\left(\frac{\partial^2}{\partial r^2} + \frac{1}{r}\frac{\partial}{\partial r}\right)\left(\frac{\partial^2\Phi}{\partial r^2} + \frac{1}{r}\frac{\partial\Phi}{\partial r}\right) = 0$$

and the general solution can be written in the following form:

$$\Phi = A\log(r) + Br^2\log(r) + Cr^2 + D \tag{10.66}$$

With the aid of Eq. (10.66), the stress components in the stress state (B) can be calculated by the following equations:

$$\sigma_{rr} = \frac{1}{r}\frac{\partial\Phi}{\partial r} = 2C + \frac{A}{r^2} + B[1 + 2\log(r)] \tag{10.67}$$

$$\sigma_{\theta\theta} = \frac{\partial^2\Phi}{\partial r^2} = 2C - \frac{A}{r^2} + B[3 + \log(r)] \tag{10.68}$$

The parameters A, B, C, D can be determined from the boundary conditions.

Because the term log(r) in Eqs. (10.67) and (10.68) cannot be determined for $r \to \infty$, the parameter B should be equal to zero:

$$B = 0 \tag{10.69}$$

Then, Eqs. (10.67) and (10.68) can be simplified:

$$\sigma_r = 2C + \frac{A}{r^2} \tag{10.70}$$

$$\sigma_\theta = 2C - \frac{A}{r^2} \tag{10.71}$$

In the above equations, the parameters C and A can be calculated with the aid of boundary conditions 2:

For $r = a$: $\sigma_r = 0$ (see Fig. 10.6)

Then, Eq. (10.70) yields:

$$2C + \frac{A}{a^2} = 0$$

or

$$C = -\frac{A}{2a^2} \tag{10.72}$$

For $r = b$: $\sigma_r = \frac{\sigma_o}{2}$ (see Fig. 10.6)

Then, Eq. (10.70) yields:

$$2C + \frac{A}{b^2} = \frac{\sigma_o}{2}$$

With the aid of Eq. (10.72), the above equation can provide the value of the parameter A:

$$A\left(-\frac{1}{a^2}+\frac{1}{b^2}\right)=\frac{\sigma_o}{2}$$

Because $b \to \infty$ for remote stresses, the ratio $1/b^2 \to 0$ can be eliminated. Then:

$$A=-\frac{\sigma_o}{2}a^2 \tag{10.73}$$

With the aid of Eq. (10.73), Eq. (10.72) yields:

$$C=\frac{\sigma_o}{4} \tag{10.74}$$

Considering the values of the parameters A and C, Eqs. (10.70) and (10.71) provide the solution for the stress state (B):

$$\sigma_{rr}=\frac{\sigma_o}{2}-\frac{\sigma_o}{2}\left(\frac{a}{r}\right)^2 \tag{10.75}$$

$$\sigma_{\theta\theta}=\frac{\sigma_o}{2}+\frac{\sigma_o}{2}\left(\frac{a}{r}\right)^2 \tag{10.76}$$

Superposition of the solutions of the stress states (B) and (C) yields the stress field of the stress state (A). Adding the stress components of Eqs. (10.63)–(10.65) with the corresponding stress components of Eqs. (10.75) and (10.76), the stress field near a hole of an infinite plate under remote uniaxial tension can be obtained by the following equations:

$$\sigma_{rr}=\frac{\sigma_o}{2}\left[1-\left(\frac{a}{r}\right)^2\right]+\frac{\sigma_o}{2}\left[1+3\left(\frac{a}{r}\right)^4-4\left(\frac{a}{r}\right)^2\right]\cos 2\theta \tag{10.77}$$

$$\sigma_{\theta\theta}=\frac{\sigma_o}{2}\left[1+\left(\frac{a}{r}\right)^2\right]+\frac{\sigma_o}{2}\left[-1-3\left(\frac{a}{r}\right)^4\right]\cos 2\theta \tag{10.78}$$

$$\tau_{r\theta}=\frac{\sigma_o}{2}\left[-1+3\left(\frac{a}{r}\right)^4-2\left(\frac{a}{r}\right)^2\right]\sin 2\theta \tag{10.79}$$

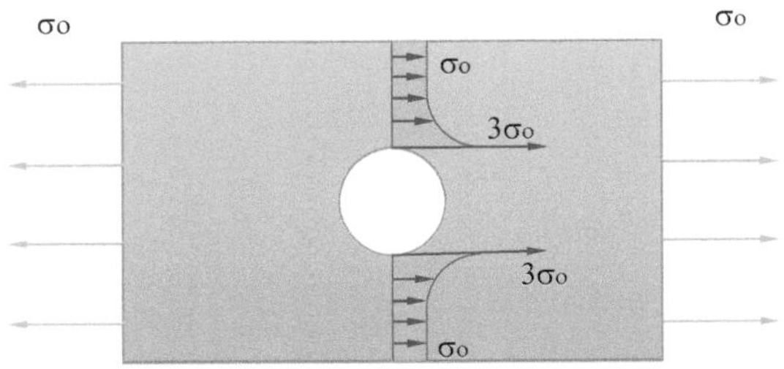

FIG. 10.7

Hole-induced stress concentration in an infinite plane under uniaxial stress.

10.3.4 STRESS CONCENTRATION FACTOR

Eqs. (10.77)–(10.79) show that the distribution of stresses versus r depends on the ratio $1/r$. In the location $r = a$ and $\theta = \pi/2$ on the border of the hole, the local tensile stress $\sigma_{\theta\theta}$ can be calculated by Eq. (10.78):

$$\sigma_{\theta\theta} = \frac{\sigma_o}{2}[1 + 1^2] + \frac{\sigma_o}{2}[-1 - 3 \times 1^4]\cos\pi$$

or

$$\sigma_{\theta\theta} = 3\sigma_o \qquad (10.80)$$

Eq. (10.80) shows that a stress concentration of a factor

$$k_t = \frac{\sigma_{\theta\theta}}{\sigma_o} = 3 \qquad (10.81)$$

is observed on the root of the hole. Fig. 10.7 demonstrates the distribution of $\sigma_{\theta\theta}$ versus r on the plane $\theta = \pi/2$ and the stress concentration on the location $(r, \theta) = (a, \pi/2)$.

The local tensile stress at the border of the hole is three times higher than the remote stress σ_o, and there is no influence from the radius of the hole.

REFERENCES

[1] Timoshenko SP, Goodier JN. Theory of elasticity. McGraw-Hill; 1970.
[2] Fenner RT. Engineering elasticity. Ellis Horwood ltd; 1986.
[3] Boresi Arthur P, Schmidt Richard J. Advanced Mechanics of Materials. Wiley; 2002.
[4] Armenakas Anthony. Advanced mechanics of materials and applied elasticity. CRC press; 2006.
[5] Bittencourt Marco. Computational solid mechanics. CRC press; 2015.

[6] Pavlou DG, et al. Stress Concentration Analysis of Interfacial Micro-structural Cracks Under Internal Singular Loading Sources. Journal of Computational Methods in Sciences and Engineering 2010;10(1-2):3–12. https://doi.org/10.3233/JCM-2010-0287

[7] Pavlou DG. Green's function for the bimaterial elastic solid containing interface annular crack. Engineering Analysis with Boundary Elements 2002;26(10):845–53. https://doi.org/10.1016/S0955-7997(02)00052-8

[8] Mavrothanasis FI, Pavlou FI, G D. Green's function for KI determination of axisymmetric elastic solids containing external circular crack. Engineering Fracture Mechanics 2008;75(8):1891–905. https://doi.org/10.1016/j.engfracmech.2007.08.014

[9] Ugural Ansel, Fenster Saul. Advanced Mechanics of Materials and Applied Elasticity. Pearson; 2019.

[10] Solecki Roman, Jay Conant R. Advanced Mechanics of Materials. Oxford University Press; 2003.

[11] Cook Robert. Advanced Mechanics of Materials. Pearson; 1998.

[12] Bickford William. Advanced Mechanics of Materials. Longman Pub Group; 1998.

[13] Anand Lallit, Govindjee Sanjay. Continuum Mechanics of Solids. OUP Oxford; 2020.

CHAPTER

Thick-Wall and Compound Cylinders

11

Chapter Outline

Thick-wall cylinders have many industrial applications (pressure vessels, industrial tanks, high-pressure pipelines, internal combustion engines, cannon tubes, etc.). The aim of this chapter is to determine the stress and displacement distribution in the area $a < r < b$ and $0 < \theta < 2\pi$ of a cross section of thick-wall cylinders under radial pressure on the inner and outer surface (Fig. 11.1). The stress analysis of thick-wall cylinders is convenient with the aid of Airy stress function.

11.1 STRESS ANALYSIS

The stress distribution within the wall of thick-wall cylinders can be achieved by the solution of the biharmonic equation of the Airy stress function $\Phi(r, \theta)$:

$$\left(\frac{\partial^2}{\partial r^2} + \frac{1}{r}\frac{\partial}{\partial r} + \frac{1}{r^2}\frac{\partial^2}{\partial \theta^2}\right)\left(\frac{\partial^2 \Phi}{\partial r^2} + \frac{1}{r}\frac{\partial \Phi}{\partial r} + \frac{1}{r^2}\frac{\partial^2 \Phi}{\partial \theta^2}\right) = 0 \tag{11.1}$$

Both the geometry of the cylinder and the load distribution (pressure) are axisymmetric; therefore all the partial derivatives of θ in the above equation should be eliminated:

$$\left(\frac{d^2}{dr^2} + \frac{1}{r}\frac{d}{dr}\right)\left(\frac{d^2 \Phi}{dr^2} + \frac{1}{r}\frac{d\Phi}{dr}\right) = 0 \tag{11.2}$$

or

$$\frac{d^4 \Phi}{dr^4} + \frac{2}{r}\frac{d^3 \Phi}{dr^3} - \frac{1}{r^2}\frac{d^2 \Phi}{dr^2} + \frac{1}{r^3}\frac{d\Phi}{dr} = 0 \tag{11.3}$$

Advanced Mechanics of Solids and Structures. DOI: http://doi.org/10.1016/B978-0-443-23778-2.00011-1

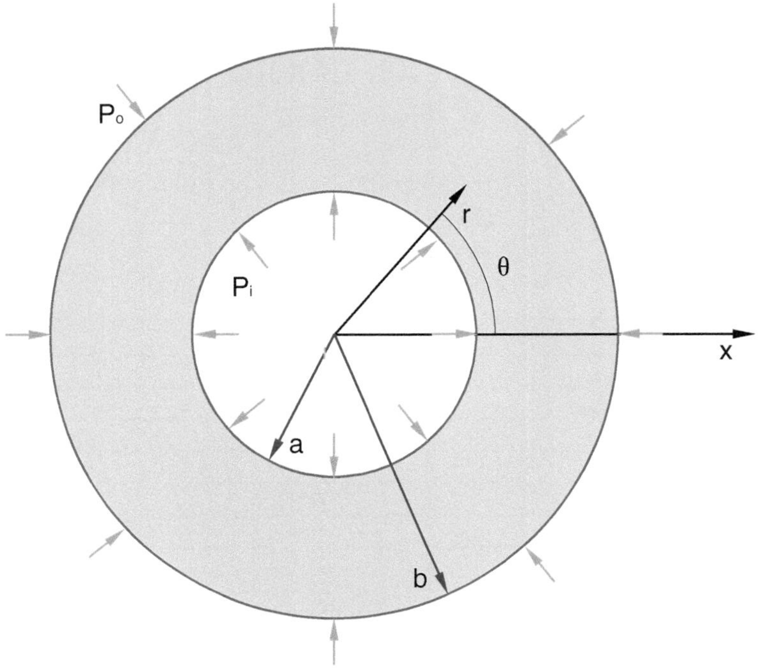

FIG. 11.1

Thick-walled cylinder under radial pressure on the inner and outer surface.

The general solution of the above ordinary differential equation is

$$\Phi = A\log(r) + Br^2\log(r) + Cr^2 + D \tag{11.4}$$

and the corresponding stresses are given by the following formulae (see Chapter 10):

$$\sigma_r = \frac{1}{r}\frac{\partial\Phi}{\partial r} = 2C + \frac{A}{r^2} + B[1 + 2\log(r)] \tag{11.5}$$

$$\sigma_\theta = \frac{\partial^2\Phi}{\partial r^2} = 2C - \frac{A}{r^2} + B[3 + \log(r)] \tag{11.6}$$

A. Determination of constants A, B, and C

For deriving the distribution of the stress components σ_r and σ_θ, the determination of the constants A, B, and C in Eqs. (11.5) and (11.6) is necessary. The constant B can be determined with the aid of the strain-displacement equations in polar coordinates:

$$\varepsilon_\theta = \frac{u_r}{r} \tag{11.7}$$

$$\varepsilon_r = \frac{\partial u_r}{\partial r} \tag{11.8}$$

Combining Eqs. (11.7) and (11.8), the following result can be obtained:

$$\varepsilon_r = \frac{\partial (r\varepsilon_\theta)}{\partial r} = \varepsilon_\theta + r\frac{\partial \varepsilon_\theta}{\partial r}$$

or

$$\varepsilon_r - \varepsilon_\theta - r\frac{\partial \varepsilon_\theta}{\partial r} = 0 \tag{11.9}$$

The strain components ε_r and ε_θ can be correlated with the stress components σ_r and σ_θ with the aid of Hooke's law:

$$\varepsilon_r = \frac{1}{E}(\sigma_r - \nu\sigma_\theta) \tag{11.10}$$

$$\varepsilon_\theta = \frac{1}{E}(\sigma_\theta - \nu\sigma_r) \tag{11.11}$$

Considering Eqs. (11.10) and (11.11), Eq. (11.9) yields:

$$\frac{\nu + 1}{E}(\sigma_r - \sigma_\theta) - \frac{1}{E}\frac{\partial}{\partial r}(\sigma_\theta - \nu\sigma_r) = 0 \tag{11.12}$$

The stress components σ_r and σ_θ in the above equation can be replaced by the ones given in Eqs. (11.5) and (11.6). Then, the following formula can be obtained:

$$(\nu + 1)\left(\frac{2A}{r^2} - 2B\right) - r\left(-(\nu + 1)\frac{A2r}{r^4} + 2(1 - \nu)B\frac{1}{r}\right) = 0$$

or

$$B = 0 \tag{11.13}$$

The result $B = 0$ was expected because $\log(r)$ in the terms $B[1 + 2\log(r)]$ and $B[3 + \log(r)]$ of Eqs. (11.5) and (11.6) cannot be determined for $r = 0$.

With the aid of Eq. (11.13), Eqs. (11.5) and (11.6) can now be simplified

$$\sigma_r = 2C + \frac{A}{r^2} \tag{11.14}$$

$$\sigma_\theta = 2C - \frac{A}{r^2} \tag{11.15}$$

and the constants A and C can be determined by the following boundary conditions:

$$r = a:\ \sigma_r = -P_i \tag{11.16}$$

$$r = b:\ \sigma_r = -P_o \tag{11.17}$$

The corresponding equations

$$-P_i = 2C + \frac{A}{a^2} \tag{11.18}$$

$$-P_o = 2C + \frac{A}{b^2} \tag{11.19}$$

yield:

$$A = \frac{a^2b^2}{b^2 - a^2}(-P_i + P_o) \tag{11.20}$$

and

$$2C = \frac{-b^2P_o + a^2P_i}{b^2 - a^2} \tag{11.21}$$

With the aid of Eqs. (11.14), (11.15), (11.20), and (11.21), the following formulae can be obtained for the stress distribution within the wall of a thick-wall cylinder:

$$\sigma_r = \frac{-b^2P_o + a^2P_i}{b^2 - a^2} + \frac{1}{r^2}\frac{a^2b^2}{b^2 - a^2}(-P_i + P_o) \tag{11.22}$$

$$\sigma_\theta = \frac{-b^2P_o + a^2P_i}{b^2 - a^2} - \frac{1}{r^2}\frac{a^2b^2}{b^2 - a^2}(-P_i + P_o) \tag{11.23}$$

11.2 DISPLACEMENT ANALYSIS

Due to the axisymmetric geometry and pressure distribution, only radial displacements u_r take place within the wall of the cylinders. The radial displacement can easily be determined from Eq. (11.7):

$$u_r = r\varepsilon_\theta = r\frac{1}{E}(\sigma_\theta - \nu\sigma_r) \tag{11.24}$$

Considering Eqs. (11.22) and (11.23), the above equation yields:

$$Eu_r = r(1-\nu)\frac{-b^2P_o + a^2P_i}{b^2 - a^2} - \frac{1}{r}(1+\nu)\frac{a^2b^2}{b^2 - a^2}(-P_i + P_o) \qquad (11.25)$$

11.3 COMPOUND CYLINDERS

In many engineering applications, compound cylinders consisting of two (or more) cylinders where the inner cylinder has different material properties than the outer one are used.

Usually, the inner cylinder has special tribological properties or is impermeable to special gases (e.g., hydrogen) or has anticorrosive properties for liquids, and the outer one should withstand the pressure. The compound cylinder is usually made by shrink fitting the smaller cylinder into the larger one (Fig. 11.2). For effective shrink

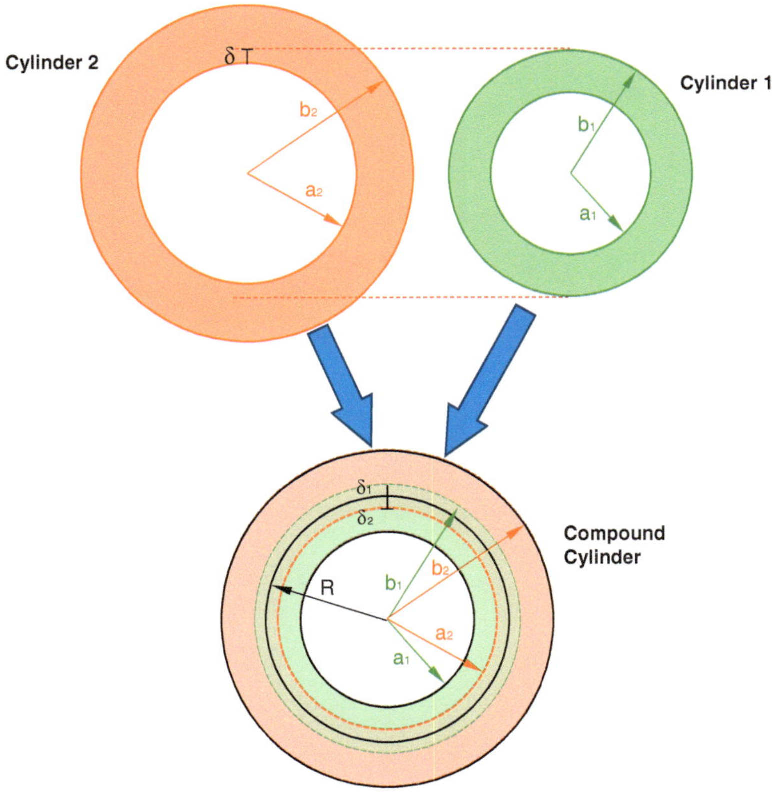

FIG. 11.2

Compound cylinder after shrink fitting of two cylinders.

fitting, the radius b_1 of the external surface of the cylinder 2 should be slightly larger than the radius a_2 of the hole of the cylinder 1 (see Fig. 11.2). We can define the shrink-fitting clearance as

$$\delta = b_1 - a_2 \tag{11.26}$$

After making the compound cylinder, the radius of interface between cylinders 1 and 2 can be symbolized by R.

If we use the notations δ_1 and δ_2 to symbolize the decrease of b_1 and the increase of a_2, the following condition (see Fig. 11.2) can be obtained:

$$|\delta_1| + |\delta_2| = \delta \tag{11.27}$$

Because of the forced fitting of cylinder 2 into cylinder 1, residual pressure P_s takes place on the interface between the two cylinders. The magnitude of P_s is needed for determining the stress distribution in the domains of cylinders 1 and 2 in the compound cylinder.

The radial displacement of the inner and outer surface of any thick-wall cylinder of inner radius a and outer radius b can be calculated with the aid of Eq. (11.25):

For $r = a$, Eq. (11.25) yields:

$$Eu_r(a) = a\left[\nu + \frac{a^2b^2}{(b^2 - a^2)}\right]P_i - \frac{2ab^2}{(b^2 - a^2)}P_o \tag{11.28}$$

Likewise, for $r = b$ the following equation can be obtained:

$$Eu_r(b) = b\left[\nu - \frac{a^2 + b^2}{b^2 - a^2}\right]P_o + \frac{2ba^2}{b^2 - a^2}P_i \tag{11.29}$$

Utilizing Eqs. (11.28) and (11.29), the radial displacements δ_1 and δ_2 can be obtained by applying the following boundary conditions to the free-body diagram of each cylinder separately. In cylinder 1, δ_1 symbolizes the decrease of the radius b_1 and can be calculated with the aid of Eq. (11.29):

$$\left.\begin{array}{l} P_i = 0 \\ P_o = P_s \\ a = a_1 \\ b = b_1 \end{array}\right\} \Rightarrow u_r(b_1) = \delta_1 = \frac{1}{E_1}b_1\left[\nu - \frac{a_1^2 + b_1^2}{b_1^2 - a_1^2}\right]P_s \tag{11.30}$$

Likewise, for cylinder 2, the increase of radius a_2 is:

$$\left.\begin{array}{l} P_i = P_s \\ P_o = 0 \\ a = a_2 \\ b = b_2 \end{array}\right\} \Rightarrow u_r(a_2) = \delta_2 = \frac{1}{E_2}a_2\left[\nu + \frac{a_2^2 + b_2^2}{b_2^2 - a_2^2}\right]P_s \tag{11.31}$$

Because $\nu \approx 0.3$ and $\frac{a_1^2+b_1^2}{b_1^2-a_1^2} > 1$, then $\nu - \frac{a_1^2+b_1^2}{b_1^2-a_1^2} < 0$

Therefore:

$$\left| \nu - \frac{a_1^2 + b_1^2}{b_1^2 - a_1^2} \right| = -\left(\nu - \frac{a_1^2 + b_1^2}{b_1^2 - a_1^2} \right)$$

and

$$|\delta_1| = \frac{b_1}{E_1}\left[\frac{a_1^2 + b_1^2}{b_1^2 - a_1^2} - \nu\right]P_s \tag{11.32}$$

In the right-hand side of Eq. (11.31), all the quantities are positive. Therefore:

$$|\delta_2| = \frac{a_2}{E_2}\left[\frac{a_1^2 + b_1^2}{b_1^2 - a_1^2} + \nu\right]P_s \tag{11.33}$$

With the aid of Eqs. (11.32) and (11.33), Eq. (11.27) yields

$$\delta = \frac{b_1}{E_1}\left[\frac{a_1^2 + b_1^2}{b_1^2 - a_1^2} - \nu\right]P_s + \frac{a_2}{E_2}\left[\frac{a_2^2 + b_2^2}{b_2^2 - a_2^2} + \nu\right]P_s$$

or

$$P_s = \frac{1}{\frac{b_1}{E_1}\left[\frac{a_1^2+b_1^2}{b_1^2-a_1^2} - \nu\right] + \frac{a_2}{E_2}\left[\frac{a_2^2+b_2^2}{b_2^2-a_2^2} + \nu\right]}\delta \tag{11.34}$$

Usually, $\delta = 1.5 \cdot 10^{-3}b$; thus the difference $b_1 - a_2$ is very small. Therefore, in Eq. (11.34)), the nominal value of radius a_2 and b_1 can be symbolized by R, that is, $b_1 = a_2 = R$. Then Eq. (11.34) can be simplified:

$$P_s = \frac{\delta/R}{\frac{1}{E_1}\left[\frac{R^2+a_1^2}{R-a_1^2} - \nu\right] + \frac{1}{E_2}\left[\frac{b_2^2+R^2}{b_2^2-R^2} + \nu\right]} \tag{11.35}$$

For shrink fitting of a shaft into a cylinder, the inner radius of the shaft is $a_1 = 0$. According to Eq. (11.35), the residual pressure on the interface between the outer surface of the shaft and the inner surface of the cylinder is:

$$P_s = \frac{\delta/R}{\frac{1-\nu}{E_1} + \frac{1}{E_2}\left[\frac{b_2^2+R^2}{b_2^2-R^2} + \nu\right]} \tag{11.36}$$

The stress distribution within the domain of cylinder 1 and 2 in the compound cylinder can be achieved with the aid of Eqs. (11.22) and (11.23).

Stress distribution on the wall of cylinder 1:

$$\left.\begin{matrix} P_i = 0 \\ P_o = P_s \\ a = a_1 \\ b = R \end{matrix}\right\} \Rightarrow \begin{cases} \sigma_r = \dfrac{-R^2 P_s}{R^2 - a_1^2} + \dfrac{1}{r^2}\dfrac{a_1^2 R^2}{R^2 - a_1^2} P_s \\ \sigma_\theta = \dfrac{-R^2 P_s}{R^2 - a_1^2} - \dfrac{1}{r^2}\dfrac{a_1^2 R^2}{R^2 - a_1^2} P_s \end{cases}$$

or

$$\begin{cases} \sigma_r = \dfrac{R^2}{R^2 - a_1^2}\left(\dfrac{a_1^2}{r^2} - 1\right) P_s \\ \sigma_\theta = \dfrac{-R^2}{R^2 - a_1^2}\left(\dfrac{a_1^2}{r^2} + 1\right) P_s \end{cases} \tag{11.37}$$

Stress distribution on the wall of cylinder 2:

$$\left.\begin{matrix} P_i = P_s \\ P_o = 0 \\ a = R \\ b = b_2 \end{matrix}\right\} \Rightarrow \begin{cases} \sigma_r = \dfrac{R^2 P_s}{b_2^2 - R^2} + \dfrac{1}{r^2}\dfrac{R^2 b_2^2}{b_2^2 - R^2}(-P_s) \\ \sigma_\theta = \dfrac{R^2 P_s}{b_2^2 - R^2} - \dfrac{1}{r^2}\dfrac{R^2 b_2^2}{b_2^2 - R^2}(-P_s) \end{cases}$$

or

$$\begin{cases} \sigma_r = \dfrac{R^2}{b_2^2 - R^2}\left(1 - \dfrac{b_2^2}{r^2}\right) P_s \\ \sigma_\theta = \dfrac{R^2}{b_2^2 - R^2}\left(1 + \dfrac{b_2^2}{r^2}\right) P_s \end{cases} \tag{11.38}$$

EXERCISE 1

The compact cylinder 1 (shaft) is shrink fitted into the thick-wall cylinder 2 (see figure below). The shrink-fit clearance is δ = 0.15 mm. Calculate the von Mises stress on cylinders 1 and 2 on their interface in the compound cylinder.

11.4 GIVEN DATA

Elasticity modulus of cylinders 1 and 2: $E_1 = 120\,GPa$, $E_2 = 69\,GPa$
Poisson's ratio of cylinders 1 and 2: $v_1 = v_2 = v = 0.3$

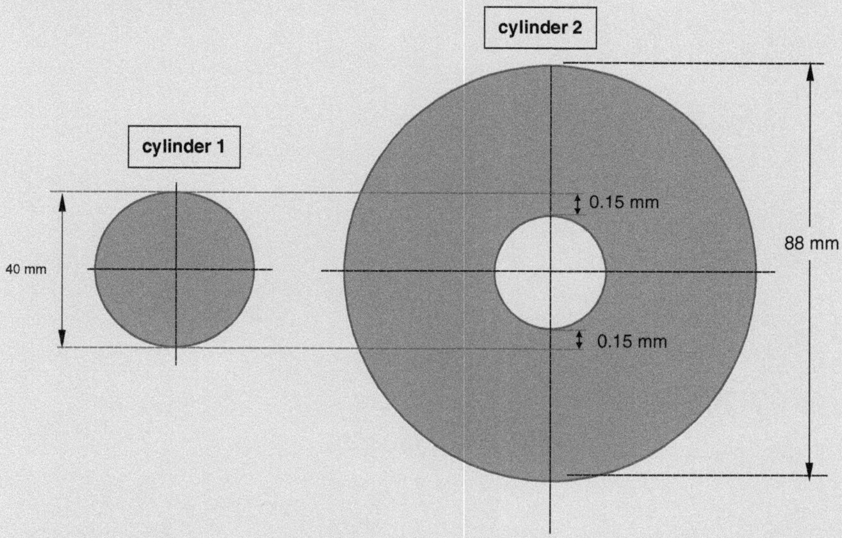

Solution

The residual pressure on the interface between the shaft and the cylinder 2 can be calculated by Eq. (11.36). To this end, the following data should be used:

$$\delta = 0.15\text{ mm}$$
$$a_1 = 0$$
$$R \approx b_1 \approx a_2 = 20\text{ mm}$$
$$b_2 = 44\text{ mm}$$
$$E_1 = 120\text{ GPa}$$
$$E_2 = 69\text{ GPa}$$
$$v = 0.3$$

Then, Eq. (11.36) yields:

$$P_s = \frac{\delta/R}{\frac{1-v}{E_1} + \frac{1}{E_2}\left(\frac{b_2^2+a_2^2}{b_2^2-a_2^2} + v\right)} = \frac{0.15/20}{\frac{1-0.3}{120} + \frac{1}{69}\left(\frac{44^2+20^2}{44^2-20^2} + 0.3\right)}\text{ GPa}$$

(Continued)

EXERCISE 1—CONT'D

or

$$P_s = 269\,\text{MPa}$$

11.5 CYLINDER 1

For $a_1 = 0$ and $r = R$, the stress components σ_r and σ_θ of the shaft on the interface can be obtained by Eq. (11.37):

$$\sigma_r = \frac{R^2}{R^2 - a_1^2}\left(\frac{a_1^2}{R^2} - 1\right)P_s = -P_s$$

$$\sigma_\theta = -\frac{R^2}{R^2 - a_1^2}\left(\frac{a_1^2}{R^2} + 1\right)P_s = P_s$$

Then, the von Mises stress for the shaft (cylinder 1) is:

$$\sigma_{VM_1} = \sqrt{\frac{1}{2}\left[(\sigma_r - \sigma_\theta)^2 + \sigma_\theta^2 + (-\sigma_r)^2\right]} = P_s\sqrt{3}$$

11.6 CYLINDER 2

Following the same procedure, the stresses of the cylinder 2 on the interface $r = R$ can be obtained by Eq. (11.38):

$$\sigma_r = \frac{R^2}{{b_2}^2 - R^2}\left(1 - \frac{{b_2}^2}{R^2}\right)P_s = \frac{20^2}{44^2 - 20^2}\left(1 - \frac{44^2}{20^2}\right)P_s = -P_s$$

$$\sigma_\theta = \frac{R^2}{{b_2}^2 - R^2}\left(1 + \frac{{b_2}^2}{R^2}\right)P_s = \frac{20^2}{44^2 - 20^2}\left(1 + \frac{44^2}{20^2}\right)P_s = 1.52P_s$$

The von Mises stress on the interface of cylinder 2 is:

$$\sigma_{VM_2} = \sqrt{\frac{1}{2}\left[(\sigma_r - \sigma_\theta)^2 + \sigma_\theta + (-\sigma_r)^2\right]} = 2.2P_s$$

EXERCISE 2

The dimensions of the cross section of three thick-wall cylinders, *C*1, *C*2, and *C*3, are shown in the following figure. All the cylinders are made by steel of elasticity modulus E = 210 GPa. The cylinder *C*1 is shrink fitted into cylinder *C*2 causing enlargement of the exterior diameter of the cylinder *C*2. Then, the compound cylinder *C*1-*C*2 is shrink fitted into the cylinder *C*3. Derive the functions of the stress distribution $\sigma_\theta(r)$, $\sigma_r(r)$ in the layer *C*3 of the compound cylinder *C*1-*C*2-*C*3.

EXERCISE 2—CONT'D

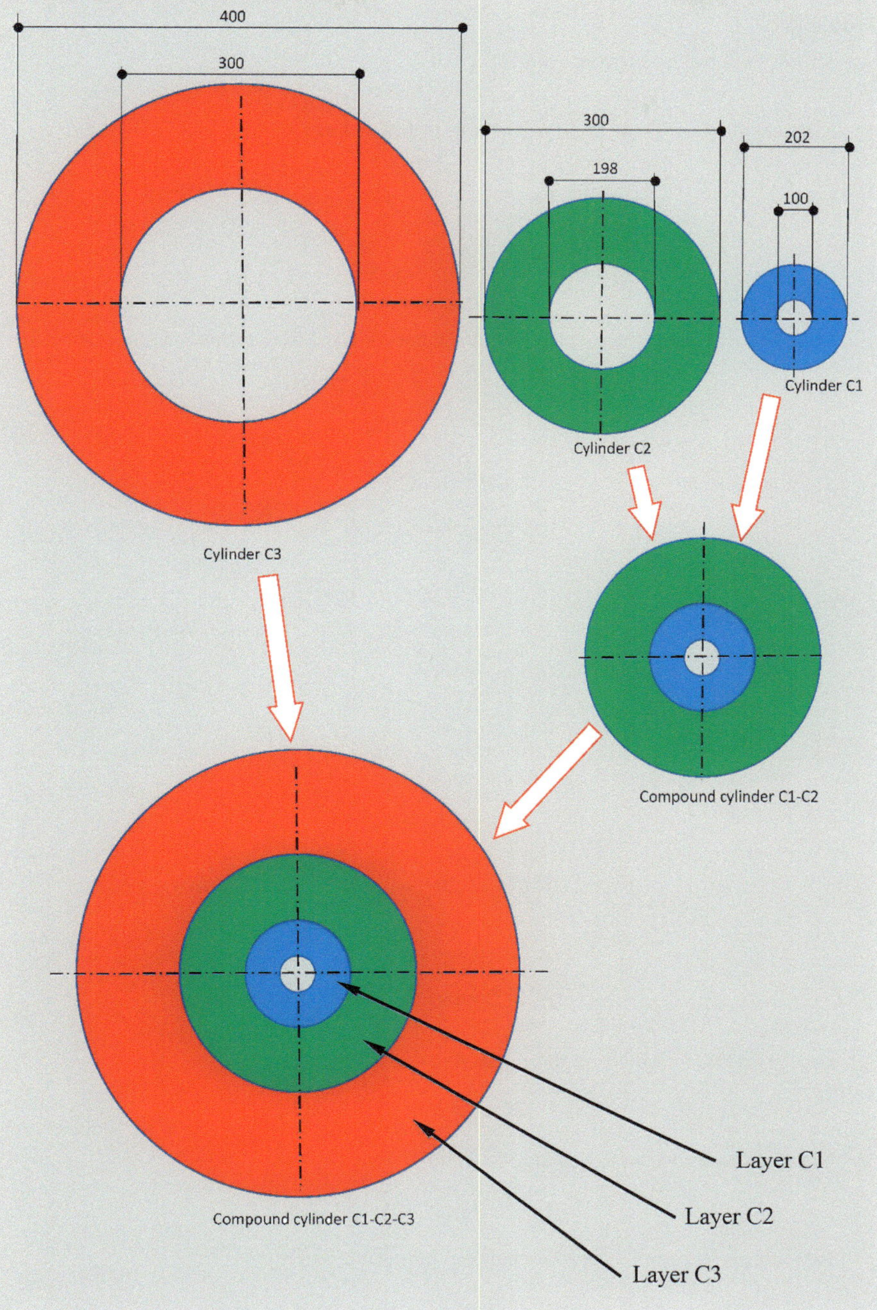

(Continued)

EXERCISE 2—CONT'D

Solution

- **Shrink-fit clearance between cylinders 1 and 2:**

$$\delta_{1,2} = 101 - 99 = 2$$

- **Residual pressure on the interface between cylinders 1 and 2:**

$$P_{S_{1,2}} = \frac{E \cdot \delta_{1,2}}{R\left[\frac{R^2 + a_1{}^2}{R^2 - a_1{}^2} + \frac{b_2{}^2 + R^2}{b_2{}^2 - R^2}\right]} = \frac{2\text{E}}{R\left[\frac{100^2 + 50^2}{100^2 - 50^2} + \frac{150^2 + 100^2}{150^2 - 100^2}\right]}$$

or

$$P_{S_{1,2}} = 0.00468E$$

- **Radial displacement on the exterior surface of the cylinder 2 after shrink fitting of cylinder 1:**

Data:

$P_o = 0$
$P_i = P_{S_{1,2}}$
$a = 100$
$b = 150$

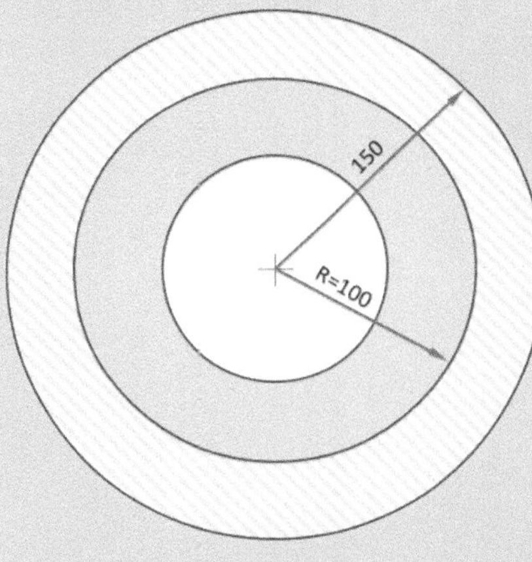

$$Eu_r(150) = b\left[v - \frac{a^2 + b^2}{b^2 - a^2}\right]P_o + \frac{2ba^2}{b^2 - a^2}P_i = 0 + \frac{2 \times 150 \times 100^2}{150^2 - 100^2}0.00468E$$

or

$$u_r(150) = 1.123\text{mm}$$

Therefore the exterior cylinder 2 after shrink fitting of cylinder 1 is:

$$b_2 = 150 + 1.123 = 151.123\text{mm}$$

- **Shrink-fit clearance between cylinders 2 and 3**

$$\delta_{2,3} = 151.123 - 150 = 1.123\text{mm}$$

- **Residual pressure on the interface between cylinders 2 and 3**

$$P_{S2,3} = \frac{E \cdot \delta_{2,3}}{R_{23}\left[\frac{R_{23}{}^2 + a_2{}^2}{R_{23}{}^2 - a_2{}^2} + \frac{b_3{}^2 + R_{23}{}^2}{b_3{}^2 - R_{23}{}^2}\right]} = \frac{1.123E}{150\left[\frac{150^2 + 100^2}{150^2 - 100^2} + \frac{200^2 + 150^2}{200^2 - 150^2}\right]}$$

EXERCISE 2—CONT'D

or

$$P_{S2,3} = 0.001213E$$

- **Stress distribution on cylinder *C*3**

The distribution of stresses σ_r and σ_θ within the domain of the cylinder $C3$ in the compound cylinder can be obtained by Eq. (11.38).

Using the following data

$$P_i = P_{S2,3}$$
$$P_o = 0$$
$$a = R = 150$$
$$b = 200$$

Eq. (11.38) yields:

$$\sigma_r = \frac{R^2}{b^2 - R^2}\left(1 - \frac{b^2}{r^2}\right)P_{S2,3}$$
$$\sigma_\theta = \frac{R^2}{b^2 - R^2}\left(1 + \frac{b^2}{r^2}\right)P_{S2,3}$$

or

$$\sigma_r = \frac{150^2}{200^2 - 150^2}\left(1 - \frac{200^2}{r^2}\right)(0.001213E) = 327\left(1 - \frac{200^2}{r^2}\right)$$
$$\sigma_\theta = \frac{150^2}{200^2 - 150^2}\left(1 + \frac{200^2}{r^2}\right)(0.001213E) = 327\left(1 + \frac{200^2}{r^2}\right)$$

REFERENCES

[1] Timoshenko SP, Goodier JN. Theory of elasticity. McGraw-Hill; 1970.
[2] Fenner RT. Engineering elasticity. Ellis Horwood ltd; 1986.
[3] Boresi Arthur P, Schmidt Richard J. Advanced Mechanics of Materials. Wiley; 2002.
[4] Armenakas Anthony. Advanced mechanics of materials and applied elasticity. CRC press; 2006.
[5] Bittencourt Marco. Computational solid mechanics. CRC press; 2015.
[6] Ugural Ansel, Fenster Saul. Advanced Mechanics of Materials and Applied Elasticity. Pearson; 2019.
[7] Solecki Roman, Conant RJay. Advanced Mechanics of Materials. Oxford University Press; 2003.
[8] Cook Robert. Advanced Mechanics of Materials. Pearson; 1998.
[9] Bickford William. Advanced Mechanics of Materials. Longman Pub Group; 1998.
[10] Anand Lallit, Govindjee Sanjay. Continuum Mechanics of Solids. OUP Oxford; 2020.

CHAPTER

Contact Stresses 12

Chapter Outline

12.1 THE ESSENCE OF CONTACT STRESSES

In basic courses of mechanics the equilibrium equations of structures and components and their internal forces are examined but no attention is paid on how the forces are transferred from one body to another. The main mechanism of force transfer is body contact via small contact surface that causes very high local stresses. Even if the structural members are able to withstand external forces, the contact stresses on the joints, supports, and other contact areas may be extremely high and can cause local yielding. Local stresses on the contact surface between locomotive wheel and rail, ball bearings in machines and engines, tooths of a pair of gears, and cam and valve in internal combustion engines (Fig. 12.1) are typical engineering examples of the importance of calculating contact stresses.

12.2 THE HERTZ MODEL (1895)

The theoretical model for calculating contact stresses is complex and involves advanced engineering mathematics. Satisfactory solution of contact stress can be achieved with the aid of the Hertz model (1895). The steps for implementing the Hertz model for calculating the contact stresses in the vicinity of the contact surface of bodies of double radius of curvature (Fig. 12.2) are briefly summarized below. In this model, the contact area has an ellipsoidal shape (Fig. 12.3), where a and b are the length of the ellipse's semi-axes.

Advanced Mechanics of Solids and Structures. DOI: http://doi.org/10.1016/B978-0-443-23778-2.00012-3

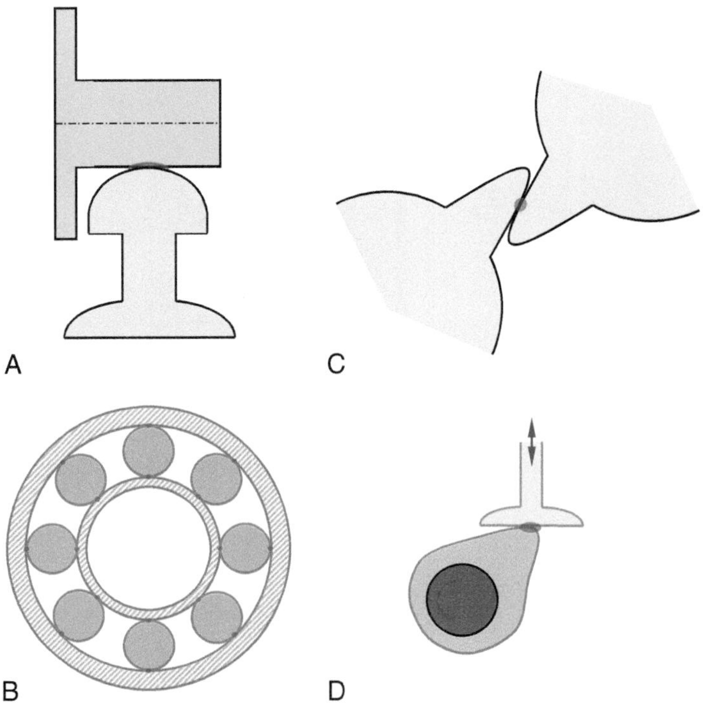

FIG. 12.1

Examples of contact stresses in engineering.

Important parameters in the Hertz model are the following:

$$k = \frac{b}{a} = \cos\theta \tag{12.1}$$

$$k' = \sqrt{1 - k^2} = \sin\theta \tag{12.2}$$

According to the Hertz model, the stresses on any material element located in a distance x, y, z from the contact point (Fig. 12.4) can be obtained by the following equations:

$$\sigma_{xx} = \left[M\left(\Omega_x + \nu\Omega'_x\right)\right]\frac{b}{\Delta} \tag{12.3}$$

$$\sigma_{yy} = \left[M\left(\Omega_y + \nu\Omega'_y\right)\right]\frac{b}{\Delta} \tag{12.4}$$

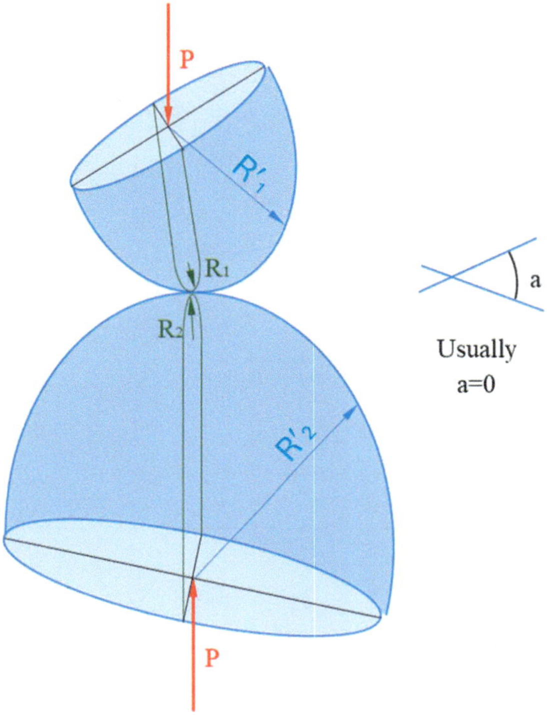

FIG. 12.2

Bodies of two radius of curvature in contact.

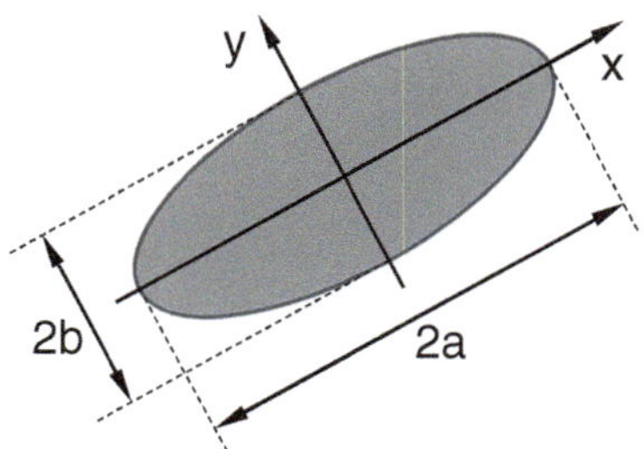

FIG. 12.3

Shape of the contact surface of two bodies of double radius of curvature.

$$\sigma_{xx} = \left[\frac{M}{2}\left(\frac{1}{\eta} - \eta\right)\right]\frac{b}{\Delta} \tag{12.5}$$

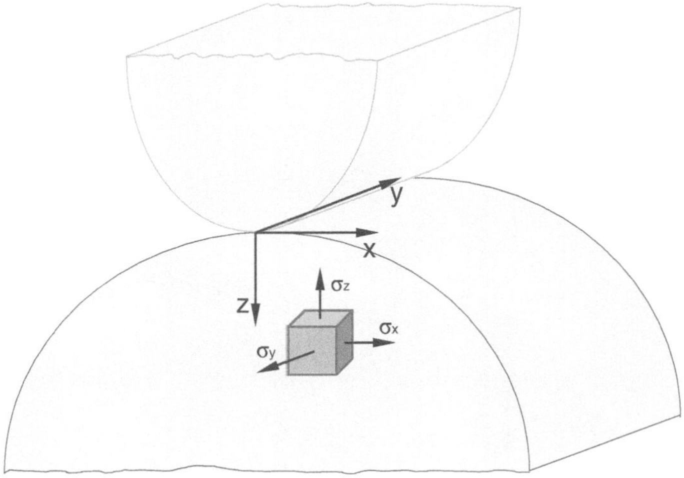

FIG. 12.4

Coordinate system and material element in the vicinity of the contact point.

The parameters of the above equations can be calculated by the following formulae:

$$\Delta = \frac{1}{A+B}\left[\frac{1-\nu_1^2}{E_1} + \frac{1-\nu_2^2}{E_2}\right] \tag{12.6}$$

$$b = \sqrt[3]{\frac{3kE(k')}{2\pi}} \tag{12.7}$$

$$M = \frac{2k}{k'^2 E(k')} \tag{12.8}$$

$$\eta = \sqrt{\frac{k^2 + k^2\left(\frac{z}{b}\right)^2}{1 + k^2\left(\frac{z}{b}\right)^2}} \tag{12.9}$$

$$A = \frac{1}{4}\left(\frac{1}{R_1^{'}} + \frac{1}{R_2} + \frac{1}{R'_1} + \frac{1}{R'_2}\right) - \frac{1}{4}\left(\frac{1}{R_1} + \frac{1}{R_2} - \frac{1}{R'_1} - \frac{1}{R'_2}\right) \tag{12.10}$$

$$B = \frac{1}{4}\left(\frac{1}{R_1} + \frac{1}{R_2} + \frac{1}{R'_1} + \frac{1}{R'_2}\right) + \frac{1}{4}\left(\frac{1}{R_1} + \frac{1}{R_2} - \frac{1}{R'_1} - \frac{1}{R'_2}\right) \tag{12.11}$$

$$\frac{B}{A} = \frac{\left(\frac{1}{k^2}\right) E(k') - K(k')}{E(k') - K(k')} \tag{12.12}$$

The terms $E(k')$ and $K(k')$ in Eqs. (12.7), (12.8), and (12.12) are the following elliptic integrals.

$$E(k') = H\left(\frac{\pi}{2}, k'\right) \int_0^{\pi/2} \sqrt{1 - k'^2 \sin^2\theta}\, d\theta \tag{12.13}$$

$$K(k') = F\left(\frac{\pi}{2}, k'\right) \int_0^{\pi/2} \frac{d\theta}{\sqrt{1 - k'^2 \sin^2\theta}} \tag{12.14}$$

$$H(\phi, k') = \int_0^{\phi} \sqrt{1 - k'^2 \sin^2\theta}\, d\theta \tag{12.15}$$

$$F(\phi, k') \int_0^{\phi} \frac{d\theta}{\sqrt{1 - k'^2 \sin^2\theta}} \tag{12.16}$$

where

$$\cot\phi = k\left(\frac{z}{b}\right) \tag{12.17}$$

With the aid of the above equations, the required parameters Ω_x, Ω'_x, Ω_y, and Ω'_y for calculating the contact stresses by Eqs. (12.3)–(12.5) can be provided below. It should be noted that these parameters are given as a function of the distance z of the material element by the contact surface.

$$\Omega_x = -\frac{1-\eta}{2} + k\frac{z}{b}\left[F(\phi, k') - H(\phi, k')\right] \tag{12.18}$$

$$\Omega'_x = \frac{\eta}{k^2} + 1 + k\frac{z}{b}\left[\frac{1}{k^2}H(\phi, k') - F(\phi, k')\right] \tag{12.19}$$

$$\Omega_y = \frac{1}{2\eta} + \frac{1}{2} - \frac{\eta}{k^2} + k\frac{z}{b}\left[\frac{1}{k^2}H(\phi, k') - F(\phi, k')\right] \tag{12.20}$$

$$\Omega'_y = -1 + \eta + k\frac{z}{b}[F(\phi, k') - H(\phi, k')] \tag{12.21}$$

Analytical solution of contact stresses is obviously very challenging. However, the model can be simplified for bodies in line contact (contact surface of single radius of curvature). For bodies of double radius of curvature, numerical solution can be performed and formulae for approximate calculation of contact stresses can be obtained.

Based on the Hertz model, a schematic distribution of contact stresses versus z beneath the contact surface is demonstrated in Fig. 12.5. In the same figure, the maximum shear stress distribution is derived, indicating that it takes maximum value beneath the contact surface.

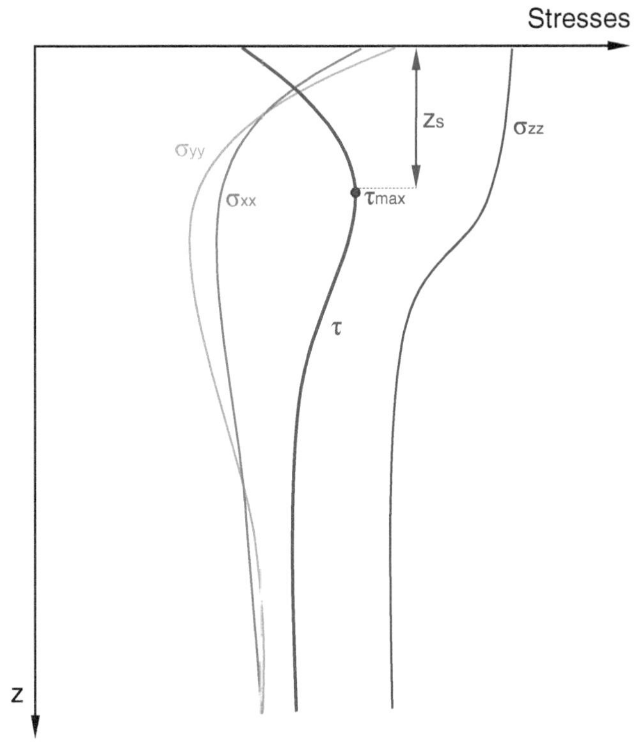

FIG. 12.5

Schematic distribution of contact stresses underneath the contact surface.

12.3 BODIES IN LINE CONTACT

Contact surfaces of a single radius of curvature is a usual case in engineering practice. Cylindrical components of cylinders and flat surfaces (Fig. 12.6) are usually in line contact. For this special case where the contact area can be defined as a line, the following simple formulae

can provide the contact stresses. To this end the following parameters should be calculated first:

$$\Delta = \frac{1}{\frac{1}{2}\left(\frac{1}{R_1} + \frac{1}{R_2}\right)}\left[\frac{1 - \nu_1^2}{E_1} + \frac{1 - \nu_2^2}{E_2}\right] \tag{12.22}$$

$$b = \sqrt{\frac{2w\Delta}{\pi}} \tag{12.23}$$

In the above equation, w is the load per unit length (line force).

With the aid of Eqs. (12.22) and (12.23), the contact stresses versus z can be obtained by the following equations:

$$\sigma_{xx} = -\left[\frac{\left(\sqrt{1 + \left(\frac{z}{b}\right)^2} - \frac{z}{b}\right)^2}{\sqrt{1 + \left(\frac{z}{b}\right)^2}}\right]\frac{b}{\Delta} \tag{12.24}$$

$$\sigma_{yy} = -2\nu\left[\sqrt{1 + \left(\frac{z}{b}\right)^2} - \frac{z}{b}\right]\frac{b}{\Delta} \tag{12.25}$$

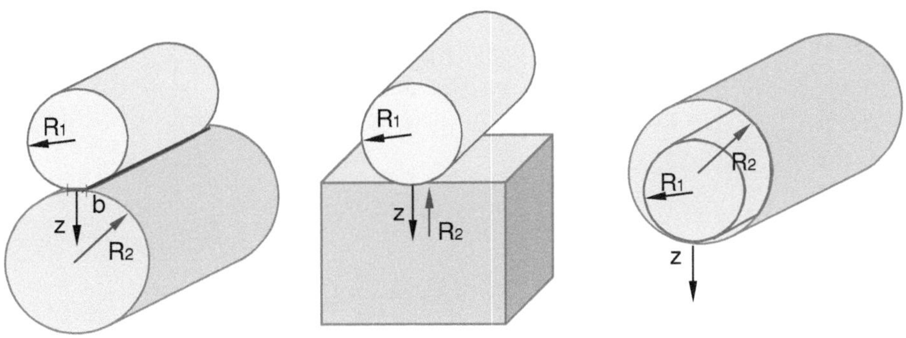

FIG. 12.6

Bodies in line contact.

$$\sigma_{zz} = -\left[\frac{1}{\sqrt{1 + \left(\frac{z}{b}\right)^2}}\right]\frac{b}{\Delta} \tag{12.26}$$

The notation b in Eqs. (12.24)–(12.26) denoted the width of the contact surface (Fig. 12.6). Considering the above equations, the maximum value of the contact stresses takes place at $z/b = 0$:

$$\sigma_{xx}^{\max} = -\frac{b}{\Delta} \tag{12.27}$$

$$\sigma_{yy}^{\max} = -2\nu\frac{b}{\Delta} \tag{12.28}$$

$$\sigma_{zz}^{\max} = -\frac{b}{\Delta} \tag{12.29}$$

According to the yield criteria, the maximum shear stress is an important parameter for designing joints for force transfer without yielding. The maximum shear stress can be calculated by the following formula:

$$\tau_{\max} = \frac{\sigma_{xx} - \sigma_{zz}}{2} \tag{12.30}$$

The location where $\tau_{\max}$ takes the maximum value can be obtained by the following condition:

$$\frac{d\tau_{\max}}{dz} = 0 \tag{12.31}$$

With the aid of Eqs. (12.27), (12.29), and (12.30), the above equation yields:

$$z = 0.7861\ b \tag{12.32}$$

Then, the following magnitude for the maximum shear stress can be obtained:

$$\tau_{\max} = 0.3\frac{b}{\Delta} \tag{12.33}$$

12.4 NUMERICAL CALCULATION OF CONTACT STRESSES FOR BODIES OF DOUBLE RADIUS OF CURVATURE

Analytic solution for contact stresses of bodies of double radius of curvature is not convenient because of the elliptic integrals in the Hertz model. However, numerical solution is performed and simple formulae can be provided for approximate calculation of contact stresses. To this end, the following approximate formulae can be used:

$$\sigma_{\max} = -\lambda_\sigma \frac{b}{\Delta} \tag{12.34}$$

$$\tau_{\max} = \lambda_\tau \frac{b}{\Delta} \tag{12.35}$$

where

$$b = \lambda_b \sqrt[3]{P\Delta} \tag{12.36}$$

In the above equations, P is the contact force (Fig. 12.2). The parameter Δ can be obtained by Eq. (12.6). For obtaining the coefficients λ_σ, λ_τ, λ_b only the geometric parameters A and B, given by Eqs. (12.10) and (12.11) are necessary. Numeric results yield the following formulae for the coefficients λ_σ, λ_τ, λ_b:

$$\lambda_\sigma = \begin{cases} 10^{0.226609 Log_{10}(B/A) - 0.187087} & for \quad 1 \le B/A \le 2.5 \\ 10^{0.123964 Log_{10}(B/A) - 0.14624} & for \quad 2.5 \le B/A \le 10 \\ 10^{0.0740006 Log_{10}(B/A) - 0.096277} & for \quad 10 \le B/A \le 20 \\ 1.000 & for \quad B/A \ge 20 \end{cases} \tag{12.37}$$

$$\lambda_\tau = \begin{cases} 10^{0.2742736771 Log_{10}(B/A) - 0.6777807053} & for \quad 1 \le B/A \le 2.5 \\ 10^{0.0760015 Log_{10}(B/A) - 0.59888} & for \quad 2.5 \le B/A \le 10 \\ 0.300 & for \quad B/A \ge 10 \end{cases} \tag{12.38}$$

$$\lambda_b = \begin{cases} 10^{-0.282666 Log_{10}(B/A) - 0.0457575} & for \quad 1 \le B/A \le 8 \\ 10^{-0.191429 Log_{10}(B/A) - 0.128152} & for \quad 8 \le B/A \le 200 \\ 10^{-0.186831 Log_{10}(B/A) - 0.138732} & for \quad 200 \le B/A \le 10^3 \end{cases} \tag{12.39}$$

EXERCISE 1

Calculate the maximum contact stresses σ_{max}, τ_{max} on a rail wheel that is in contact with a steel rail (see figure below). The wheel load is P = 150 kN, the wheel radius is $R_1' = 500$ mm, and the radius of curvature of the rail contact surface is $R_2 = 670$ mm. The material properties of the steel of the rail and the wheel are $E = 200$ GPa, $v = 0.29$, and Y = 880 MPa. Estimate whether the material on the contact surface will yield, and calculate the safety factor against yielding with the aid of the maximum shear stress criterion.

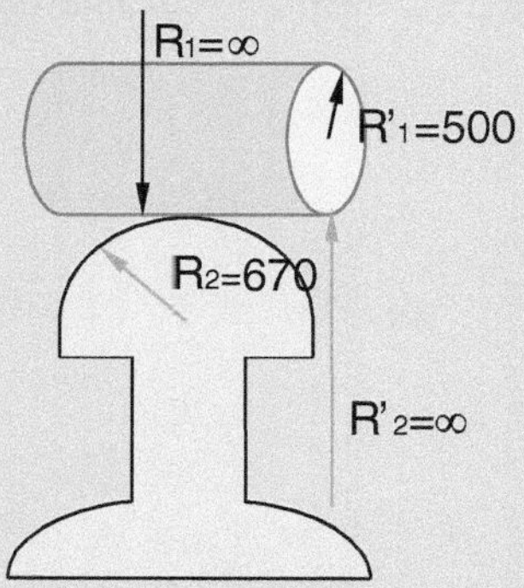

Solution

For $R_1 = \infty$, $R_1' = 500$ mm, $R_2 = 670$ mm, and $R_2' = \infty$, the following results can be obtained for the geometric parameters A and B:

$$A = \frac{1}{4}\left(\cancel{\frac{1}{R_1}} + \frac{1}{R_2} + \frac{1}{R_1'} + \cancel{\frac{1}{R_2'}}\right) - \frac{1}{4}\left(\cancel{\frac{1}{R_1}} + \frac{1}{R_2} - \frac{1}{R_1'} - \cancel{\frac{1}{R_2'}}\right) =$$

$$= \frac{1}{4}\left(\frac{1}{670 \cdot 10^{-3}} + \frac{1}{500 \cdot 10^{-3}}\right) - \frac{1}{4}\left(\frac{1}{670 \cdot 10^{-3}} - \frac{1}{500 \cdot 10^{-3}}\right) = 1.00 \text{ m}^{-1}$$

$$B = \frac{1}{4}\left(\cancel{\frac{1}{R_1}} + \frac{1}{R_2} + \frac{1}{R_1'} + \cancel{\frac{1}{R_2'}}\right) + \frac{1}{4}\left(\cancel{\frac{1}{R_1}} + \frac{1}{R_2} - \frac{1}{R_1'} - \cancel{\frac{1}{R_2'}}\right) =$$

$$= \frac{1}{4}\left(\frac{1}{670 \cdot 10^{-3}} + \frac{1}{500 \cdot 10^{-3}}\right) + \frac{1}{4}\left(\frac{1}{670 \cdot 10^{-3}} - \frac{1}{500 \cdot 10^{-3}}\right) = 0.746268 \text{ m}^{-1}$$

$$B/A = 0.746268$$

With the aid of Eqs. (12.37)–(12.39), the following coefficients can be obtained:

$$\lambda_\sigma = 10^{0.226609\, Log_{10}(B/A) - 0.187087} = 10^{0.226609\, Log_{10}(0.746268) - 0.187087} = 0.608289$$

$$\lambda_\tau = 10^{0.2742736771\, Log_{10}(B/A) - 0.677781} = 10^{0.2742736771\, Log_{10}(0.746268) - 0.677781} = 0.193802$$

$$\lambda_b = 10^{-0.282666\, Log_{10}(B/A) - 0.0457575} = 10^{-0.282666\, Log_{10}(0.746268) - 0.0457575} = 0.977621$$

$$\Delta = \frac{1}{A+B}\left[\frac{1-\nu_1^2}{E_1} + \frac{1-\nu_2^2}{E_2}\right] = \frac{1}{(1+0.746268)}\left[\frac{1-0.29^2}{200 \cdot 10^9} + \frac{1-0.29^2}{200 \cdot 10^9}\right] = 5.2449 \cdot 10^{-12}$$

$$b = \lambda_b \sqrt[3]{P\Delta} = 0.821\sqrt[3]{150 \cdot 10^3 \cdot 5.2449 \cdot 10^{-12}} = 0.009025 \text{ m}$$

EXERCISE 1—CONT'D

$$\frac{b}{\Delta} = \frac{0.009025}{5.2449 \cdot 10^{-12}} = 1.72072 \cdot 10^9 \ Pa$$

$$\sigma_{\max} = -\lambda_\sigma \frac{b}{\Delta} = -0.608289 \cdot 1.72072 \cdot 10^9 = -1046 \cdot 10^6 \ Pa$$

$$\tau_{\max} = \lambda_\tau \frac{b}{\Delta} = 0.193802 \cdot 1.72072 \cdot 10^9 = 333 \cdot 10^6 \ Pa$$

Because $Y = 880 * 10^6 \ Pa$, the safety factor is

$SF = \frac{Y/2}{\tau_{\max}} = 1.32$ (no yielding will occur).

EXERCISE 2

A cam-rod assembly (see figure below) in an internal combustion engine consists of a steel cam with material properties $E = 200\,\text{GPa}$ and $\nu = 0.29$, and a rod of a hardened steel with the same properties. The contact surface of the rod is spherical with radius $R_2 = R_2' = 5.5$ m. The contact surface of the cam is cylindrical with radius $R_1 = 0.6$ m and $R_1' = \infty$. With the aid of the yield criterion of the maximum principal stress calculate the maximum contact force. The compressive yield stress of the material of the rod is Y = −1270 MPa.

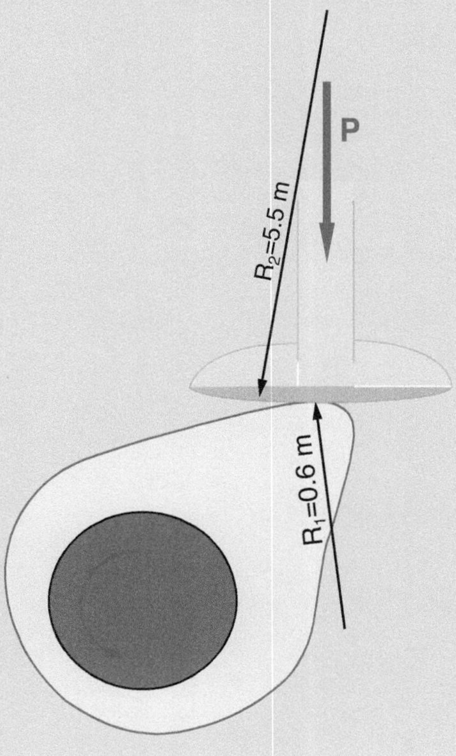

(Continued)

EXERCISE 2—CONT'D

Solution

$$A=\frac{1}{4}\left(\frac{1}{R_1}+\frac{1}{R_2}+\frac{1}{R_1'}+\frac{1}{R_2'}\right)-\frac{1}{4}\left(\frac{1}{R_1}+\frac{1}{R_2}-\frac{1}{R_1'}-\frac{1}{R_2'}\right)=\frac{1}{4}\left(\frac{1}{0.6}+\frac{1}{5.5}+\frac{1}{5.5}\right)-\frac{1}{4}\left(\frac{1}{0.6}+\frac{1}{5.5}-\frac{1}{5.5}\right)=0.091$$

$$B=\frac{1}{4}\left(\frac{1}{R_1}+\frac{1}{R_2}+\frac{1}{R_1'}+\frac{1}{R_2'}\right)+\frac{1}{4}\left(\frac{1}{R_1}+\frac{1}{R_2}-\frac{1}{R_1'}-\frac{1}{R_2'}\right)=\frac{1}{4}\left(\frac{1}{0.6}+\frac{1}{5.5}+\frac{1}{5.5}\right)+\frac{1}{4}\left(\frac{1}{0.6}+\frac{1}{5.5}-\frac{1}{5.5}\right)=0.9242$$

$$\frac{B}{A}=10.1667$$

With the aid of Eqs. (12.37) and (12.39), the following results can be obtained:

$$\begin{aligned}\lambda_\sigma &= 10^{0.0740006 Log_{10}(B/A)-0.096277}\\ &= 10^{0.0740006 Log_{10}(10.1667)-0.096277}\\ &= 0.951163\lambda_b\\ &= 10^{-0.191429 Log_{10}(B/A)-0.128152}\\ &= 10^{-0.191429 Log_{10}(10.1667)-0.128152}\\ &= 0.477579\end{aligned}$$

Parameter Δ can be calculated by Eq. (12.6):

$$\Delta=\frac{1}{A+B}\left[\frac{1-\nu_1^2}{E_1}+\frac{1-\nu_2^2}{E_2}\right]=\frac{1}{0.091+0.9242}\left[\frac{1-0.29^2}{210\cdot 10^9}+\frac{1-0.29^2}{210\cdot 10^9}\right]=8.59267\times 10^{-12}$$

According to the yield criterion of the maximum principal stress, the following equation can be formulated:

$$\sigma_{max}=Y$$

or

$$-\lambda_\sigma\frac{b}{\Delta}=-Y$$

Because

$$b=\lambda_b\sqrt[3]{P\Delta}$$

the following equation can be obtained:

$$-\lambda_\sigma\frac{\lambda_b\sqrt[3]{P\Delta}}{\Delta}=-Y$$

or

$$P=\frac{1}{\Delta}\left(\frac{Y\cdot\Delta}{\lambda_\sigma\lambda_b}\right)^3=\frac{1}{8.59267\times 10^{-12}}\left(\frac{1270\times 8.59267\times 10^{-12}}{0.951163\times 0.477579}\right)^3=1613\ kN$$

REFERENCES

[1] Boresi Arthur P, Schmidt Richard J. Advanced Mechanics of Materials. Wiley; 2002.
[2] Jaeger J. New solutions in contact mechanics. WIT press; 2005.
[3] Timoshenko SP, Goodier JN. Theory of elasticity. McGraw-Hill; 1970.
[4] Pavlou DG. Boundary element analysis of a plate on an elastic foundation loaded by a twisted annular rigid stamp. Proceedings of the Institution of Mechanical Engineers, Part J: Journal of Engineering Tribology 2004;218(6):549–56. https://journals.sagepub.com/doi/10.1243/1350650042794761
[5] Pavlou D, et al. An analytical solution of the annular plate on elastic foundation. Structural Engineering and Mechanics 2005;Volume 20(Number 2):209–23. https://doi.org/10.12989/sem.2005.20.2.209
[6] Armenakas Anthony. Advanced mechanics of materials and applied elasticity. CRC press; 2006.
[7] Bittencourt Marco. Computational solid mechanics. CRC press; 2015.
[8] Ugural Ansel, Fenster Saul. Advanced Mechanics of Materials and Applied Elasticity. Pearson; 2019.
[9] Solecki Roman, Conant RJay. Advanced Mechanics of Materials. Oxford University Press; 2003.
[10] Cook Robert. Advanced Mechanics of Materials. Pearson; 1998.
[11] Anand Lallit, Govindjee Sanjay. Continuum Mechanics of Solids. OUP Oxford; 2020.

CHAPTER

Elastic Stability

13

Chapter Outline

In most cases of elastic loading of structural members, the value of the displacement is proportional to the value of the applied force. An exception to this general rule is the compressive loading of slender and/or thin-walled structural members, that is, slender columns, thin-walled plates, shells, etc. In such cases, the initiation of failure of the structural member is not due to exceeding the yield stress but to the initiation of transverse to the direction of the force deformation causing instability. That is, the loss of stability of the member may occur even if the maximum stresses of the member are less than the yield stress. The modulus of elasticity of the material and the geometry of the loading plays a dominant role in the stability of structures. The main subject of the theory of elastic stability is to determine the critical values of the forces for which the members of a structure can be in a slightly deformed state without being led to collapse.

The elastic instability of structural members is also known as buckling and can be elastic bifurcation buckling (Fig. 13.1) or limit load buckling (Fig. 13.2). In bifurcation buckling, the axial deformation (Fig. 13.1) of a straight member (e.g., column) changes suddenly to a lateral deformation. The load at which this bifurcation occurs is called critical buckling load and is typically observed in bars and beams. The bifurcation buckling can be stable where the load deflection curve rises above the critical buckling load (Fig. 13.1a) or unstable where the load-deflection curve falls below the critical buckling load (Fig. 13.1b). Buckling where the member attains the maximum load without bifurcation (Fig. 13.2) is called limit load buckling and is typically observed in arches.

Advanced Mechanics of Solids and Structures. DOI: http://doi.org/10.1016/B978-0-443-23778-2.00006-8

13.1 CRITICAL BUCKLING LOAD OF COLUMNS

13.1.1 COLUMN FIXED-SUPPORTED IN ONE END AND FREE IN THE OTHER END

The fixed supported column in Fig. 13.3 is subjected to an axial force *P*. The column has elastic behavior, that is, the compressive stresses do not exceed the material yield stress. We consider a progressive increase of the magnitude of the axial force *P*.

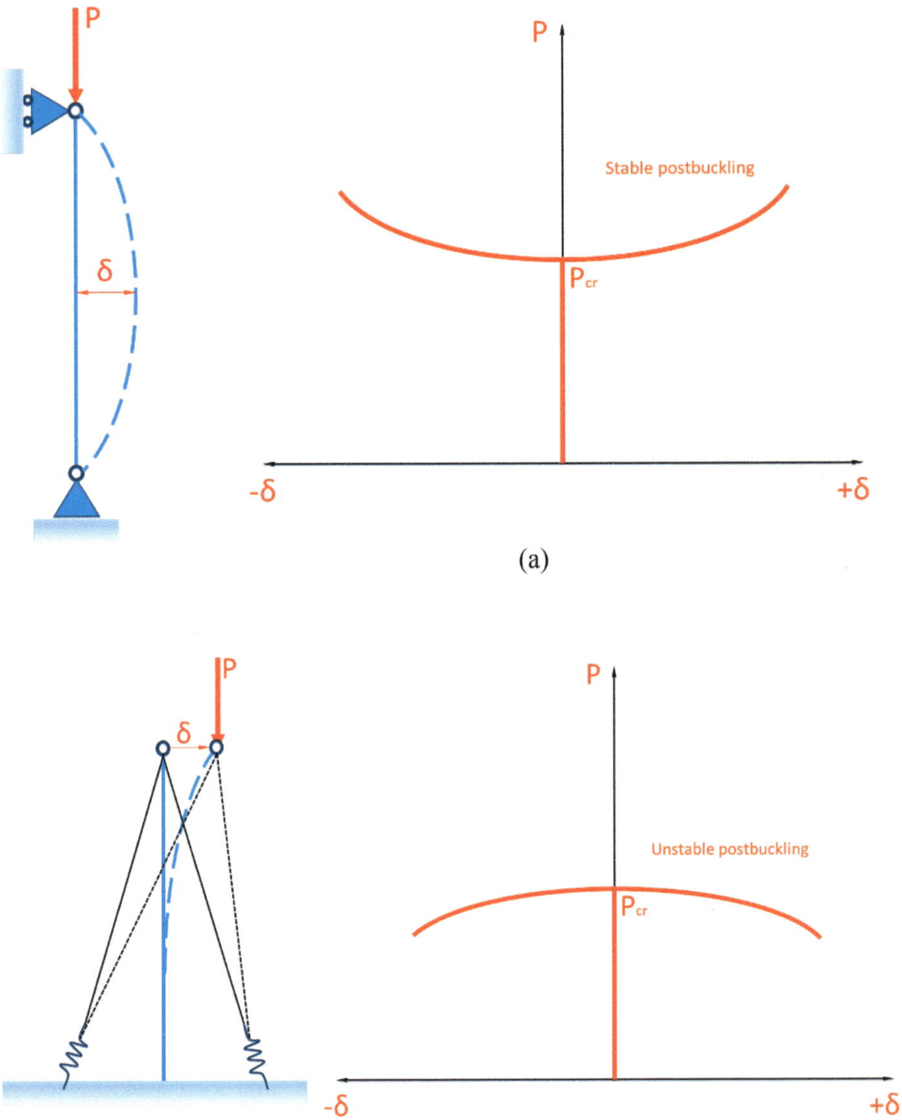

Fig. 13.1 Typical elastic bifurcation buckling examples: (a) stable postbuckling and (b) unstable postbuckling.

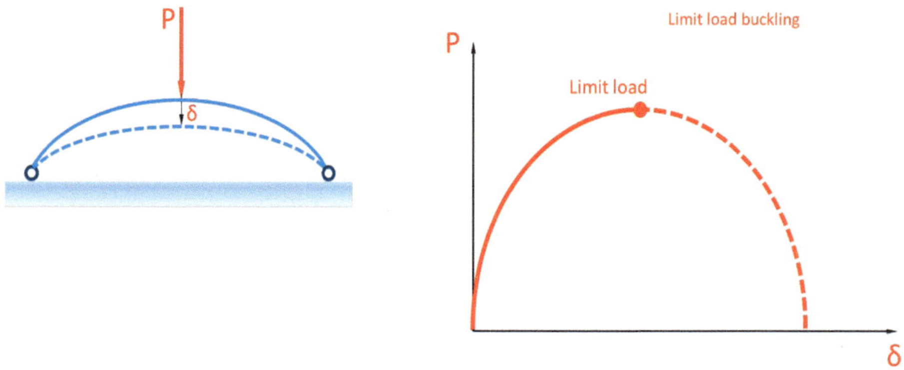

FIG. 13.2 Typical elastic limit load buckling.

For small values of P, any possible lateral force can cause a lateral deflection that can be removed when removing the lateral force. That means the elastic equilibrium of the column is stable. For a larger magnitude of P an irreversible lateral deflection δ can take place, leading to column instability. The magnitude of P causing transition from stable to unstable state is called critical buckling load P_{cr} and can be determined with the aid of the differential equation of bending. For any cross section in a distance x where a lateral deflection y takes place (Fig. 13.3), the differential equation of bending (see bending theory) is:

$$-\frac{M}{EI_z} = \frac{d^2y}{dx^2} \tag{13.1}$$

where the bending moment on the cross section is

$$-M = P(\delta - y) \tag{13.2}$$

With the aid of Eq. (13.2), Eq. (13.1) yields:

$$\frac{d^2y}{dx^2} + \frac{P}{EI_{\min}}y = \frac{P}{EI_{\min}}\delta \tag{13.3}$$

In the above equation, I_{min} is the minimum second moment of inertia with respect to the principal axes of the cross section. Using the notation

$$k^2 = \frac{P}{EI_{\min}} \tag{13.4}$$

Eq. (13.3) can be written in the following form:

$$\frac{d^2y}{dx^2} + k^2y = k^2\delta \tag{13.5}$$

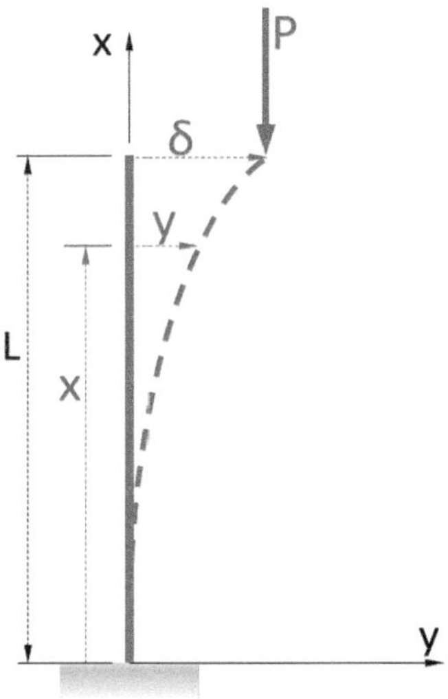

FIG. 13.3 Column under buckling loading.

The above equation has the following general solution:

$$y = A\cos(kx) + B\sin(kx) + \delta \tag{13.6}$$

where A and B are constants that can be determined by the conditions that the deflection and the slope of the beam at the fixed supported end ($x = 0$) are null:

$$y(0) = 0 \tag{13.7}$$

$$y'(0) = 0 \tag{13.8}$$

and the deflection at the free end ($x = L$) is

$$y(L) = \delta \tag{13.9}$$

With the aid of Eq. (13.6), Eqs. (13.7) and (13.8) yield:

$$A = -\delta \tag{13.10}$$

$$B = 0 \tag{13.11}$$

Then, Eq. (13.9) yields:

$$\delta\cos(kL) = 0 \tag{13.12}$$

The above equation has two solutions:

$$\delta = 0 \tag{13.13}$$

$$\cos(kL) = 0 \tag{13.14}$$

The solution $\delta = 0$ corresponds to a stable equilibrium (no buckling takes place). Eq. (13.14) corresponds to buckling state and is valid for

$$kL = (2n - 1)\frac{\pi}{2} \tag{13.15}$$

where n is a positive integer parameter. Eq. (13.15) specifies the magnitude of $k = \left(P/EI_{\text{min}}\right)^{1/2}$ when buckling takes place. For the minimum value $n = 1$, Eq. (13.15) yields the minimum magnitude of P, causing buckling, that is, the critical buckling load:

$$P_{cr} = \frac{\pi^2 EI_{\text{min}}}{4L^2} \tag{13.16}$$

The above result is known as Euler's rule for buckling.

13.1.2 COLUMN WITH FREE ENDS UNDER BUCKLING

An elastic column of length L where both ends are free is subjected to buckling load P (Fig. 13.4). Because of the symmetry, the critical buckling load can be determined by analyzing the half column. As Fig. 13.4 shows, the half column has the same conditions of support and loading like the column in Fig. 13.3. Only the

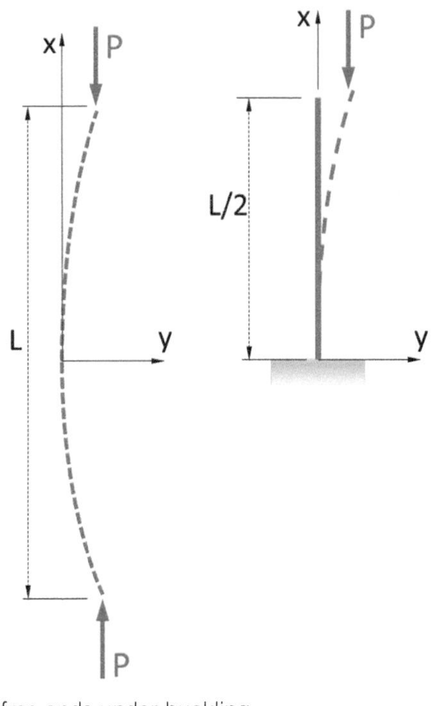

FIG. 13.4 Column with free ends under buckling.

length is half. Therefore the critical buckling load can be determined with the aid of Eq. (13.16) if we replace the length L by $(L/2)$. Therefore the critical buckling load of an elastic column of length L where both ends are free is:

$$P_{cr} = \frac{\pi^2 EI_{\min}}{L^2} \tag{13.17}$$

13.1.3 COLUMN WITH FIXED-SUPPORTED ENDS

Fig. 13.5 shows a column with fixed-supported ends under buckling. For calculating the critical buckling load, the column can be decomposed into the parts BA, AA′, and A′B′ with lengths $(L/4)$, $(L/2)$, and $(L/4)$, respectively. Points A and A′ are turning points of the curvature of the deformed shape of the column, that is, there is a change of the sign of the curvature. Therefore the second derivative of the deflection on A and A′ is null:

$$\frac{d^2y}{dx^2} = \frac{M}{EI_{\min}} = 0 \tag{13.18}$$

That means the points A and A′ are free of bending moment (see Eq. 13.18), and the segment AA′ of the beam in Fig. 13.5 has the same conditions of support and loading like the column in Fig. 13.4. Only the length is half. Therefore the critical

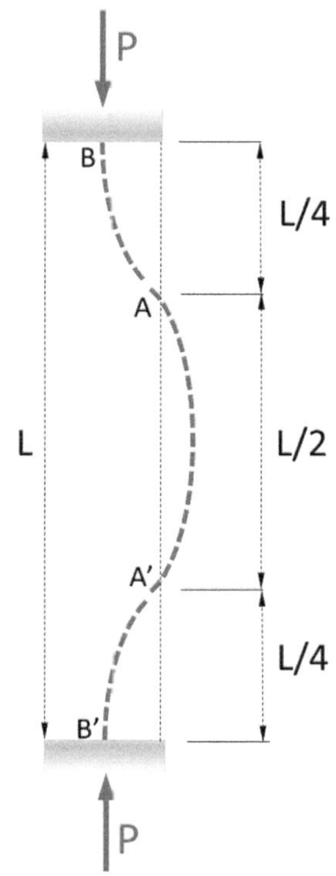

FIG. 13.5 Column with fixed-supported ends under buckling.

buckling load can be determined with the aid of Eq. (13.17) by replacing the length L by $(L/2)$. Therefore the critical buckling load of an elastic column of length L where both ends are fixed-supported is:

$$P_{cr} = \frac{4\pi^2 EI_{\min}}{L^2} \tag{13.19}$$

The critical buckling load of columns can be provided by the following general formula in terms of the equivalent buckling length L_a:

$$P_{cr} = \frac{\pi^2 EI_{\min}}{L_a^2} \tag{13.20}$$

where L_a is given in Table 13.1.

Table 1 Equivalent buckling length.

Supports of column	Fixed-supported in one end and free in the other end (Fig. 13.3)	Free ends (Fig. 13.4)	Fixed-supported ends (Fig. 13.5)
L_a	$2L$	L	$L/2$

13.2 CRITICAL BUCKLING STRESS OF COLUMNS

The critical buckling load on a column of cross-section area A, causes a stress:

$$\sigma_{cr} = \frac{P_{cr}}{A} \tag{13.21}$$

The stress given in the above equation is called critical stress.

Eq. (13.20) can be expressed in terms of the slenderness

$$\lambda = \frac{L_a}{i_{\min}} \tag{13.22}$$

where $i_{\min}$ is the radius of gyration

$$i_{\min} = \sqrt{\frac{I_{\min}}{A}} \tag{13.23}$$

Combining Eqs. (13.20), (13.22), and (13.23), Eq. (13.21) yields:

$$\sigma_{cr} = \frac{\pi^2 E}{\lambda^2} \tag{13.24}$$

The slenderness λ depends on geometric parameters of the column (cross-section area, moment of inertia, and length). Therefore, apart from the geometry of the column, the critical buckling stress is dominated by the elasticity modulus E. As Eq. (13.24) shows, the capability of a column to withstand buckling load is independent from the yield or the ultimate stress. The critical buckling stress versus λ for several materials is demonstrated in Fig. 13.6.

13.2 1 VALIDITY RANGE OF EULER'S FORMULA FOR BUCKLING

It has been mentioned that Euler's theory is valid for elastic deformation of bars and beams. Therefore the limit of the stress in Eq. (13.24) is:

$$\sigma_{cr} \leq Y \tag{13.25}$$

where Y is the yield stress of the material.

Combining Eqs. (13.24) and (13.25) the following range for λ where Euler's theory is valid (Fig. 13.7) can be obtained:

$$\lambda \geq \pi\sqrt{\frac{E}{Y}} \tag{13.26}$$

However, the existence of the knee point B in Fig. 13.7 has not been confirmed by experimental results. Experiments indicated that near point B, buckling occurs for

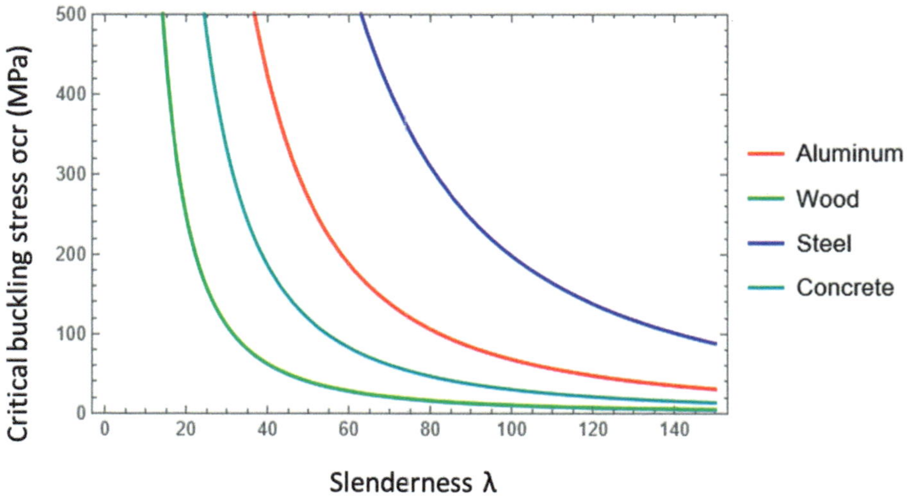

FIG. 13.6 Critical buckling stress versus slenderness for several materials.

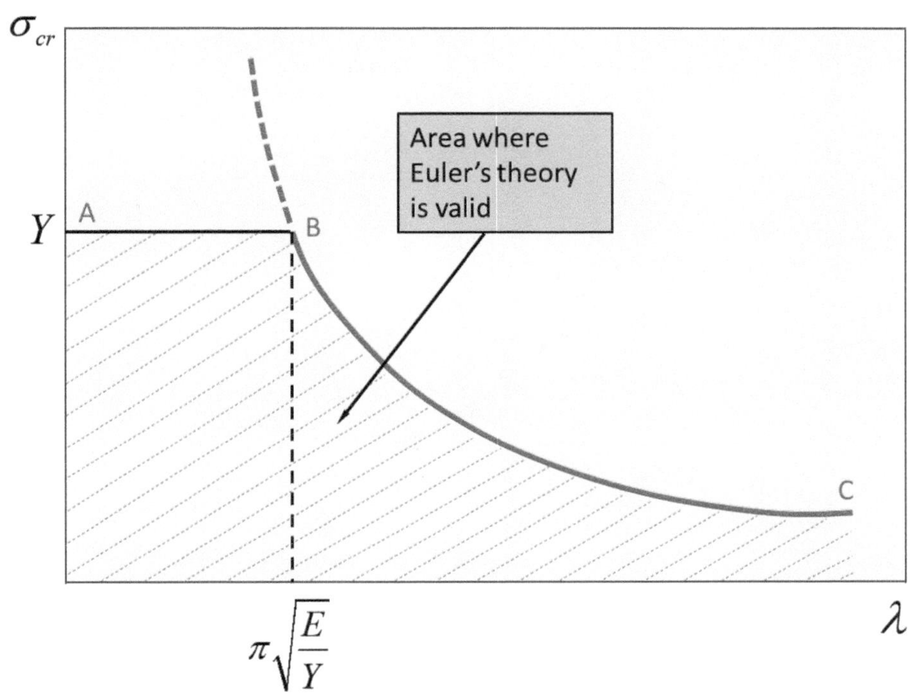

FIG. 13.7 Validity range of Euler's theory.

smaller stress that the predicted one by Euler's theory. Semi-empirical formulae have been proposed for improving Euler's theory. A suitable one is Johnson's formula

$$\sigma_{cr} = Y - \frac{Y^2}{4\pi^2 E}\lambda^2 \tag{13.27}$$

Graphical representation of Johnson's formula is a parabola that intersects with Euler's curve (Fig. 13.8) at the point

$$(\sigma_{cr}, \lambda) = \left(\frac{Y}{2}, \sqrt{\frac{2\pi^2 E}{Y}}\right) \tag{13.28}$$

Considering Johnson's and Euler's curves, we can classify the columns (Fig. 13.8) into three categories: "Long" columns" that have high slenderness ratio, "Intermediate" columns having intermediate slenderness ratio, and "Short" columns with low slenderness ratio. The formulae for calculating the critical buckling stress for each area are given below:

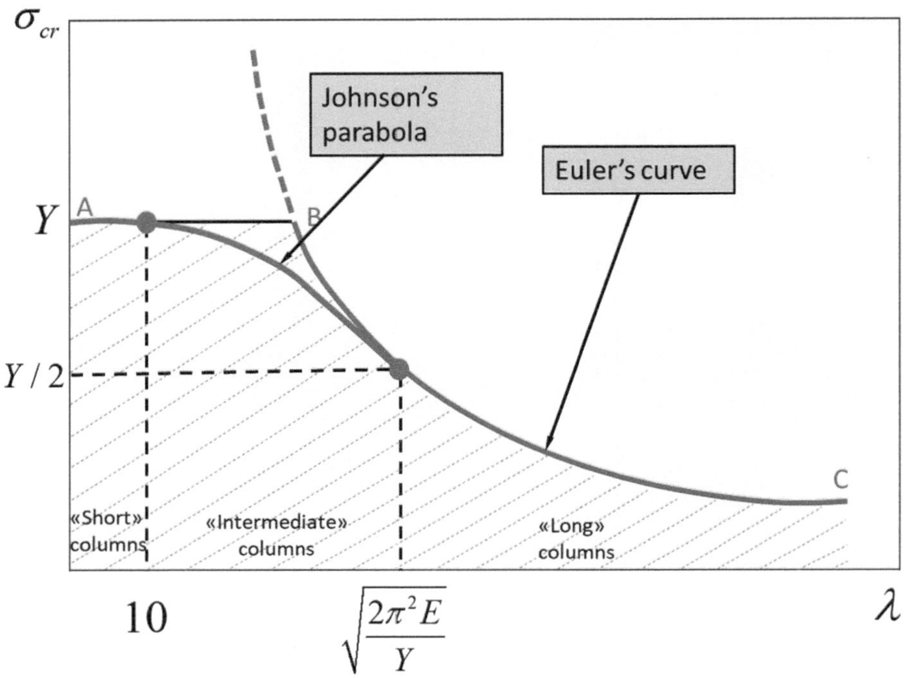

FIG. 13.8 Johnson's parabola for buckling.

Slenderness ratio λ	Column category	Critical buckling stress σ_{cr}
$\lambda > \sqrt{\frac{2\pi^2 E}{Y}}$	"Long" columns	$\sigma_{cr} = \frac{\pi^2 E}{\lambda^2}$ (Euler)
$10 < \lambda < \sqrt{\frac{2\pi^2 E}{Y}}$	"Intermediate" columns	$\sigma_{cr} = Y - \frac{Y^2}{4\pi^2 E}\lambda^2$ (Johnson)
$\lambda < 10$	"Short" columns	$\sigma_{cr} = Y$ (Hooke)

EXERCISE 1

A column of rectangular cross section with dimensions $10 \times 2\,\text{cm}^2$ has length $L = 6\,\text{m}$. The column is fixed-supported at the bottom and is free at the top. The column is made by steel with elasticity modulus $E = 210\,\text{GPa}$ and yield stress $Y = 240\,\text{MPa}$. The column is subjected to an axial force $N = 1.5\,\text{kN}$. Estimate whether the column will buckle.

Solution

$$I_{min} = \frac{(10 \times 10^{-2}) \times (2 \times 10^{-2})^3}{12} = 6.666 \times 10^{-8} m^4$$

$$i_{\min} = \sqrt{\frac{I_{\min}}{A}} = \sqrt{\frac{6.666 \times 10^{-8}}{(2 \times 10^{-2}) \times (8 \times 10^{-2})}} = 6.454 \times 10^{-3} m$$

$L_a = 2L = 12m$ (fixed-supported at one end and free at the other end)

$$\lambda = \frac{L_a}{i_{\min}} = \frac{12}{6.454 \times 10^{-3}} = 1859$$

$$\sqrt{\frac{2\pi^2 E}{Y}} = \sqrt{\frac{2\pi^2 \times 210 \times 10^9}{240 \times 10^6}} = 131$$

Obviously,

$$\lambda > \sqrt{\frac{2\pi^2 E}{Y}}$$

Therefore Euler's theory is valid for this buckling case. Then, the critical buckling stress can be calculated by the following equation:

$$\sigma_{cr} = \frac{\pi^2 E}{\lambda^2} = \frac{\pi^2 \times 210 \times 10^9}{1859^2} = 599128\ Pa$$

The critical buckling load is:

$$N_{cr} = \sigma_{cr} A = 599128 \times 20 \times 10^{-4} = 1198\ N$$

Because $N\,(=1500N) > (N_{cr} = 1198N)$, the column will be subjected to elastic buckling.

13.2.2 BUCKLING OF COLUMNS UNDER ECCENTRIC LOADING

Let's assume a column under eccentric loading (Fig. 13.9). The bending moment at each cross section is

$$M_z = Py \tag{13.29}$$

Then, the differential equation of bending

$$-\frac{M_z}{EI_z} = \frac{d^2 y}{dx^2} \tag{13.30}$$

or

$$\frac{d^2 y}{dx^2} - \frac{P}{EI_z} y = 0 \tag{13.31}$$

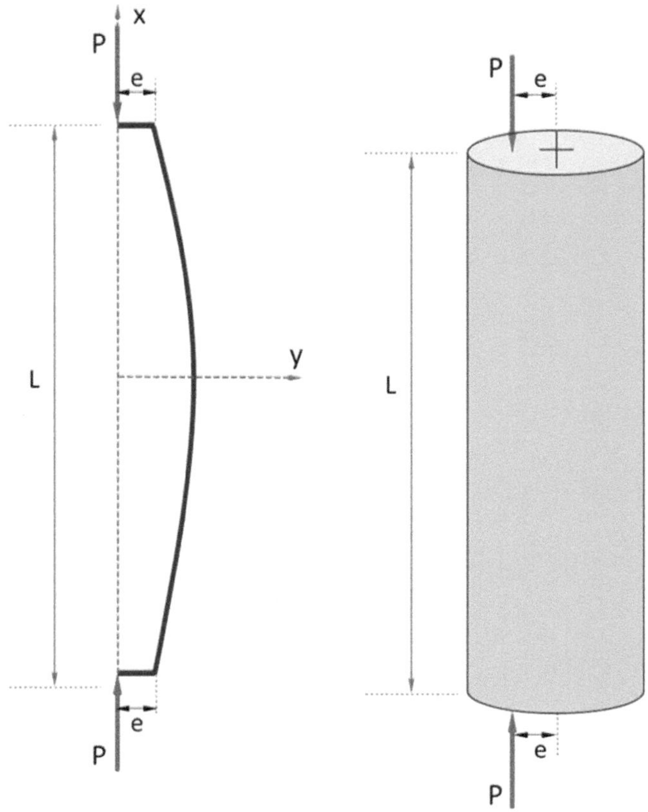

FIG. 13.9 Column under eccentric loading

has the following general solution:

$$y = A \sin\left(\sqrt{\frac{P}{EI_z}}x\right) + B \cos\left(\sqrt{\frac{P}{EI_z}}x\right) \tag{13.32}$$

With the aid of the following boundary conditions

$$y = e \quad for \;\; x = -\frac{L}{2} \tag{13.33}$$

$$y = e \quad for \;\; x = \frac{L}{2} \tag{13.34}$$

the following results for the parameters A and B can be obtained:

$$A = 0 \tag{13.35}$$

$$B = \frac{e}{\cos\left(\sqrt{\frac{P}{EI_z}}\frac{L}{2}\right)} \tag{13.36}$$

Then, the general solution of the differential equation is:

$$y = \frac{\cos\left(\sqrt{\frac{P}{EI_z}}x\right)e}{\cos\left(\sqrt{\frac{P}{EI_z}}\frac{L}{2}\right)} \text{ then } y_{max} = y(0) = \frac{e}{\cos\left(\sqrt{\frac{P}{EI_z}}\frac{L}{2}\right)} \text{ and } M_{max} = Py_{max} = \frac{P\,e}{\cos\left(\sqrt{\frac{P}{EI_z}}\frac{L}{2}\right)} \tag{13.37}$$

$$\textit{The condition } \sigma_{max} = \frac{P_{cr}}{A} + \frac{M_{max}}{I_{min}}C = Y \text{ or } \frac{P_{cr}}{A} + \frac{P_{cr}\,e\,c}{I_{min}\cos\left(\sqrt{\frac{P_{cr}}{EI_z}}\frac{L}{2}\right)} = Y$$

yields the critical buckling load P_{cr}.

13.3 FLEXURAL-TORSIONAL BUCKLING OF BEAMS

Beams and columns with cross section consisting of a web and flanges (I-beams) have wide use in steel structures. When the web and the flanges are thin, a bending moment at the ends of the beam can cause flexural-torsional buckling. For the sake of understanding the flexural-torsional buckling, a beam of a thin rectangular cross section can be analyzed firstly.

13.4 BEAM OF THIN RECTANGULAR CROSS SECTION

Fig. 13.10 demonstrates the deformed shape of a beam of thin rectangular cross section under bending moment M_x. The beam has a length L. Apart from the bending in the y-z plane, the beam is bended in the x-z plane and rotated in the z axis. Due to the beam double deformation by slopes φ and du/dz (Fig. 13.10), the bending component M_x can be analyzed into two components: $M_x \sin\varphi \approx M_x\varphi$ and $M_x \sin(du/dz) \approx M_x(du/dz)$ (because both slopes are small). The component $M_x\varphi$ dominates the bending in the x-z plane:

$$EI_y\frac{d^2u}{dz^2} = -M_x\varphi \tag{13.38}$$

and the component $M_x(du/dz)$ is a torsional moment causing torsion along the longitudinal direction:

$$\vartheta = \frac{M_x(du/dz)}{GI_o} \tag{13.39}$$

where ϑ is the angle of twist per unit length

$$\vartheta = \frac{d\varphi}{dz} \tag{13.40}$$

and GI_o is the torsional rigidity (see torsion theory)

$$GI_o = G\frac{1}{3}hb^3 \tag{13.41}$$

Taking the derivative $d\{\bullet\}/dz$ in Eq. (13.39), the following result can be obtained:

$$GI_o\frac{d^2\varphi}{dz^2} = M_x\frac{d^2u}{dz^2} \tag{13.42}$$

Combining Eqs. (13.38) and (13.42) we get the following equation:

$$\frac{d^2\varphi}{dz^2} + \frac{M_x^2}{EI_yGI_o}\varphi = 0 \tag{13.43}$$

The general solution of the above homogeneous differential equation is:

$$\varphi(z) = c_1\cos\left(\sqrt{\frac{M_x^2}{EI_yGI_o}}\,z\right) + c_2\sin\left(\sqrt{\frac{M_x^2}{EI_yGI_o}}\,z\right) \tag{13.44}$$

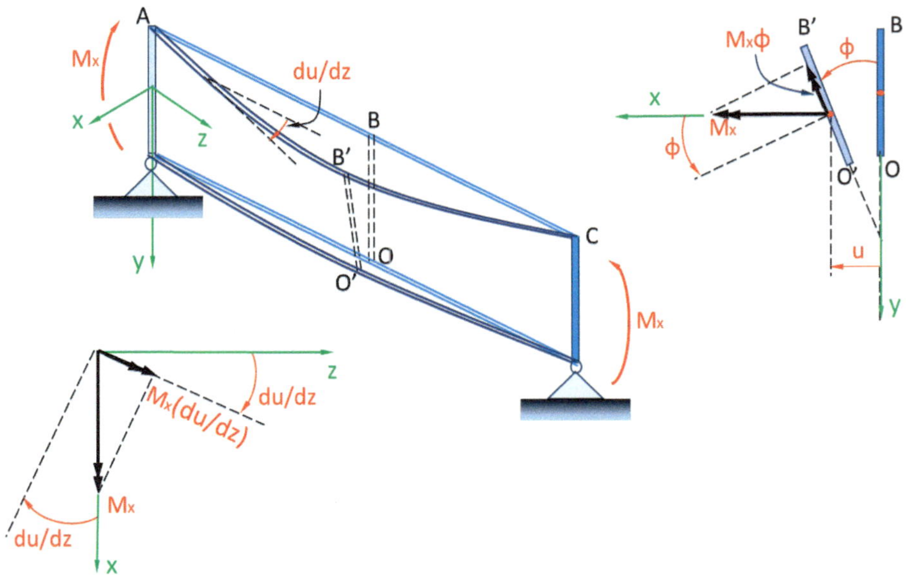

FIG. 13.10 Flexural-torsional buckling of a beam of thin rectangular cross section.

The boundary condition $\varphi(0) = 0$ on the support on $z = 0$ yields:

$$c_1 = 0 \tag{13.45}$$

Then

$$\varphi(z) = c_2 \sin\left(\sqrt{\frac{M_x^2}{EI_y GI_o}}\, z\right) \tag{13.46}$$

The boundary condition $\varphi(L) = 0$ on the support $z = L$ yields:

$$\sin\left(\sqrt{\frac{M_x^2}{EI_y GI_o}}\, L\right) = 0 \tag{13.47}$$

or

$$\frac{M_x}{\sqrt{EI_y GI_o}}\, L = n\pi \tag{13.48}$$

For $n = 0$ we get the trivial solution (undeformed shape of the beam). The smallest value of n for having buckling deformation is $n = 1$. Then, the critical bending moment is

$$M_{cr} = \frac{\pi\sqrt{EI_y GI_o}}{L} \tag{13.49}$$

EXERCISE 2

The ends of a simply supported beam of thin rectangular cross section are subjected to bending moments $M_x = 930\,Nm$ (Fig. 13.11). The beam is made by structural steel with elasticity and shear modulus $E = 210\ GPa$ and $G = 90\ GPa$, respectively. The dimensions of the cross section are $b = 0.5\ cm$ and $h = 10\,cm$. Calculate the maximum length L for which the beam will not undergo flexural-torsional buckling.

Solution

Calculation of the second moment of inertia and the torsional rigidity:

$$I_y = \frac{hb^3}{12} = \frac{(10 \times 10^{-2})(0.5 \times 10^{-2})}{12} = 0.104 \times 10^{-8} m^4$$

$$GI_o = G\frac{hb^3}{3} = (80 \times 10^9)\frac{(10 \times 10^{-2})(0.5 \times 10^{-2})}{3} = 133.333 \times 10^5 Nm^2$$

(Continued)

EXERCISE 2—CONT'D

FIG. 13.11 Simply supported beam under bending.

With the aid of Eq. (13.49), the maximum length for which the beam will not undergo flexural-torsional buckling is:

$$L = \frac{\pi\sqrt{EI_y GI_o}}{M_{cr}} = \frac{\pi\sqrt{(210 \times 10^9)(0.104 \times 10^{-8})(133.333 \times 10^5)}}{930} = 0.91\,m$$

REFERENCES

[1] Bazant ZP, Cedolin L. Stability of structures. Dover; 2003.
[2] Wang CM, Wang CY, Reddy JN. Exact solutions for buckling of structural members. CRC press; 2005.
[3] Kitipornchai S, Wang CM. Out of plane buckling formulae for beam-columns and tie-beams. Journal of Structural Engineering 1988;114(12):2773–89.
[4] Timoshenko SP, Gere JM. Theory of elastic stability. New York: McGraw Hill; 1961.

CHAPTER

Elements of the Mechanics of Composite Materials 14

Chapter Outline

Composite materials consist of a polymeric matrix reinforced with fibers. Composite materials are advantageous because of their superior mechanical and physical properties, that is, high strength, low density, corrosion resistance, excellent behavior in fatigue loading, and low maintenance cost. Because of their advantages, the application of composite materials in structural and mechanical engineering has a rapid growth in the last decades. The analysis and design of structures made by composite materials is based on the classical lamination theory of laminates.

14.1 CLASSICAL LAMINATION THEORY

Composite laminates are composed from multiple fiber reinforced laminae (Fig. 14.1). Each lamina has very small thickness and usually different fiber orientation. The stacking sequence of layers, their fiber orientation, and number influence the mechanical behavior of the multilayered medium. A multilayered composite plate consisting of N layers (Fig. 14.2) is considered for describing the stress analysis with the aid of Kirchhoff's assumption. The x axis of the global coordinate system is located in the geometric midplane of the laminate, whereas the z and y axes are aligned with the directions of the thickness and the width, respectively (Fig. 14.2). The specification of the fiber angles of the various layers starts with layer 1.

- ***The Kirchhoff assumption***

  The simplest lamination theory is an extension of the Kirchhoff isotropic plate theory to laminated composite plates. According to this theory, the straight lines normal to the xy plane (Fig. 14.2) before deformation remain straight and normal to the midsurface $z = 0$ after deformation. Therefore, due to this assumption, both transverse shear and transverse normal effects are neglected.

Advanced Mechanics of Solids and Structures. DOI: http://doi.org/10.1016/B978-0-443-23778-2.00014-7

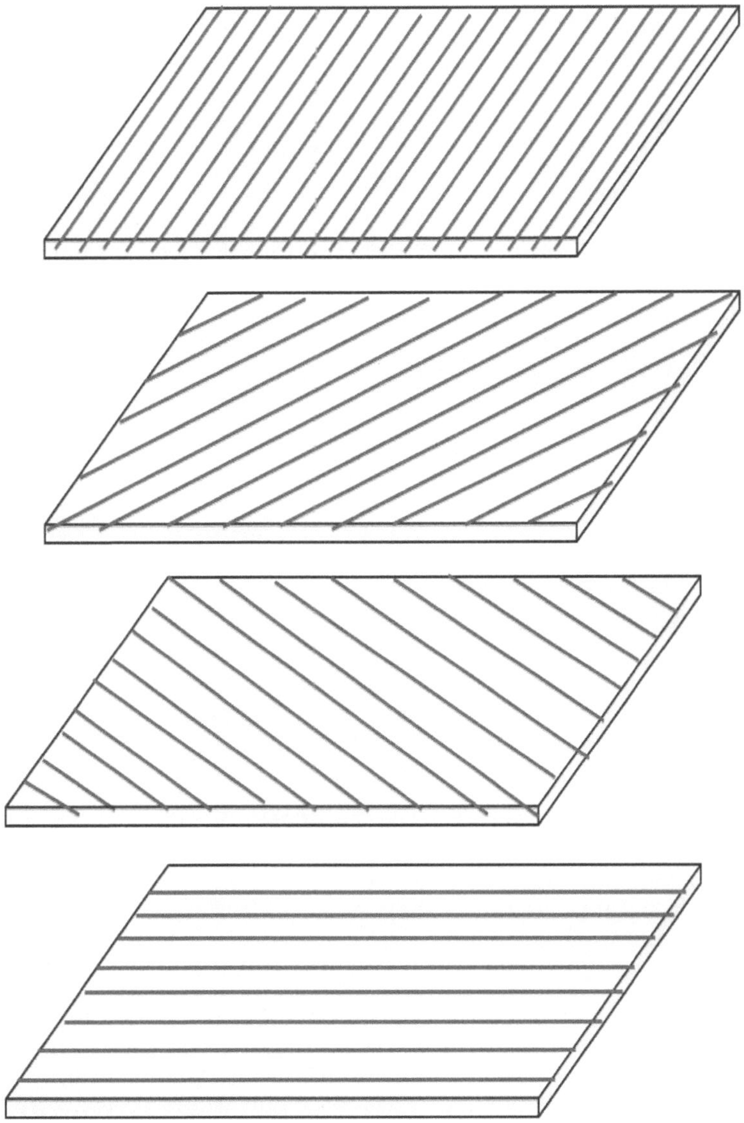

FIG. 14.1

Multiple laminae composing a laminate.

According to Fig. 14.3, the horizontal and vertical displacement of a point A due to deformation in x-z plane is given by the following formulae:

$$u = u^0 - z\frac{\partial w^0}{\partial x}, \qquad w = w^0 \tag{14.1}$$

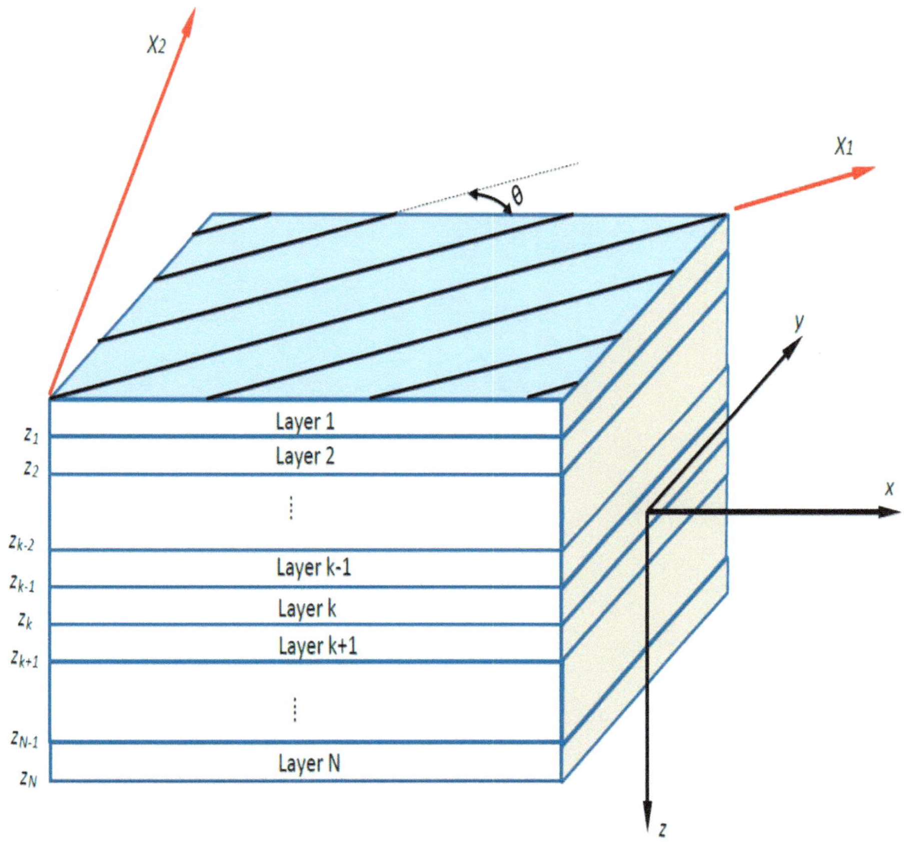

FIG. 14.2

Definition of the global coordinate system for a multilayered FRP laminate.

In the above equations, u_0 represents the horizontal movement of the reference point O located in geometric midplane, w^0 represents its vertical movement, and $\partial w^0/\partial x$ represents the slope in the x-z plane of the x-axis at point O after the deformation.

Following the same concept, the movement of point A in the direction of the y-axis (Fig. 14.2) can be given by

$$v = v^0 - z\frac{\partial w^0}{\partial y} \tag{14.2}$$

where v^0 is the deflection of reference point O in the direction of the y-axis and $\partial w^o/\partial x$ is the slope in the x-y plane of the y-axis at point O after deformation.

- **Laminate strains**

  Taking into account Eqs. (14.1) and (14.2), the strain-displacement relations

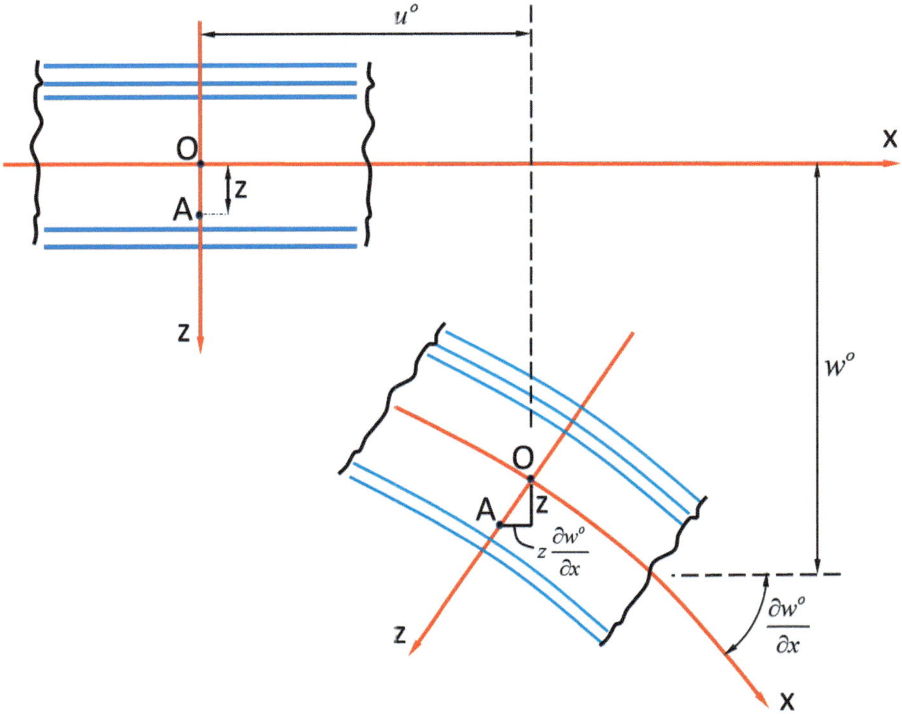

FIG. 14.3

Laminated plate deformation as viewed in the *x-z* plane.

$$\varepsilon_x = \frac{\partial u}{\partial x} \tag{14.3}$$

$$\varepsilon_y = \frac{\partial v}{\partial y} \tag{14.4}$$

$$\varepsilon_z = \frac{\partial w}{\partial z} \tag{14.5}$$

$$\gamma_{xy} = \frac{\partial v}{\partial x} + \frac{\partial u}{\partial y} \tag{14.6}$$

$$\gamma_{xz} = \frac{\partial w}{\partial x} + \frac{\partial u}{\partial z} \tag{14.7}$$

$$\gamma_{yz} = \frac{\partial w}{\partial y} + \frac{\partial v}{\partial z} \tag{14.8}$$

yield:

$$\varepsilon_x = \varepsilon_x^0 + zk_x^0 \tag{14.9}$$

$$\varepsilon_y = \varepsilon_y^0 + zk_y^0 \tag{14.10}$$

$$\varepsilon_z = 0 \tag{14.11}$$

$$\gamma_{xy} = \gamma_{xy}^0 + zk_{xy}^0 \tag{14.12}$$

$$\gamma_{xz} = 0 \tag{14.13}$$

$$\gamma_{yz} = 0 \tag{14.14}$$

In the above equations, the following definitions were used:

$$\varepsilon_x^0 = \frac{\partial u}{\partial x} \tag{14.15}$$

$$\varepsilon_y^0 = \frac{\partial v}{\partial y} \tag{14.16}$$

$$\gamma_{xy}^0 = \frac{\partial v^0}{\partial x} + \frac{\partial u}{\partial y} \tag{14.17}$$

$$k_x^0 = -\frac{\partial^2 w^0}{\partial x^2} \tag{14.18}$$

$$k_y^0 = -\frac{\partial^2 w^0}{\partial y^2} \tag{14.19}$$

$$k_{xy}^0 = -2\frac{\partial^2 w^0}{\partial x \partial y} \tag{14.20}$$

The strains ε_x^0, ε_y^0, and γ_{xy}^0 are called reference surface extensional strain in the x direction, reference surface extensional strain in the y direction, and reference surface in-plane shear strain, respectively. The quantities k_x^0 and k_y^0 are the curvatures of the reference surface in x and y directions, respectively, whereas the quantity k_{xy}^0 is the reference twisting curvature.

- ***Laminate stresses***

If we adopt plane stresses assumption for each lamina, the stresses of each point of a lamina can be determined by the already known (Chapter 3) stresse-strain relation in global coordinate system:

$$\begin{Bmatrix} \sigma_x \\ \sigma_y \\ \tau_{xy} \end{Bmatrix} = \begin{bmatrix} \overline{Q_{11}} & \overline{Q_{12}} & \overline{Q_{16}} \\ \overline{Q_{12}} & \overline{Q_{22}} & \overline{Q_{26}} \\ \overline{Q_{16}} & \overline{Q_{25}} & \overline{Q_{66}} \end{bmatrix} \begin{Bmatrix} \varepsilon_x^0 + zk_x^0 \\ \varepsilon_y^0 + zk_y^0 \\ \gamma_{xy}^0 + zk_{xy}^0 \end{Bmatrix} \tag{14.21}$$

Because the strains and the reduced stiffnesses $\overline{Q}_{ij}$ are functions of the location z of each lamina, and because the material properties controlling the value of $\overline{Q}_{ij}$ are generally different for adjacent layers, the distribution of stresses through the thickness of the laminate is expected to be stepwise.

- ***Laminate stiffness matrix***

Input data for the design of composite structures are often the forces and moments acting on bounding surfaces of laminates. The aim of the designer is to calculate the stress field on each layer in order to apply a failure criterion. To achieve this target, formulae correlating the stress resultants—that is, the loads and moments required to produce the specified midplane deformations—with the midplane strains and curvatures should be derived. Using the notations given in Fig. 14.4, the force and moment resultants are given by the following definitions:

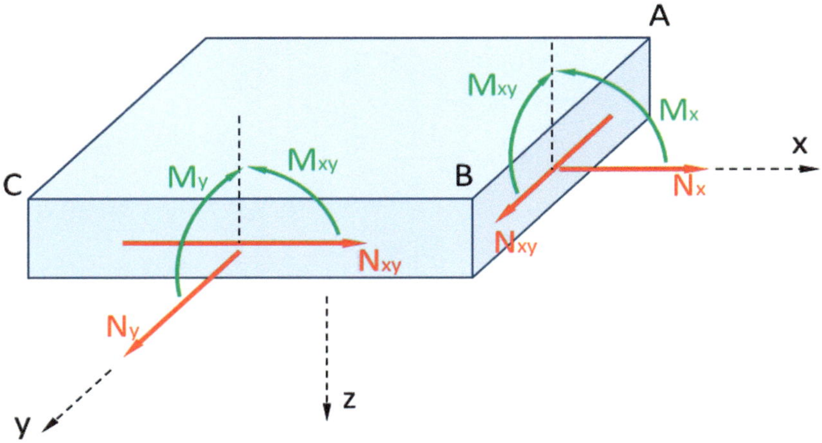

FIG. 14.4

Nomenclature for (**a**) force resultants and (**b**) moment resultants.

$$N_x \equiv \int_{-\frac{H}{2}}^{\frac{H}{2}} \sigma_x dz \tag{14.22}$$

$$N_y \equiv \int_{-\frac{H}{2}}^{\frac{H}{2}} \sigma_y dz \tag{14.23}$$

$$N_{xy} \equiv \int_{-\frac{H}{2}}^{\frac{H}{2}} \tau_{xy} dz \tag{14.24}$$

$$\mathrm{M}_x \equiv \int_{-\frac{H}{2}}^{\frac{H}{2}} \sigma_x z dz \tag{14.25}$$

$$\mathrm{M}_y \equiv \int_{-\frac{H}{2}}^{\frac{H}{2}} \sigma_y z dz \tag{14.26}$$

$$\mathrm{M}_{xy} \equiv \int_{-\frac{H}{2}}^{\frac{H}{2}} \tau_{xy} z dz \tag{14.27}$$

The unit of the force resultants is force per unit length. whereas the unit of the moment resultants is moment per unit length. For example, the resultants N_x and M_χ are

$$N_x = \frac{\hat{N}_x}{\overline{AB}} \tag{14.28}$$

$$M_x = \frac{\hat{M}_x}{\overline{AB}} \tag{14.29}$$

where $\hat{N}_x, \hat{M}_x$ are the force and moment acting on the segment AB (Fig. 14.4), and $\overline{AB}$ is the length of the segment AB.

With the aid of Eq. (14.21), the Eqs.(14.22)–(14.24) can be written as:

$$\begin{aligned} N_x = &\left\{\int_{-\frac{H}{2}}^{\frac{H}{2}} \overline{Q_{11}}dz\right\}\varepsilon_x^0 + \left\{\int_{-\frac{H}{2}}^{\frac{H}{2}} \overline{Q_{11}}zdz\right\}k_x^0 + \left\{\int_{-\frac{H}{2}}^{\frac{H}{2}} \overline{Q_{12}}dz\right\}\varepsilon_y^0 \\ &+ \left\{\int_{-\frac{H}{2}}^{\frac{H}{2}} \overline{Q_{12}}zdz\right\}k_y^0 + \left\{\int_{-\frac{H}{2}}^{\frac{H}{2}} \overline{Q_{16}}dz\right\}\gamma_{xy}^0 + \left\{\int_{-\frac{H}{2}}^{\frac{H}{2}} \overline{Q_{16}}zdz\right\}k_{xy}^0 \end{aligned} \tag{14.30}$$

$$\begin{aligned} N_y = &\left\{\int_{-\frac{H}{2}}^{\frac{H}{2}} \overline{Q_{12}}dz\right\}\varepsilon_x^0 + \left\{\int_{-\frac{H}{2}}^{\frac{H}{2}} \overline{Q_{12}}zdz\right\}k_x^0 + \left\{\int_{-\frac{H}{2}}^{\frac{H}{2}} \overline{Q_{22}}dz\right\}\varepsilon_y^0 \\ &+ \left\{\int_{-\frac{H}{2}}^{\frac{H}{2}} \overline{Q_{22}}zdz\right\}k_y^0 + \left\{\int_{-\frac{H}{2}}^{\frac{H}{2}} \overline{Q_{26}}dz\right\}\gamma_{xy}^0 + \left\{\int_{-\frac{H}{2}}^{\frac{H}{2}} \overline{Q_{26}}zdz\right\}k_{xy}^0 \end{aligned} \tag{14.31}$$

$$\begin{aligned} N_{xy} = &\left\{\int_{-\frac{H}{2}}^{\frac{H}{2}} \overline{Q_{16}}dz\right\}\varepsilon_x^0 + \left\{\int_{-\frac{H}{2}}^{\frac{H}{2}} \overline{Q_{16}}zdz\right\}k_x^0 + \left\{\int_{-\frac{H}{2}}^{\frac{H}{2}} \overline{Q_{26}}dz\right\}\varepsilon_y^0 \\ &+ \left\{\int_{-\frac{H}{2}}^{\frac{H}{2}} \overline{Q_{26}}zdz\right\}k_y^0 + \left\{\int_{-\frac{H}{2}}^{\frac{H}{2}} \overline{Q_{66}}dz\right\}\gamma_{xy}^0 + \left\{\int_{-\frac{H}{2}}^{\frac{H}{2}} \overline{Q_{66}}zdz\right\}k_{xy}^0 \end{aligned} \tag{14.32}$$

For a laminate with N layers, each integral $\int_{-\frac{H}{2}}^{+\frac{H}{2}} \overline{Q}_{ij}dz$ has the form

$$\begin{aligned} \int_{-\frac{H}{2}}^{\frac{H}{2}} \overline{Q_{ij}}dz = &\overline{Q}_{ij1}(z_1 - z_0) + \overline{Q}_{ij2}(z_2 - z_1) + \overline{Q}_{ij3}(z_3 - z_2) + \cdots \\ &+\overline{Q}_{ijk}(z_k - z_{k-1}) + \cdots + \overline{Q}_{ijN}(z_N - z_{N-1}) \end{aligned} \tag{14.33}$$

Using the notation

$$A_{ij} = \int_{-\frac{H}{2}}^{-\frac{H}{2}} \overline{Q}_{ij}dz \tag{14.34}$$

Eq. (14.33) can be written in the following simple form:

$$A_{ij} = \sum_{k=1}^{N} \overline{Q}_{ijk}(z_k - z_{k-1}) \tag{14.35}$$

Following the same concept, the integrals of the form

$$B_{ij} = \int_{-H/2}^{+H/2} \overline{Q}_{ij} z dz \tag{14.36}$$

yield

$$B_{ij} = \frac{1}{2}\sum_{k=1}^{N} \overline{Q}_{ijk}(z_k^2 - z_{k-1}^2) \tag{14.37}$$

Therefore the force resultants can be written in the following matrix format:

$$\begin{Bmatrix} N_x \\ N_y \\ N_{xy} \end{Bmatrix} = \begin{bmatrix} A_{11} & A_{12} & A_{16} \\ A_{12} & A_{22} & A_{26} \\ A_{16} & A_{26} & A_{66} \end{bmatrix} \begin{Bmatrix} \varepsilon_x^0 \\ \varepsilon_y^0 \\ \gamma_{xy}^0 \end{Bmatrix} + \begin{bmatrix} B_{11} & B_{12} & B_{16} \\ B_{12} & B_{22} & B_{26} \\ B_{16} & B_{26} & B_{66} \end{bmatrix} \begin{Bmatrix} k_x^0 \\ k_y^0 \\ k_{xy}^0 \end{Bmatrix} \tag{14.38}$$

Combining Eq. (14.21) with Eqs. (14.25)–(14.27), the following formulas for the moment resultants can be obtained:

$$\begin{aligned} M_x = {} & \left\{ \int_{-\frac{H}{2}}^{\frac{H}{2}} \overline{Q_{11}} z dz \right\} \varepsilon_x^0 + \left\{ \int_{-\frac{H}{2}}^{\frac{H}{2}} \overline{Q_{11}} z^2 dz \right\} k_x^0 + \left\{ \int_{-\frac{H}{2}}^{\frac{H}{2}} \overline{Q_{12}} z dz \right\} \varepsilon_y^0 \\ & + \left\{ \int_{-\frac{H}{2}}^{\frac{H}{2}} \overline{Q_{12}} z^2 dz \right\} k_y^0 + \left\{ \int_{-\frac{H}{2}}^{\frac{H}{2}} \overline{Q_{16}} z dz \right\} \gamma_{xy}^0 + \left\{ \int_{-\frac{H}{2}}^{\frac{H}{2}} \overline{Q_{16}} z^2 dz \right\} k_{xy}^0 \end{aligned} \tag{14.39}$$

$$\begin{aligned} M_y = {} & \left\{ \int_{-\frac{H}{2}}^{\frac{H}{2}} \overline{Q_{12}} z dz \right\} \varepsilon_x^0 + \left\{ \int_{-\frac{H}{2}}^{\frac{H}{2}} \overline{Q_{12}} z^2 dz \right\} k_x^0 + \left\{ \int_{-\frac{H}{2}}^{\frac{H}{2}} \overline{Q_{22}} z dz \right\} \varepsilon_y^0 \\ & + \left\{ \int_{-\frac{H}{2}}^{\frac{H}{2}} \overline{Q_{22}} z^2 dz \right\} k_y^0 + \left\{ \int_{-\frac{H}{2}}^{\frac{H}{2}} \overline{Q_{26}} z dz \right\} \gamma_{xy}^0 + \left\{ \int_{-\frac{H}{2}}^{\frac{H}{2}} \overline{Q_{26}} z^2 dz \right\} k_{xy}^0 \end{aligned} \tag{14.40}$$

$$M_{xy} = \left\{ \int_{-\frac{H}{2}}^{\frac{H}{2}} \overline{Q_{16}} z dz \right\} \varepsilon_x^0 + \left\{ \int_{-\frac{H}{2}}^{\frac{H}{2}} \overline{Q_{16}} z^2 dz \right\} k_x^0 + \left\{ \int_{-\frac{H}{2}}^{\frac{H}{2}} \overline{Q_{26}} z dz \right\} \varepsilon_y^0$$
$$+ \left\{ \int_{-\frac{H}{2}}^{\frac{H}{2}} \overline{Q_{26}} z^2 dz \right\} k_y^0 + \left\{ \int_{-\frac{H}{2}}^{\frac{H}{2}} \overline{Q_{66}} z dz \right\} \gamma_{xy}^0 + \left\{ \int_{-\frac{H}{2}}^{\frac{H}{2}} \overline{Q_{66}} z^2 dz \right\} k_{xy}^0 \tag{14.41}$$

Using the notation

$$D_{ij} = \int_{-H/2}^{H/2} \overline{Q}_{ij} z^2 dz \tag{14.42}$$

Eqs. (14.39)–(14.41) can now be written

$$\begin{Bmatrix} M_x \\ M_y \\ M_{xy} \end{Bmatrix} = \begin{bmatrix} B_{11} & B_{12} & B_{16} \\ B_{12} & B_{22} & B_{26} \\ B_{16} & B_{26} & B_{66} \end{bmatrix} \begin{Bmatrix} \varepsilon_x^0 \\ \varepsilon_y^0 \\ \gamma_{xy}^0 \end{Bmatrix} + \begin{bmatrix} D_{11} & D_{12} & D_{16} \\ D_{12} & D_{22} & D_{26} \\ D_{16} & D_{26} & D_{66} \end{bmatrix} \begin{Bmatrix} k_x^0 \\ k_y^0 \\ k_{xy}^0 \end{Bmatrix} \tag{14.43}$$

where

$$D_{ij} = \frac{1}{3} \sum_{k=1}^{N} \overline{Q_{ijk}} (z_k^3 - z_{k-1}^3) \tag{14.44}$$

Superposition of Eqs. (14.38), (14.43) yields

$$\begin{Bmatrix} N_x \\ N_y \\ N_{xy} \\ M_x \\ M_y \\ M_{xy} \end{Bmatrix} = \begin{bmatrix} A_{11} & A_{12} & A_{16} & B_{11} & B_{12} & B_{16} \\ A_{12} & A_{22} & A_{26} & B_{12} & B_{22} & B_{26} \\ A_{16} & A_{26} & A_{66} & B_{16} & B_{26} & B_{66} \\ B_{11} & B_{12} & B_{16} & D_{11} & D_{12} & D_{16} \\ B_{12} & B_{22} & B_{26} & D_{12} & D_{22} & D_{26} \\ B_{16} & B_{26} & B_{66} & D_{16} & D_{26} & D_{66} \end{bmatrix} \begin{Bmatrix} \varepsilon_x^0 \\ \varepsilon_y^0 \\ \gamma_{xy}^0 \\ k_x^0 \\ k_y^0 \\ k_{xy}^0 \end{Bmatrix} \tag{14.45}$$

In the above equation, the 6 × 6 matrix consisting of components A_{ij}, B_{ij}, and D_{ij} is called laminate stiffness matrix, or *ABD* matrix. Inversion of the above equation provides the relation of the strains and curvatures with the force and moment resultants:

$$\begin{Bmatrix} \varepsilon_x^0 \\ \varepsilon_y^0 \\ \gamma_{xy}^0 \\ k_x^0 \\ k_y^0 \\ k_{xy}^0 \end{Bmatrix} = \begin{bmatrix} a_{11} & a_{12} & a_{16} & b_{11} & b_{12} & b_{16} \\ a_{12} & a_{22} & a_{26} & b_{21} & b_{22} & b_{26} \\ a_{16} & a_{26} & \alpha_{66} & b_{61} & b_{62} & b_{66} \\ b_{11} & b_{12} & b_{61} & d_{11} & d_{12} & d_{16} \\ b_{12} & b_{22} & b_{62} & d_{12} & d_{22} & d_{26} \\ b_{16} & b_{26} & b_{66} & d_{16} & d_{26} & d_{66} \end{bmatrix} \begin{Bmatrix} N_x \\ N_y \\ N_{xy} \\ M_x \\ M_y \\ M_{xy} \end{Bmatrix} \tag{14.46}$$

The 6 × 6 matrix consisting of the components α_{ij} is given by:

$$\begin{bmatrix} a_{11} & a_{12} & a_{16} & b_{11} & b_{12} & b_{16} \\ a_{12} & a_{22} & a_{26} & b_{21} & b_{22} & b_{26} \\ a_{16} & a_{26} & \alpha_{66} & b_{61} & b_{62} & b_{66} \\ b_{11} & b_{12} & b_{61} & d_{11} & d_{12} & d_{16} \\ b_{12} & b_{22} & b_{62} & d_{12} & d_{22} & d_{26} \\ b_{16} & b_{26} & b_{66} & d_{16} & d_{26} & d_{66} \end{bmatrix} = \begin{bmatrix} A_{11} & A_{12} & A_{16} & B_{11} & B_{12} & B_{16} \\ A_{12} & A_{22} & A_{26} & B_{12} & B_{22} & B_{26} \\ A_{16} & A_{26} & A_{66} & B_{16} & B_{26} & B_{66} \\ B_{11} & B_{12} & B_{16} & D_{11} & D_{12} & D_{16} \\ B_{12} & B_{22} & B_{26} & D_{12} & D_{22} & D_{26} \\ B_{16} & B_{26} & B_{66} & D_{16} & D_{26} & D_{66} \end{bmatrix}^{-1} \tag{14.47}$$

and is called laminate compliance matrix, or *abd* matrix. Knowing the force and moment resultants acting on the bounding surface of a laminate, the principal stresses σ_1, σ_2, and τ_{12} of each layer can now be calculated by following the steps shown in Fig. 14.5.

- **The Tsai-Wu failure criterion**

Models taking into account the applied stress components in principal coordinate system and the corresponding tensile and compressive strengths have been developed for estimation of the limits of stresses yielding failure. These models are called "failure criteria."

Generally speaking, the main target of a designer aiming to estimate the safety factor of a composite structure is the determination (e.g., according to the steps of Fig. 14.5) of the principal stresses. In analogy to the yielding criteria for isotropic materials, a number of criteria for composites have been developed.

Among existing failure criteria for composites, the Tsai-Wu criterion seems to be the most popular. Based on a concept similar to that followed for the derivation of the von Mises criterion for isotropic materials, the Tsai-Wu criterion for composites, assuming the plane-stresses assumption, is given by the following formula:

$$F_1\sigma_1 + F_2\sigma_2 + F_{11}\sigma_1^2 + F_{22}\sigma_2^2 + F_{66}\tau_{12}^2 - \sqrt{F_{11}F_{22}}\sigma_1\sigma_2 \leq 1 \tag{14.48}$$

Parameters F_1, F_2, F_{66}, F_{11}, and F_{22} are material constants and involve failure stresses in the principal directions:

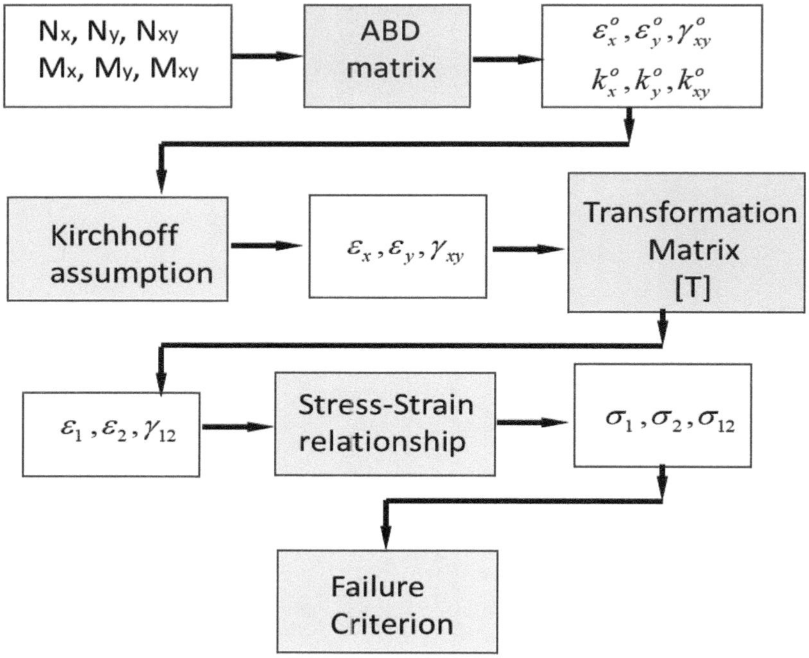

FIG. 14.5

Steps for calculating the principal stresses on each layer of a laminate.

$$F_1 = \left(\frac{1}{\sigma_1^T} + \frac{1}{\sigma_1^C} \right) \tag{14.49}$$

$$F_2 = \left(\frac{1}{\sigma_2^T} + \frac{1}{\sigma_2^C} \right) \tag{14.50}$$

$$F_{66} = \left(\frac{1}{\tau_{12}^F} \right)^2 \tag{14.51}$$

$$F_{11} = -\frac{1}{\sigma_1^T \sigma_1^C} \tag{14.52}$$

$$F_{22} = -\frac{1}{\sigma_2^T \sigma_2^C} \tag{14.53}$$

where:

σ_1^C: compressive failure stress in the x_1 direction
σ_1^T: tensile failure stresses in the x_1 direction
σ_2^C: compressive failure stresses in the x_2 direction
σ_2^T: tensile failure stresses in the x_2 direction
τ_{12}^F: shear failure stresses in the x_1-x_2 plane

In Fig. 14.6, the bounding surface including the allowable values of σ_1, σ_2, and τ_{12} in σ_1-σ_2-τ_{12} Cartesian system is schematically represented, whereas Fig. 14.7 shows the corresponding area of σ_1 and σ_2 values for the case of $\tau_{12} = 0$. In both figures, it can be shown that the corresponding 3D and 2D ellipsoids are very long (in the direction of σ_1 axis) and very slender (in the direction of σ_2 and/or τ_{12} axis), indicating the dependency of the high strength of the fibers.

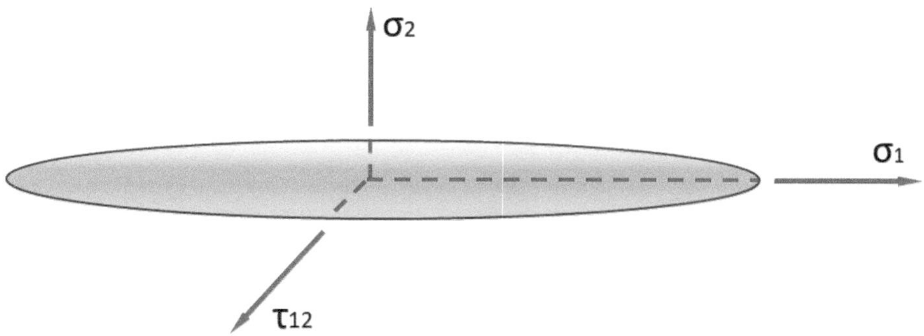

FIG. 14.6

Schematic 3D graphical representation of Tsai-Wu failure criterion.

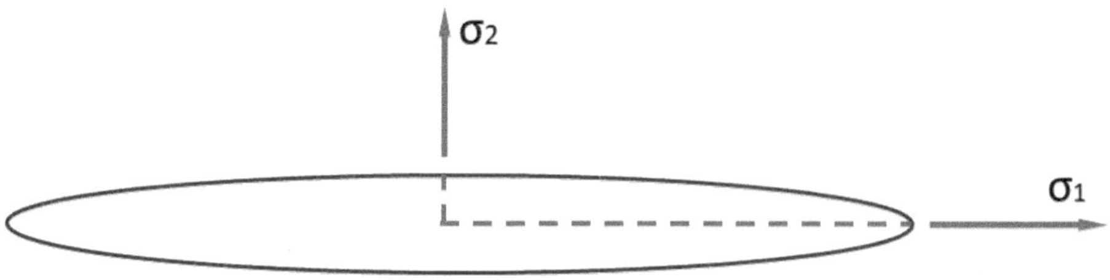

FIG. 14.7

Schematic 2D graphical representation of Tsai-Wu criterion for the case of τ_{12}= 0.

REFERENCES

[1] Pavlou D. Composite materials in piping applications. Destech publications; 2013.
[2] Hyer M. Stress analysis of fiber reinforced composite materials. Destech Publications; 2009.
[3] Reddy JN. Mechanics of laminated composite plates and shelle. CRC presss; 2004.
[4] Timoshenko SP, Goodier JN. Theory of elasticity. McGraw-Hill; 1970.
[5] Tsai SW, Wu EM. A general theory of strength for anisotropic materials. Journal of Composite Materials 1971;5:58–80.

Appendix
Solid Mechanics Problems With Solutions

PROBLEM 1

A short cantilever beam of a circular cross section is subjected to forces P_1, P_2, and torque M_T on the free end.

Calculate:

1. The principal stresses and the direction cosines of the principal planes on the point M located on the cross section of the fixed support "A" (see Fig. A.1a,b).
2. The stress components σ_{Px}, σ_{Py}, and σ_{Pz} of the plane OAB (Fig. A.1c) of the material element on the point "M"

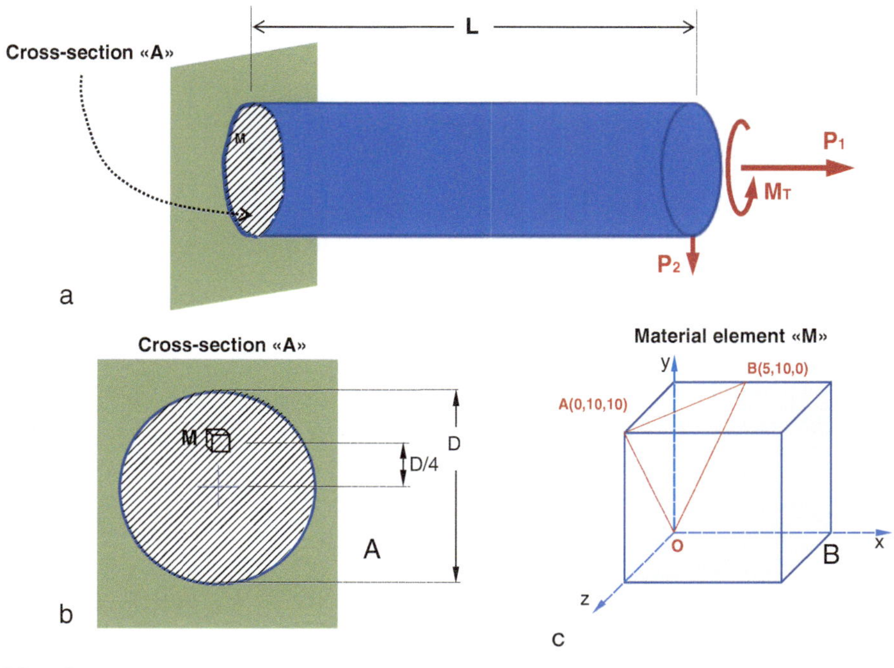

FIG. A.1

SOLUTION

The forces P_1, P_2, and the torque M_T cause the following stress distribution on the cross section "A":

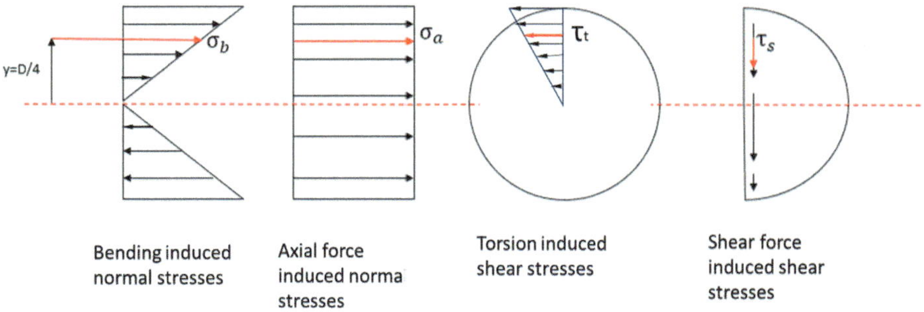

The stresses on the material element "M" are shown in the following figure:

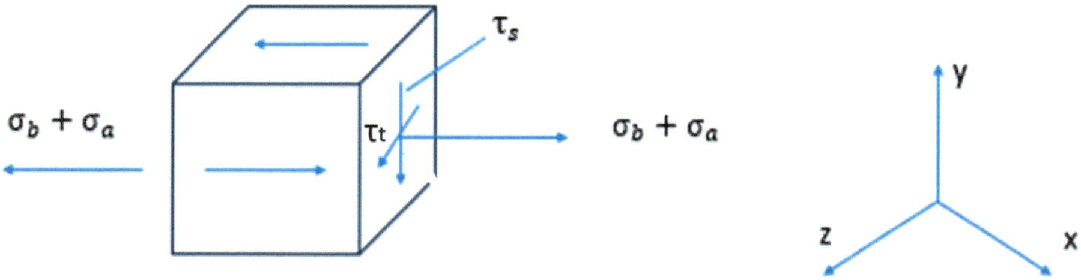

Stresses on the material element M

Normal stresses on the material element "M" due to bending and axial force:

$$\sigma_{xx} = \sigma_b + \sigma_a = \frac{P_2 L}{I}\frac{D}{4} + \frac{P1}{A} \qquad \sigma_y = 0,\ \sigma_z = 0$$

Shear stresses on the material element "M" due to shear force:

$$\tau_{xy} = -\tau_s = \frac{4P_2}{3\pi (D/2)^4}\left(\frac{D^4}{4} - \frac{D^2}{16}\right) \qquad \tau_{yz} = 0$$

Shear stresses on the material element "M" due to torsion:

$$\tau_{xz} = \tau_t = \frac{M_T}{I_p}\frac{D}{4}$$

Calculation of the principal stresses:

$$I_1 = \sigma_x + \sigma_y + \sigma_z = \frac{P_2 LD}{4I} + \frac{P_1}{A}$$

$$I_2 = \cancel{\sigma_x \sigma_y} + \cancel{\sigma_y \sigma_z} + \cancel{\sigma_x \sigma_z} - \tau_{xy}^{\ 2} - \cancel{\tau_{yz}^{\ 2}} - \tau_{xz}^{\ 2} = -\tau_s^{\ 2}$$

$$I_3 = \cancel{\sigma_x \sigma_y \sigma_z} + 2\tau_{xy}\cancel{\tau_{yz}}\tau_{xz} - \sigma_x \cancel{\tau_{yz}^{\ 2}} - \cancel{\sigma_y}\tau_{xz}^{\ 2} - \cancel{\sigma_z}\tau_{xy}^{\ 2} = 0$$

$$\sigma^3 - I_1\sigma^2 + I_2\sigma - I_3 = 0$$

$$\sigma^3 - I_1\sigma^2 + I_2\sigma = 0 \Rightarrow$$

$$\sigma^2 - I_1\sigma + I_2 = 0 \Rightarrow$$

$$\Rightarrow \begin{cases} \sigma_1 = 0 \\ \sigma_{2,3} = \dfrac{I_1 \pm \sqrt{I_1^{\ 2} - 4I_2}}{2} \end{cases}$$

Determination of the principal planes:
For $\sigma = \sigma_{2,3}$

$$\begin{aligned} &l_1(\sigma_x - \sigma_{2,3}) + m_1\tau_{xy} + n_1\tau_{xz} = 0 \\ &l_1\tau_{xy} + m_1(\cancel{\sigma_y} - \sigma_{2,3}) + n_1\cancel{\tau_{yz}} = 0 \\ &l_1\tau_{xz} + m_1\tau_{yz} + n_1(\cancel{\sigma_z} - \sigma_{2,3}) = 0 \\ &l_1^{\ 2} + m_1^{\ 2} + n_1^{\ 2} = 1 \end{aligned} \quad \Rightarrow \quad \begin{aligned} &l_1(\sigma_x - \sigma_{2,3}) + m_1\tau_{xy} + n_1\tau_{xz} = 0 && (1) \\ &l_1\tau_{xy} - m_1\sigma_{2,3} = 0 && (2) \\ &l_1\tau_{xz} - n_1 - \sigma_{2,3} = 0 && (3) \\ &l_1^{\ 2} + m_1^{\ 2} + n_1^{\ 2} = 1 && (4) \end{aligned}$$

$$(2) \Rightarrow m_1 = \frac{\tau_{xy}}{\sigma_{2,3}} l_1$$

$$(3) \Rightarrow n_1 = \frac{\tau_{xz}}{\sigma_{2,3}} l_1$$

$$(1) \Rightarrow l_1(\sigma_x - \sigma_{2,3}) + \frac{\tau_{xy}^{\ 2}}{\sigma_{2,3}} l_1 + \frac{\tau_{xz}^{\ 2}}{\sigma_{2,3}} l_1 = 0 \Rightarrow l_1 = 0$$

$$\text{Then:} \quad \begin{aligned} &(2) \Rightarrow m_1 = 0 \\ &(3) \Rightarrow n_1 = 0 \end{aligned}$$

For $\sigma = \sigma_1 = 0$

$$\left\{\begin{array}{c} l_1(\sigma_x - 0) + m_1\tau_{xy} + n_1\tau_{xz} = 0 \\ l_1\tau_{xy} = 0 \\ l_1\tau_{xz} = 0 \\ l_1^{\,2} + m_1^{\,2} + n_1^{\,2} = 1 \end{array}\right\} \Rightarrow \left\{\begin{array}{c} l_1 = 0 \\ m_1\tau_{xy} + n_1\tau_{xz} = 0 \\ m_1^{\,2} + n_1^{\,2} = 1 \end{array}\right\}$$

$$\Rightarrow \left\{\begin{array}{c} l_1 = 0 \\ n_1 = -\dfrac{\tau_{xy}}{\tau_{xz}} m_1 \\ m_1^{\,2} + \left(\dfrac{\tau_{xy}}{\tau_{xz}}\right)^2 m_1^{\,2} = 1 \end{array}\right\} \Rightarrow \left\{\begin{array}{c} l_1 = 0 \\ m_1 = \pm\dfrac{1}{\sqrt{1+\left(\frac{\tau_{xy}}{\tau_{xz}}\right)^2}} \\ n_1 = \mp\dfrac{\tau_{xy}}{\tau_{xz}}\dfrac{1}{\sqrt{1+\left(\frac{\tau_{xy}}{\tau_{xz}}\right)^2}} \end{array}\right\}$$

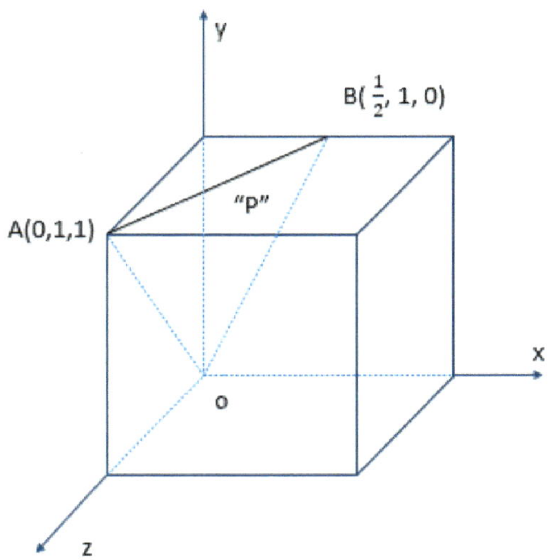

$$\overrightarrow{OA} = \vec{j} + \vec{k}$$

$$\overrightarrow{OB} = \frac{1}{2}\vec{i} + \vec{j}$$

$$\overrightarrow{OB}\times\overrightarrow{OA}=\begin{vmatrix}\vec{i} & \vec{j} & \vec{k}\\ \frac{1}{2} & 1 & 0\\ 0 & 1 & 1\end{vmatrix}=\vec{i}-\frac{1}{2}\vec{j}+\frac{1}{2}\vec{k}$$

$$l=\frac{1}{\sqrt{1+\left(\frac{1}{2}\right)^2+\left(\frac{1}{2}\right)^2}}$$

$$m=\frac{-\frac{1}{2}}{\sqrt{1+\left(\frac{1}{2}\right)^2+\left(\frac{1}{2}\right)^2}}$$

$$n=\frac{\frac{1}{2}}{\sqrt{1+\left(\frac{1}{2}\right)^2+\left(\frac{1}{2}\right)^2}}$$

$$\begin{Bmatrix}\sigma_{px}\\ \sigma_{py}\\ \sigma_{pz}\end{Bmatrix}=\begin{bmatrix}\sigma_x & \tau_{xy} & \tau_{xz}\\ \tau_{xy} & \sigma_y & \tau_{yz}\\ \tau_{xz} & \tau_{yz} & \sigma_z\end{bmatrix}\begin{Bmatrix}l\\ m\\ n\end{Bmatrix}$$

Because $\sigma_y = 0, \sigma_z = 0$, and $\tau_{yz} = 0$, the above equation yields:

$$\begin{Bmatrix}\sigma_{px}\\ \sigma_{py}\\ \sigma_{pz}\end{Bmatrix}=\begin{bmatrix}\sigma_x & \tau_{xy} & \tau_{xz}\\ \tau_{xy} & 0 & 0\\ \tau_{xz} & 0 & 0\end{bmatrix}\begin{Bmatrix}l\\ m\\ n\end{Bmatrix}$$

or

$$\sigma_{px} = \sigma_x l + \tau_{xy} m + \tau_{xz} n$$

$$\sigma_{py} = \tau_{xy} l$$

$$\sigma_{pz} = \tau_{xz} l$$

PROBLEM 2

The fixed supported beam in Fig. A.2 is subjected to forces $\mathbf{F_x}$, $\mathbf{F_y}$, and $\mathbf{F_z}$. On point **A** of the cross section **n-n′**, a plane **OAB** oriented as shown in Fig. A.3 has the following stress tensor:

$$\begin{Bmatrix} \sigma_{Px} \\ \sigma_{Py} \\ \sigma_{Pz} \end{Bmatrix} = \begin{Bmatrix} -21.3 \\ -63.9 \\ -132.5 \end{Bmatrix} MPa$$

Considering that the stress components σ_{xx}, σ_{yy}, and τ_{xy} on point **A** are null, that is, $\sigma_{xx} = 0$,

$\sigma_{yy} = 0$, and $\tau_{xy} = 0$, calculate the forces $\mathbf{F_x}$, $\mathbf{F_y}$, and $\mathbf{F_z}$.

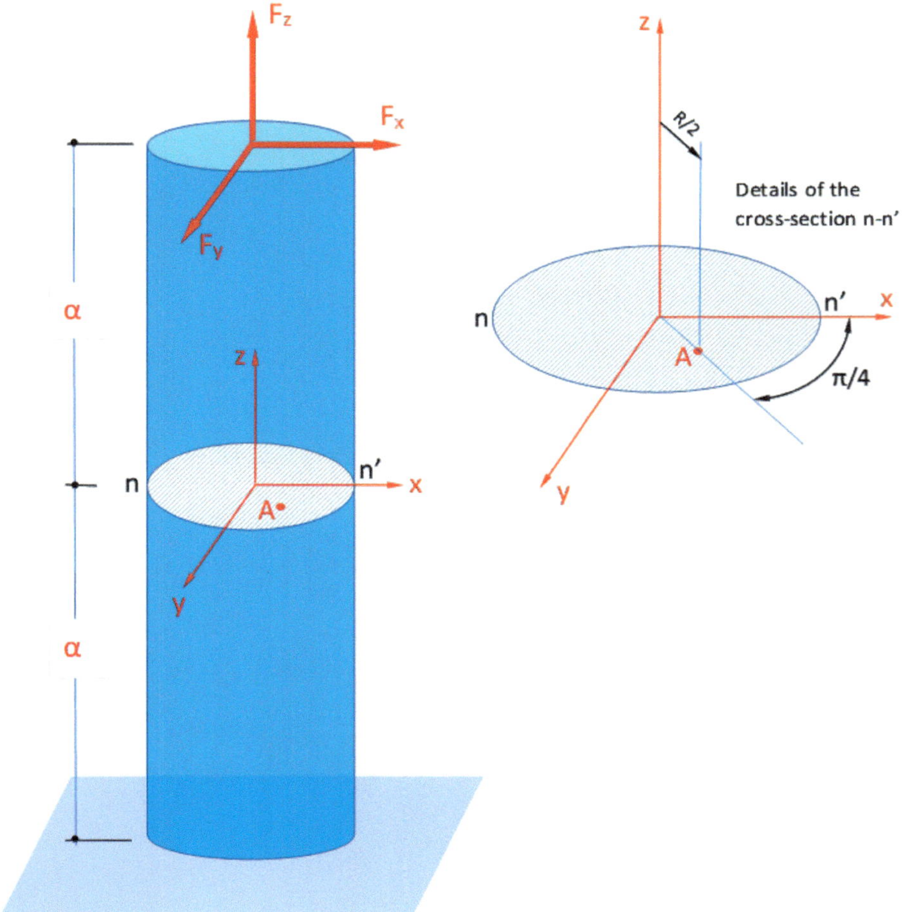

FIG. A.2

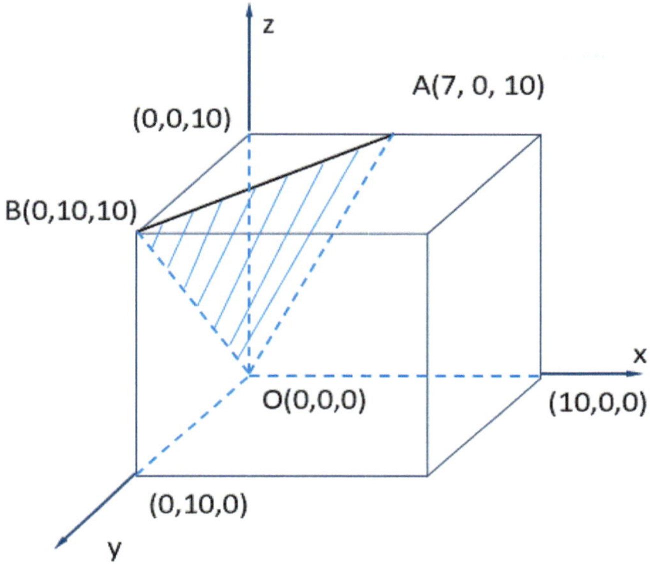

FIG. A.3

Given data:

Beam length $2\alpha = 4\,\text{m}$
Cross section radius $R = 10\,\text{cm}$

SOLUTION

Calculation of the direction cosines:

$$\overrightarrow{OA} \times \overrightarrow{OB} = \begin{vmatrix} \vec{i} & \vec{j} & \vec{k} \\ 7 & 0 & 10 \\ 0 & 10 & 10 \end{vmatrix} = -100\vec{i} - 70\vec{j} + 70\vec{k}$$

$$l = \frac{-100}{\sqrt{100^2 + 70^2 + 70^2}} = -0.71$$

$$m = \frac{-70}{\sqrt{100^2 + 70^2 + 70^2}} = -0.50$$

$$n = \frac{70}{\sqrt{100^2 + 70^2 + 70^2}} = 0.50$$

The components σ_{Px}, σ_{Py}, and σ_{Pz} on the plane OAB are given by the following formula:

$$\begin{Bmatrix} \sigma_{px} \\ \sigma_{py} \\ \sigma_{pz} \end{Bmatrix} = \begin{bmatrix} \sigma_{xx} & \sigma_{xy} & \sigma_{xz} \\ \sigma_{xy} & \sigma_{yy} & \sigma_{yz} \\ \sigma_{xz} & \sigma_{yz} & \sigma_{zz} \end{bmatrix} \begin{Bmatrix} l \\ m \\ n \end{Bmatrix}$$

or

$$\sigma_{px} = \sigma_{xz} \cdot n$$
$$\sigma_{py} = \sigma_{yz} \cdot n$$
$$\sigma_{pz} = \sigma_{xz} \cdot l + \sigma_{yz} \cdot m + \sigma_{zz} \cdot n$$

The above equations yield:

$$\sigma_{xz} = \frac{\sigma_{px}}{n} = \frac{-21.3}{-0.71} = 30MPa \quad (1)$$

$$\sigma_{yz} = \frac{\sigma_{py}}{n} = \frac{-63.9}{-0.71} = 90MPa \quad (2)$$

$$\begin{aligned} \sigma_{zz} &= \frac{1}{n}\left(\sigma_{pz} - \sigma_{xz} l - \sigma_{yz} m\right) \\ &= \frac{1}{-0.71}\left(\sigma_{pz} + 30 \times 0.71 - 90 \times 0.50\right) \\ &= 220MPa \end{aligned} \quad (3)$$

The normal stress due to the axial force F_z and the bending moments M_x, M_y (caused by the forces F_x and F_y) is given by the following equation:

$$\sigma_{zz} = \frac{F_z}{A} - \frac{M_y I_x + M_x I_{xy}}{I_x I_y - I_{xy}^2} x_0 + \frac{M_x I_y + M_y I_{xy}}{I_x I_y - I_{xy}^2} y_0 \quad (4)$$

The shear stress due to the force F_x is:

$$\tau_{zx} = \frac{F_x}{2B(x_0) I_y} \int_{x_o}^{R} x dA$$

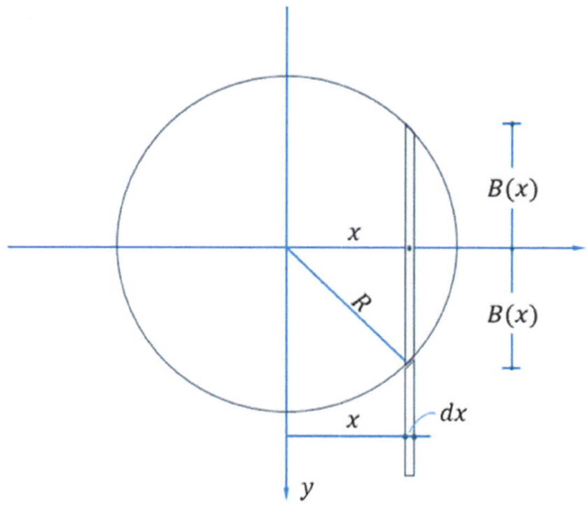

where

$$dA = 2 \cdot B(x) \cdot dx$$

$$B(x) = \sqrt{R^2 - x^2}$$

$$\int x dA = \int 2x\sqrt{R^2 - x^2}\, dx = \frac{2}{3}(R^2 - x^2)^{\frac{3}{2}}$$

$$x_0 = \frac{R}{2} \cos \frac{\pi}{4} = \frac{\sqrt{2}}{2}\frac{R}{2}$$

Then,

$$\tau_{zx}(x_0) = \frac{F_x}{2B(x_0)I_y} \int_{x_0}^{R} x dA \Rightarrow \tau_{zx} = \frac{F_x}{2\sqrt{R^2 - \left(\frac{\sqrt{2}}{2}\frac{R}{2}\right)^2} I_y} \frac{2}{3}\left[R^2 - \left(\frac{\sqrt{2}}{2}\frac{R}{2}\right)^2\right]^{\frac{3}{2}}$$

or

$$\tau_{zx} = \frac{7R^2}{24I_y} F_x \tag{5}$$

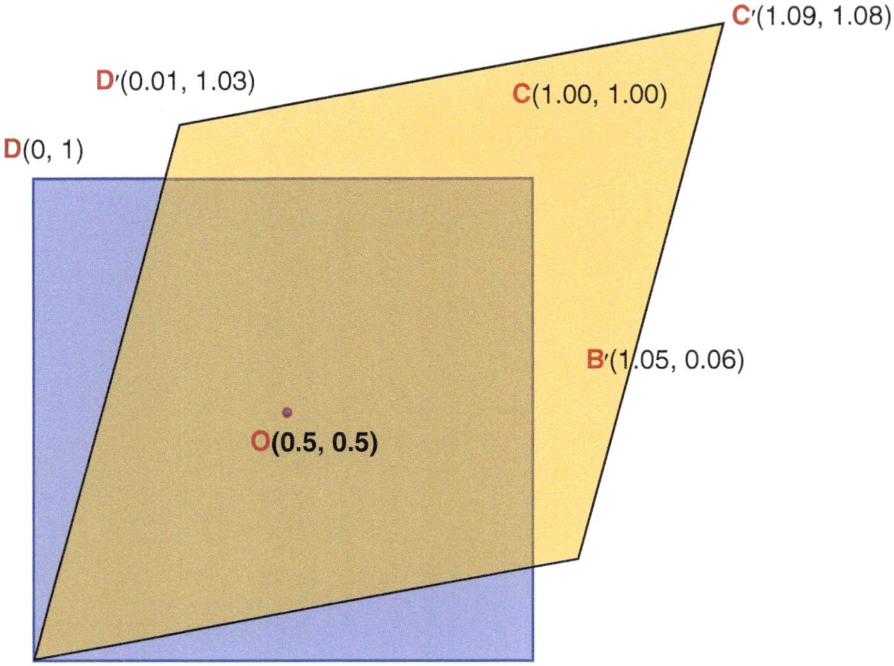

FIG. A.4

Following the same steps, the following equation for τ_{zy} can be obtained:

$$\tau_{zy} = \frac{7R^2}{24I_y}F_y \tag{6}$$

Because $M_x = F_y a$ and $M_y = -F_x a$ and considering Eqs. (1)–(3), Eqs. (4)–(6) yield:

$$\frac{F_z}{A} - \frac{-F_x a I_x + F_y a I_{xy}}{I_x I_y - I_{xy}{}^2}\frac{\sqrt{2}}{2}R + \frac{F_y a I_y}{I_x I_y - I_{xy}{}^2}\frac{\sqrt{2}}{2}R = 220 \tag{7}$$

$$\frac{7R^2}{24I_y}F_x = 30 \tag{8}$$

$$\frac{7R^2}{24I_y}F_y = 90 \tag{9}$$

where

$$I_x = I_y = \frac{\pi R^4}{4}$$

$$I_{xy} = 0$$

Eqs. (7)–(9) yield the forces F_z, F_x, and F_y.

PROBLEM 3

Fig. A.4 shows the elastically deformed shape AB′C′D′ of a thin plate subjected to plane stress conditions. The elasticity modulus and Poisson's ratio are E = 210 GPa and $\nu = 0.33$, respectively. Determine the direction (angle θ) of the maximum shear stress on the point O (0.5, 0.5).

SOLUTION

Because the sides of the plate remain straight, the following linear functions for displacements can be used:

$$u = a_1 + b_1 x + c_1 y + d_1 xy \tag{1}$$

$$v = a_2 + b_2 x + c_2 y + d_2 xy \tag{2}$$

Boundary conditions:

Point A:

$$x = 0,\ \ y = 0,\ \ u = 0,\ \ v = 0$$

$$\begin{aligned} eq.\ (1) &\Rightarrow a_1 = 0 \\ eq.\ (2) &\Rightarrow a_2 = 0 \end{aligned}$$

Point B:

$$x = 1,\ \ y = 0,\ \ u = 1.05 - 1.00 = 0.05,\ \ v = 0.06 - 0.00 = 0.06$$

$$\left.\begin{aligned} eq.\ (1) &\Rightarrow b_1 \cdot 1 + c_1 \cdot 0 + d_1 \cdot 1 \cdot 0 = 0.05 \\ eq.\ (2) &\Rightarrow b_2 \cdot 1 + c_2 \cdot 0 + d_1 \cdot 1 \cdot 0 = 0.06 \end{aligned}\right\} \Rightarrow \begin{cases} b_1 = 0.05 \\ b_2 = 0.06 \end{cases}$$

Point D:

$$x = 0,\ \ y = 1,\ \ u = 0.01 - 0 = 0.01,\ \ v = 1.03 - 1.00 = 0.03$$

$$\left.\begin{array}{l} eq.\ (1) \Rightarrow 0.05 \times 0 + c_1 \cdot 1 + d_1 \cdot 0 = 0.01 \\ eq.\ (2) \Rightarrow 0.06 \times 0 + c_2 \cdot 1 + d_2 \cdot 0 = 0.03 \end{array}\right\} \Rightarrow \left\{\begin{array}{l} c_1 = 0.01 \\ c_2 = 0.03 \end{array}\right.$$

Point C:

$$x = 1,\ \ y = 1,\ \ u = 1.08 - 1.00 = 0.08,\ \ v = 1.09 - 1.00 = 0.09$$

$$\left.\begin{array}{l} eq.\ (1) \Rightarrow 0.05 \times 1 + 0.01 \times 1 + d_1 \times 1 \times 1 = 0.08 \\ eq.\ (2) \Rightarrow 0.06 \times 1 + 0.03 \times 1 + d_2 \times 1 \times 1 = 0.09 \end{array}\right\}$$

$$\Rightarrow \left\{\begin{array}{l} d_1 = 0.08 - 0.05 - 0.01 \\ d_2 = 0.09 - 0.06 - 0.03 \end{array}\right.$$

$$\Rightarrow \left\{\begin{array}{l} d_1 = 0.02 \\ d_2 = 0 \end{array}\right.$$

With the aid of the calculated coefficients, the displacement functions in Eqs. (1) and (2) are:

$$\begin{array}{l} u(x, y) = 0.05x + 0.01y + 0.02xy \\ v(x, y) = 0.06x + 0.03y \end{array}$$

Strain components

$$\varepsilon_{xx} = \frac{\partial u}{\partial x} = 0.05 + 0.02y$$

$$\varepsilon_{yy} = \frac{\partial v}{\partial y} = 0.03$$

$$\varepsilon_{xy} = \frac{1}{2}\left(\frac{\partial v}{\partial x} + \frac{\partial u}{\partial y}\right) = \frac{1}{2}(0.06 + 0.01 + 0.02x) = 0.035 + 0.01x$$

Stress components

$$\sigma_{xx} = \frac{E}{1 - \nu^2}\left(\varepsilon_{xx} + \nu\varepsilon_{yy}\right) \Rightarrow \sigma_{xx} = \frac{210}{1 - 0.33^2}(0.05 + 0.02y + 0.33 \times 0.03)$$

$$\sigma_{yy} = \frac{E}{1 - \nu^2}\left(\varepsilon_{yy} + \nu\varepsilon_{xx}\right) \Rightarrow \sigma_{yy} = \frac{210}{1 - 0.33^2}(0.03 + 0.33(0.05 + 0.02y))$$

$$\tau_{xy} = \frac{E}{1+v}\varepsilon_{xy} \Rightarrow \tau_{xy} = \frac{210}{1+0.33}(0.035+0.01x)$$

or

$$\begin{cases} \sigma_{xx} = 14.14 + 4.712y \\ \sigma_{yy} = 10.95 + 1.55y \\ \tau_{xy} = 5.53 + 1.58x \end{cases}$$

Direction of the maximum shear stress on the point O (0.5, 0.5)
The stress components on the point O (0.5, 0.5) are:

$$\sigma_{xx} = 14.14 + 4.712 \times 0.5 \Rightarrow \sigma_{xx} = 16.49GPa$$

$$\sigma_{yy} = 10.95 + 1.55 \times 0.5 \Rightarrow \sigma_{yy} = 11.72GPa$$

$$\tau_{xy} = 5.53 + 1.58 \times 0.5 \Rightarrow \tau_{xy} = 6.32GPa$$

$$\tan\ 2\theta = -\frac{\sigma_{xx} - \sigma_{yy}}{2\tau_{xy}}$$

or

$$2\theta = \arctan\left(-\frac{16.49 - 11.72}{2 \times 6.32}\right)$$

Therefore:

$$\theta = -3.3 \times 10^{-3} rad$$

PROBLEM 4

The suspension system of a truck consists of three beams as shown in Fig. A.5a. The three beams are perfectly bonded. Calculate the angle **φ** in the force-displacement diagram in Fig. A.5b. (*Tip: Ignore the effect of the shear forces on the elastic energy of the beam.*)

Given data:
Beam length **α**
Dimensions of cross sections **b, h**
Elasticity modulus **E**

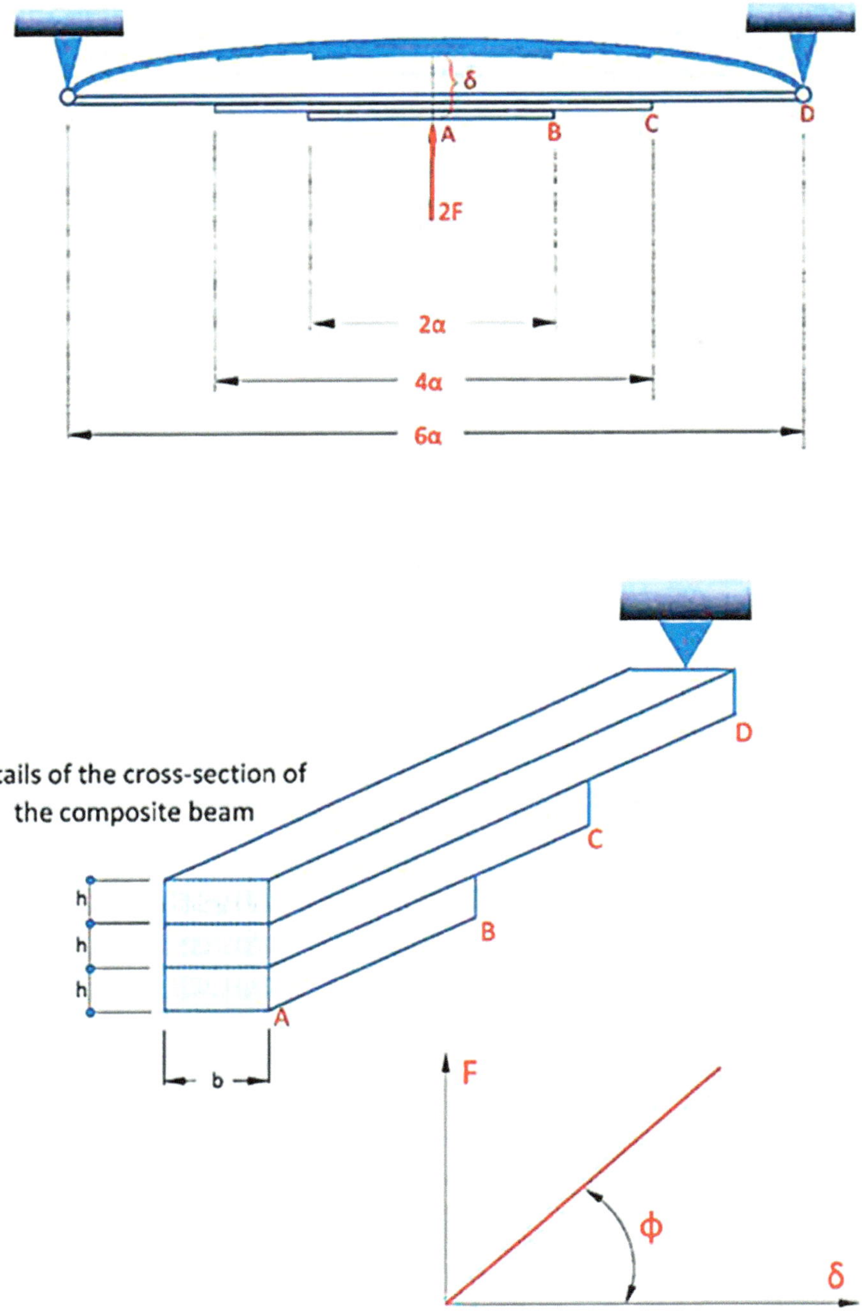
δ
A
B
C
D
2F
2α
4α
6α
Details of the cross-section of the composite beam
h
h
h
b
A
B
C
D
F
ϕ
δ

A

B

FIG. A.5

a, b

SOLUTION

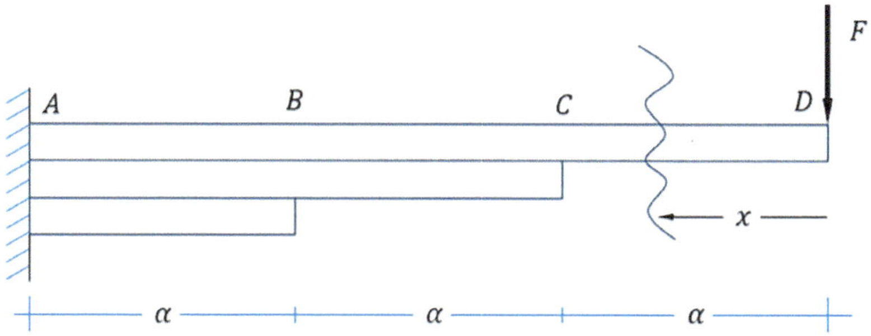

With the aid of Castigliano's theorem, we calculate the deflection on point D:

$$\delta = \frac{\partial U}{\partial F}$$

or

$$\delta = \frac{1}{EI_{CD}}\int_0^{\alpha} M_{CD}\frac{\partial M_{CD}}{\partial F}dx + \frac{1}{EI_{BC}}\int_{\alpha}^{2\alpha} M_{BC}\frac{\partial M_{BC}}{\partial F}dx + \frac{1}{EI_{AB}}\int_{2\alpha}^{3\alpha} M_{AB}\frac{\partial M_{AB}}{\partial F}dx$$

In the above equation, we replace the bending moments:

$$\delta = \frac{1}{EI_{CD}}\int_0^{\alpha}(F_x)xdx + \frac{1}{EI_{BC}}\int_{\alpha}^{2\alpha}(F_x)xdx + \frac{1}{EI_{AB}}\int_{2\alpha}^{3\alpha}(F_x)xdx$$

or

$$E\frac{\delta}{F} = \frac{1}{I_{CD}}\int_0^{\alpha}x^2dx + \frac{1}{I_{BC}}\int_{\alpha}^{2\alpha}x^2dx + \frac{1}{I_{AB}}\int_{2\alpha}^{3\alpha}x^2dx$$

The above equation yields:

$$E\frac{\delta}{F} = \frac{1}{3I_{CD}}(\alpha^3 - 0^3) + \frac{1}{3I_{BC}}[(2\alpha)^3 - \alpha^3] + \frac{1}{3I_{AB}}[(3\alpha)^3 - (2\alpha)^3]$$

or

$$\frac{F}{\delta} = \frac{3E}{\alpha^3}\frac{I_{CD}}{I_{BC}I_{AB}(19I_{CD}I_{BC} + 7I_{CD}I_{AB} + I_{BC}I_{AB})}$$

The angle φ in Fig. A.5b is given by the following equation:

$$\varphi = \arctan\frac{F}{\delta}$$

where:

$$I_{CD} = \frac{bh^3}{12}$$

$$I_{BC} = \frac{b(2h)^3}{12}$$

$$I_{AB} = \frac{b(3h)^3}{12}$$

PROBLEM 5

A circular ring of radius **R** is made by a beam with circular cross section of diameter **d**. The ring is subjected to forces **2F** as shown in Fig. A.6. Calculate the displacement **δ** of the points **B** and **B′**. (Tip: Ignore the effect of the axial and shear forces on the elastic energy of the beam.)

Given data:

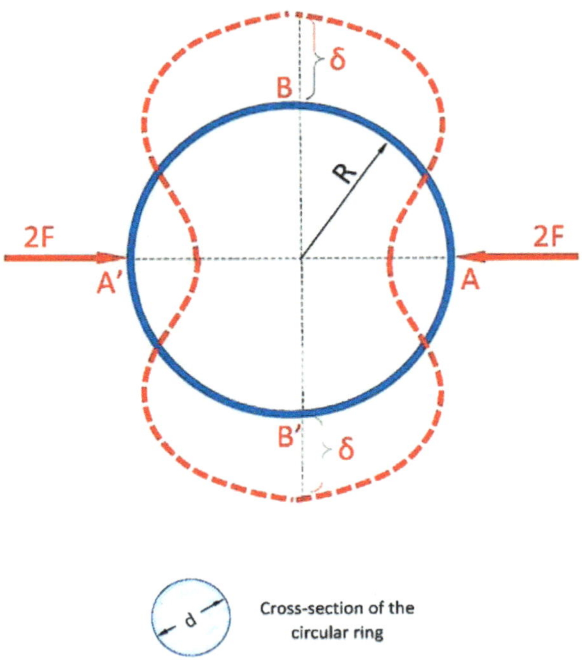

FIG. A.6

Force **F**
Elasticity modulus **E**
Ring radius **R**
Beam diameter **d**

SOLUTION

The deflection δ can be calculated with the Castigliano's theorem. Because a force on point B does not exist, a dummy load S(=0) should be considered. The structure in Fig. A.5 has symmetric geometry and loads. Therefore we can analyze the one quarter of the structure. The free body diagram and the boundary condition of the part to be analyzed is shown in the following figure:

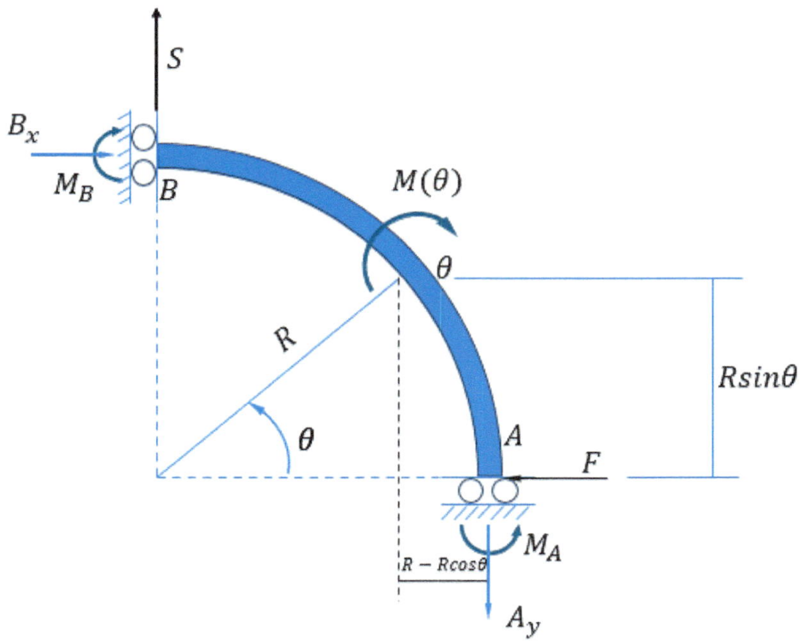

Equilibrium equations:

$$\sum F_x = 0 \Rightarrow B_x - F = 0 \Rightarrow B_x = F$$

$$\sum F_y = 0 \Rightarrow -A_y + S = 0 \Rightarrow A_y = S$$

$$\sum M_B = 0 \Rightarrow M_B + FR + A_y R - M_A = 0 \Rightarrow M_A = M_B + (F + S)R \tag{1}$$

Boundary condition on the support B:

$$\varphi_B = 0 \Rightarrow \frac{\partial U}{\partial M_B} = 0 \Rightarrow \frac{1}{EI}\int_0^{\frac{\pi}{4}} M(\theta)\frac{\partial M(\theta)}{\partial M_B}\bigg|_{S=0} Rd\theta = 0 \tag{2}$$

From the equilibrium of the free body diagram of the arch AΘ, we get:

$$M(\theta) = M_A - FR\ \sin\theta - SR(1 - \cos\theta)$$

With the aid of Eq. (1), the above equation yields:

$$M(\theta) = [M_B + (F + S)R] - FR\ \sin\theta - SR(1 - \cos\theta)$$

or

$$M(\theta) = M_B + FR(1 - \sin\theta) + SR\ \cos\theta \tag{3}$$

From the condition $\varphi_B = 0$, we can determine the moment M_B:
Considering Eq. (3), Eq. (2) yields:

$$\pi M_B + 1.97FR = 0$$

or

$$M_B = -0.63FR \tag{4}$$

Castigliano's theorem for calculating the deflection δ yields:

$$\delta = \frac{\partial U}{\partial S}\bigg|_{s=0}$$

or

$$\delta = \frac{1}{EI}\int_0^{\frac{\pi}{4}} M(\theta)\frac{\partial M(\theta)}{\partial M_B}\bigg|_{S=0} Rd\theta = 0 \tag{5}$$

With the aid of Eq. (4), the bending moment distribution $M(\theta)$ can be obtained by Eq. (3):

$$M(\theta) = -0.63FR + FR(1 - \sin\theta) + SR\ \cos\theta$$

$$\text{Then: } \frac{\partial M(\theta)}{\partial S} = R\ \cos\theta$$

Finally, Eq. (5) yields:

$$\delta = \frac{1}{EI}\int_0^{\frac{\pi}{4}} [-0.63FR + FR(1 - \sin\theta)]R\ \cos\theta \cdot Rd\theta$$

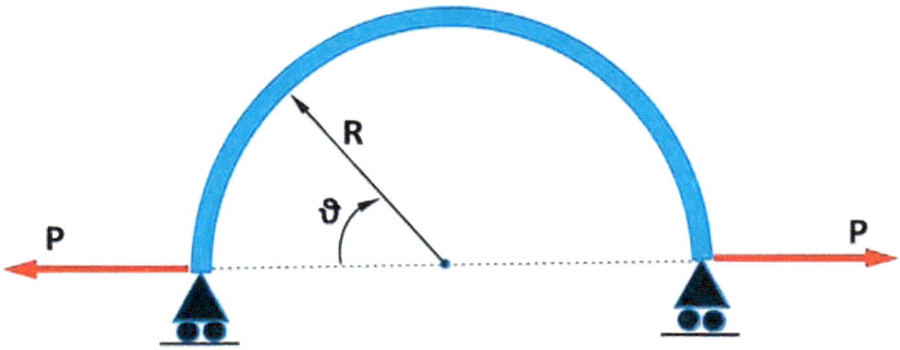

FIG. A.7

or

$$\delta = \frac{0.01163}{EI} FR^3$$

PROBLEM 6

Derive the function of the radial displacements $u_r(\theta)$ for the beam of Fig. A.7. Take into account only the bending effect.

Data: R = 2.0 m, P = 15KN, 1 = 8 × 10^{-8} m^4, E = 210 GPa

SOLUTION

The geometry of the structure and loading is symmetric. Therefore it can be simulated by the following model (Fig. a). For calculating the radial displacement to any arbitrary point on θ, a dummy load should be places (Fig. b)

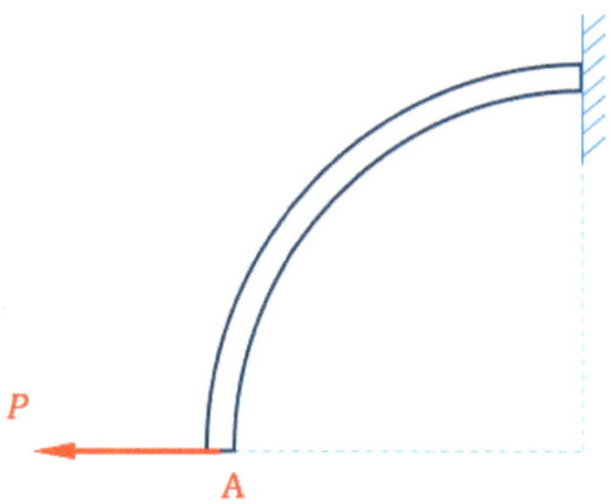

A

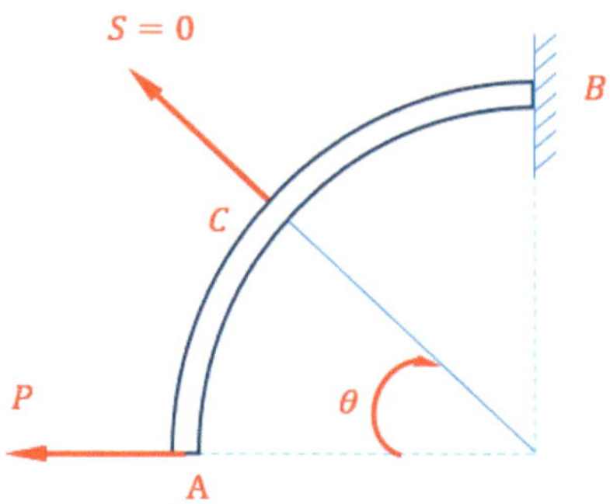

B

According to Castigliano's theorem, the radial deflection on C can be obtained by the following equation:

$$u_r(\theta) = \left.\frac{\partial U}{\partial S}\right|_{S=0}$$

or

$$u_r(\theta) = \frac{1}{EI}\int_0^{\theta} M_{AC}(\xi)\frac{\partial M_{AC}(\xi)}{\partial S}rd\xi + \int_{\theta}^{\frac{\pi}{2}} M_{CB}(\xi)\frac{\partial M_{CB}(\xi)}{\partial S}Rd\xi \tag{1}$$

Derivation of the bending moments $M_{AC}(\xi)$
In the following free body diagram for $0 \leq \xi \leq \theta$,

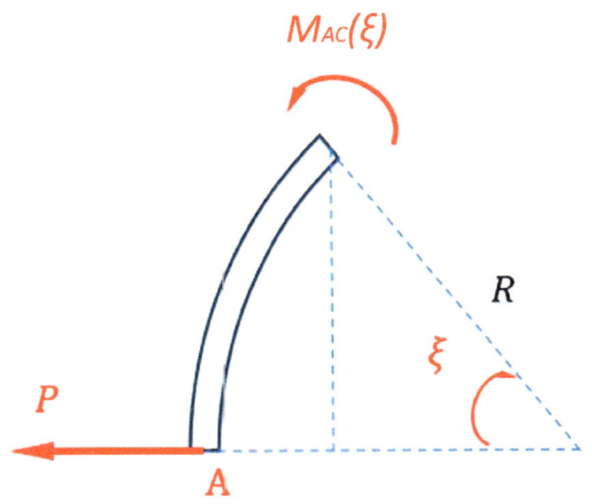

we apply the equilibrium equation of moments:

$$M_{AC}(\xi) = -PR \sin \xi = 0 \Rightarrow M_{AC}(\xi) = PR \sin \xi$$

Taking the derivative of the above equation with respect to the dummy load, we get:

$$\frac{\partial M_{AC}(\xi)}{\partial S} = 0$$

Derivation of the bending moments $M_{CB}(\xi)$
In the following free body diagram for $\theta < \xi \leq \pi/2$,

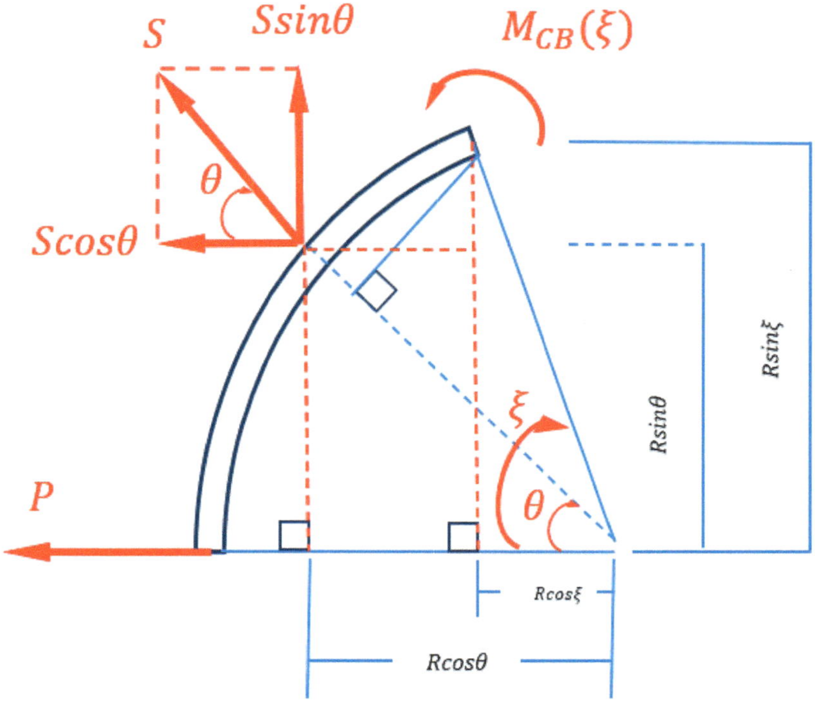

we apply the equilibrium equation of moments:

$$M_{CB}(\xi) - S\sin\theta(R\cos\theta - R\cos\xi) - S\cos\theta(R\sin\xi - R\sin\theta) - PR\sin\xi = 0$$

or

$$M_{CB}(\xi) = -SR\sin\theta(\cos\theta - \cos\xi) - SR\cos\theta(\sin\xi - \sin\theta) - PR\sin\xi$$

Taking the derivative of the above equation with respect to the dummy load, we get:

$$\frac{\partial M_{CB}(\xi)}{\partial S} = -R\sin\theta\cos\theta + R\sin\theta\cos\xi - R\cos\theta\sin\xi + R\cos\theta\sin\theta$$

Taking into account the above bending moments and the corresponding derivatives, the radial deflection can be obtained by Eq. (1), where the value of the dummy load should be taken S = 0:

$$u_r(\theta) = 0 + \int_{\theta}^{\frac{\pi}{2}} -PR\sin\xi (-R\sin\theta\cos\theta + R\sin\theta\cos\xi - R\cos\theta\sin\xi + R\cos\theta\sin\theta)Rd\xi$$

or

$$u_r(\theta) = PR^3\left[\cos\theta \int_\theta^{\frac{\pi}{2}} \sin^2\xi \, d\xi - \sin\theta \int_\theta^{\frac{\pi}{2}} \sin\xi \ \cos\xi \, d\xi\right]$$

PROBLEM 7

Point B of the semicircular beam (Fig. A.8) is subjected to a vertical force F. The curved beam is reinforced by a straight bar AB with cross-section area A_{br}. Calculate the vertical displacement of point B. (*Tip: Ignore the effect of the axial and shear forces on the elastic energy of the curved beam.*)

Given data:

Force F
Radius R
Bar cross-section area A_{br}
Beam stiffness El

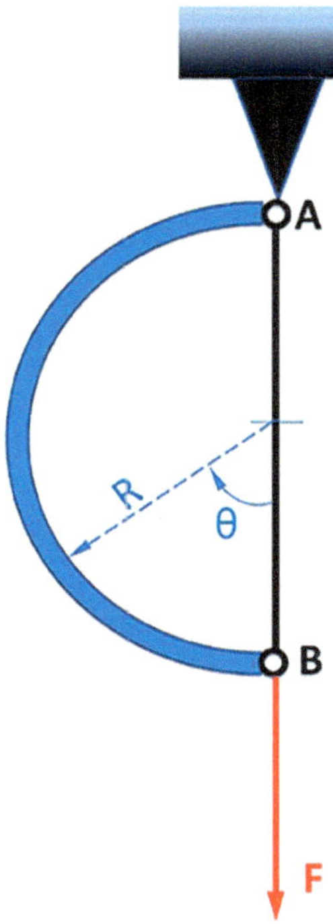

FIG. A.8

SOLUTION

We decompose the structure into the substructures (I) and (II).

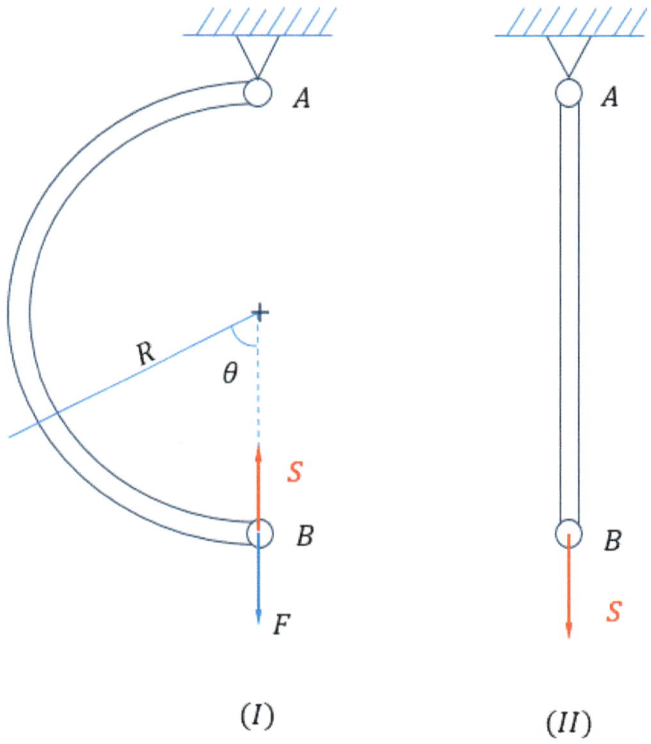

From the condition

$$u_B^{(I)} = u_B^{(II)} \tag{1}$$

we can obtain the unknown force S that acts to the straight bar AB.

The resultant force on the curved beam (substructure I) is:

$$Q = F - S \tag{2}$$

Bending moments in the curved beam:

Taking the equilibrium if moments in the free body diagram below,

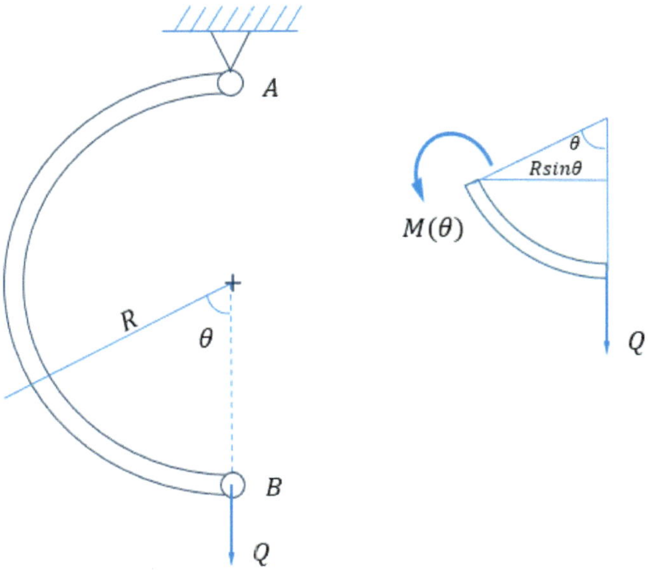

we get:

$$M(\theta) = QR \sin \theta$$

The derivative of the above equation with respect to Q is:

$$\frac{\partial M(\theta)}{\partial Q} = R \sin \theta$$

Vertical displacement of the point B in the curved beam (substructure I)

According to Castigliano's theorem, the vertical displacement of the point B in the substructure I is:

$$u_B{}^{(I)} = \frac{\partial U}{\partial Q} \Rightarrow u_B{}^{(I)} = \frac{1}{EI}\int_0^{\pi} M(\theta)\frac{\partial M(\theta)}{\partial Q} R d\theta$$

or

$$u_B{}^{(I)} = \frac{1}{EI}\int_0^{\pi} (QR \sin \theta)(R \sin \theta) R d\theta \Rightarrow u_B{}^{(I)} = \frac{QR^3}{EI}\int_0^{\pi} \sin^2 \theta d\theta$$

The above equation yields:

$$u_B^{(I)} = \frac{\pi Q R^3}{2EI}$$

Because Q = F – S, the following result can be obtained:

$$u_B^{(I)} = \frac{\pi (F - S) R^3}{2EI} \quad (3)$$

Vertical displacement of point B in the straight bar (substructure II)
Hooke's law $\sigma = E\varepsilon$ for the straight bar yields

$$\frac{S}{A_{br}} = E\frac{u_B^{(II)}}{2R}$$

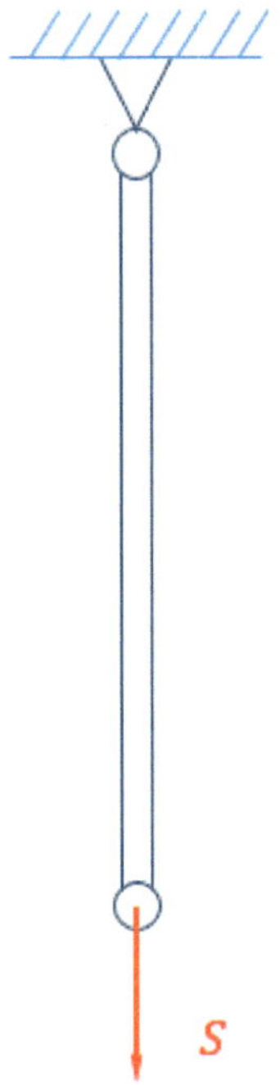

or

$$u_B^{(II)} = \frac{2RS}{EA_{br}} \quad (4)$$

With the aid of Eqs. (3) and (4), Eq. (1) yields:

$$\frac{\pi(F-S)R^3}{2EI}=\frac{2RS}{EA_{br}}$$

or

$$S=\frac{\pi A_{br}R^2}{4I+\pi A_{br}R^2}F \qquad (5)$$

Then, the vertical displacement of B can be obtained by Eq. (3) or Eq. (4). Selecting Eq. (4), the following result can be obtained:

$$u_B=\frac{2\pi A_{br}R^3}{EA_{br}\left(4I+\pi A_{br}R^2\right)}F$$

PROBLEM 8

A circular ring of radius R is made by a beam with circular cross section of diameter d. The ring is subjected to a pair of forces 2 F and supported by two springs of elastic coefficient k (Fig. A.9). Calculate the vertical displacement δ of Points B and B′. (*Tip: Ignore the effect of the axial and shear forces on the elastic energy of the beam.*)

Given data:

Force F
Spring elastic coefficient k
Beam stiffness El
Ring radius R
Beam cross section diameter d

SOLUTION

The deflection δ can be calculated with the aid of Castigliano's theorem. The structure in Fig. A.9 has symmetric geometry and loads. Therefore we can analyze one-quarter of the structure. The free body diagram and the boundary condition of the part to be analyzed is shown in the following figure:

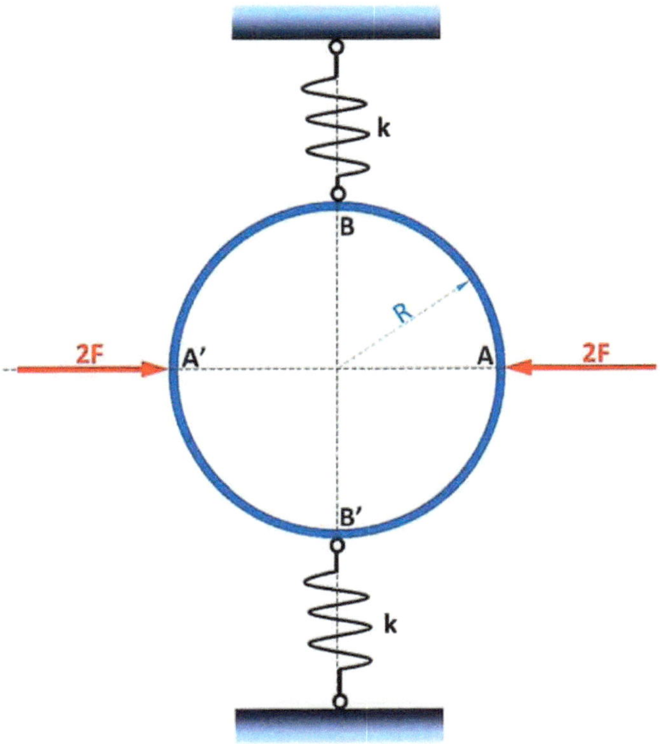
k
B
R
2F
A'
A
2F
B'
k

FIG. A.9

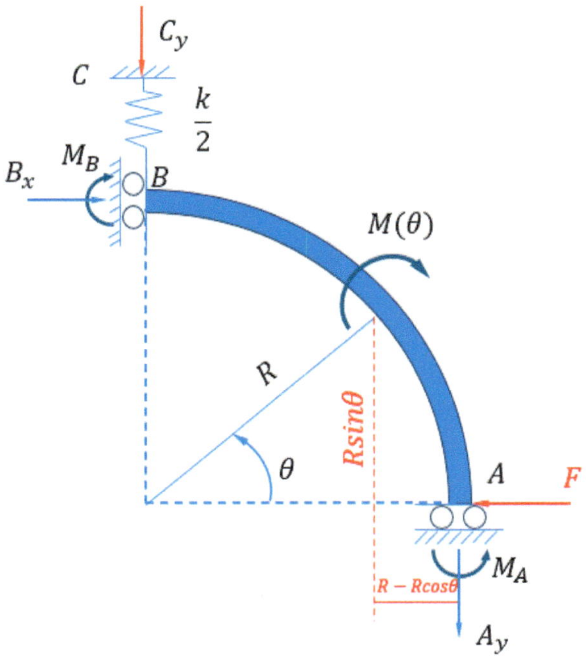
C_y
C
$\frac{k}{2}$
M_B
B_x
B
$M(\theta)$
R
$R\sin\theta$
θ
A
F
M_A
$R - R\cos\theta$
A_y

Equilibrium equations

$$\sum F_x = 0 \Rightarrow B_x - F = 0 \Rightarrow B_x = F$$

$$\sum F_y = 0 \Rightarrow -A_y - C_y = 0 \Rightarrow A_y = -C_y$$

$$\sum M_B = 0 \Rightarrow M_B + FR + A_y R - M_A \Rightarrow M_A = M_B + (F - C_y)R \tag{1}$$

Boundary condition on the support "B"

$$\varphi_B = 0 \Rightarrow \frac{\partial U}{\partial M_B} = 0 \Rightarrow \frac{1}{EI}\int_0^{\frac{\pi}{4}} M(\theta)\frac{\partial M(\theta)}{\partial M_B} R d\theta = 0 \tag{2}$$

Bending moments

$$M(\theta) = M_A - FR \sin\theta - A_y(r - R\cos\theta)$$

With the aid of Eq. (1), the above equation yields:

$$M(\theta) = M_B + (F - C_y)R - FR\sin\theta - C_y R(1 - \cos\theta)$$

or

$$M(\theta) = M_B + FR(1 - \sin\theta) - C_y R\cos\theta \tag{3}$$

With the aid of the above equation, Eq. (2) yields:

$$\int_0^{\frac{\pi}{4}} \left[M_B + FR(1 - \sin\theta) - C_y R\cos\theta\right][1]d\theta = 0$$

or

$$M_B\frac{\pi}{4} + FR\frac{\pi}{4} - FR\int_0^{\frac{\pi}{4}} \sin\theta d\theta - C_y R\int_0^{\frac{\pi}{4}} \cos\theta d\theta = 0$$

Then, the following result can be obtained:

$$M_B = -1.38FR - 0.89C_y R \tag{4}$$

Boundary condition on the support "C"
The displacement on the support "C" is null:

$$u_c = 0 \Rightarrow \frac{\partial U}{\partial C_y} = 0 \Rightarrow \int_0^{\frac{\pi}{4}} M(\theta)\frac{\partial M(\theta)}{\partial C_y} R d\theta = 0 \tag{5}$$

With the aid of Eq. (4), Eq. (3) yields:

$$M(\theta) = -1.38FR - 0.89C_y R + FR(1 - \sin\theta) - C_y R\cos\theta$$

Then

$$\frac{\partial M(\theta)}{\partial C_y} = 0.89R - R\ \cos\ \theta = -R(0.89 + \cos\ \theta)$$

With the aid of the above expressions, Eq. (5) yields:

$$\int_0^{\frac{\pi}{4}} [-1.38FR - 0.89C_yR + FR(1 - \sin\theta) - C_yR\cos\theta][0.89 + \cos\theta]d\theta = 0$$

or

$$\int_0^{\frac{\pi}{4}} \left[-1.38FR - 0.89C_yR + FR(1 - \sin\theta) - C_yR\ \cos\theta\right]d\theta +$$
$$\int_0^{\frac{\pi}{4}} \left(-1.38FR\ \cos\theta - 0.89C_yR\ \cos\theta + FR\ \cos\ \theta - FR\right.$$
$$\left.\sin\theta\ \cos\theta - C_yR\ \cos^2\theta\right)d\theta = 0$$

After some simple algebraic operations, we get

$$-\ 1.23FR\frac{\pi}{4}\ \underline{-0.79C_yR\frac{\pi}{4}} + FR\frac{\pi}{4} - 0.89FR$$
$$\left(1 - \frac{1}{\sqrt{2}}\right) - \underline{0.89C_yR\left(\frac{2+\pi}{8}\right)}$$
$$-\ 1.38FR\frac{\sqrt{2}}{2}\ \underline{-0.89C_yR\frac{\sqrt{2}}{2}} + FR\frac{\sqrt{2}}{2}$$
$$-\ FR\frac{1}{4}\ \underline{-CyR\left(\frac{2+\pi}{8}\right)} = 0$$

or

$$C_y = -\frac{1}{2.52}F \text{ compression}$$

Considering Hooke's law for the spring on the support "C,"

$$C_y = k\delta$$

the following result can be obtained:

$$\delta = \frac{C_y}{k}$$

or

$$\delta = -\frac{1}{2.52}\frac{F}{k}$$

PROBLEM 9

The curved beam of Fig. A.10 is subjected to a force **F**. Calculate the vertical displacement of the point **B**.

Given data:
Force **F**
Radius **R**

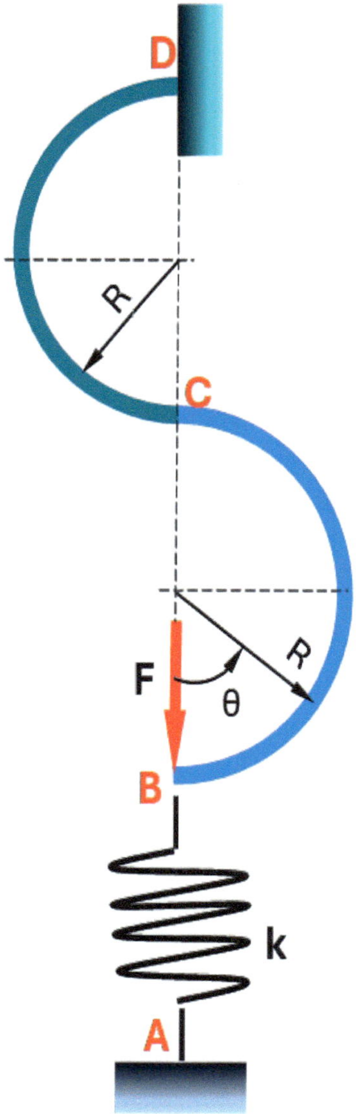

FIG. A.10

Spring constant **k**
Beam stiffness **EI**
Useful formula: $\int (\sin^2 x)dx = \frac{x}{2} - \frac{1}{4} \sin 2x$
Tip: Ignore the effect of the axial and shear forces on the elastic energy of the beam.

SOLUTION

We first apply Castigliano's theorem for the deflection on the point A to calculate the support reaction force S:

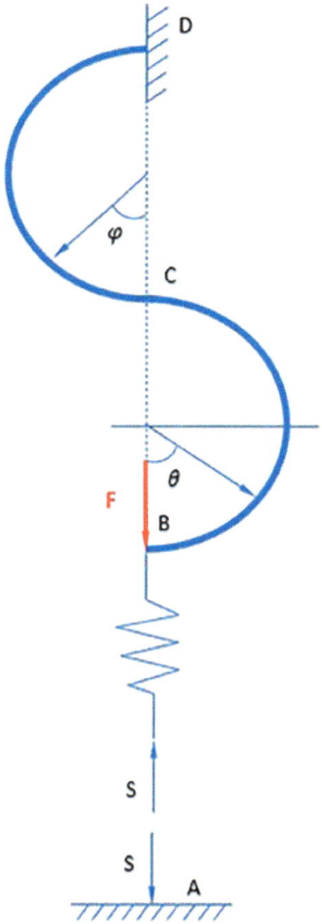

$$u_A = \frac{\partial U}{\partial S} = 0$$

or

$$\frac{1}{EI}\int_0^{\pi} M_{BC}\frac{\partial M_{BC}}{\partial S}Rd\theta + \frac{1}{EI}\int_0^{\pi} M_{CD}\frac{\partial M_{CD}}{\partial S}RD\varphi + \frac{1}{k}S\frac{\partial S}{\partial S} = 0 \qquad (1)$$

Bending moment M_{BC}

The bending moment M_{BC} can be obtained by the moment equilibrium in the following free body diagram:

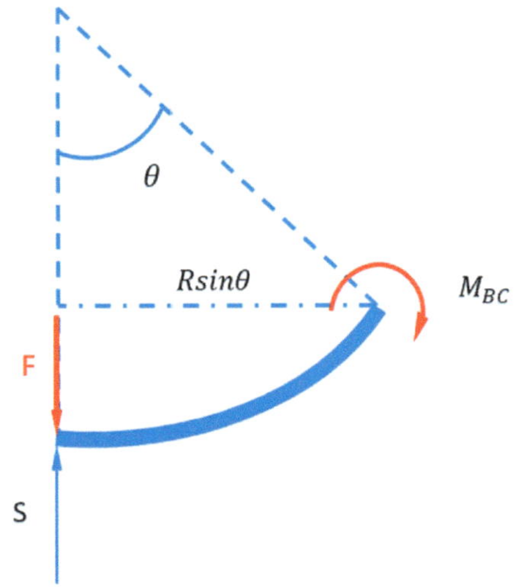

$$M_{BC} = (F - S)R \sin \theta$$

Taking the derivative of the above equation with respect to S, we get:

$$\frac{\partial M_{BC}}{\partial S} = -R \sin \theta$$

Bending moment M_{CD}

We follow a similar procedure for deriving the bending moment M_{CD}. The moment equilibrium in the following free body diagram yields:

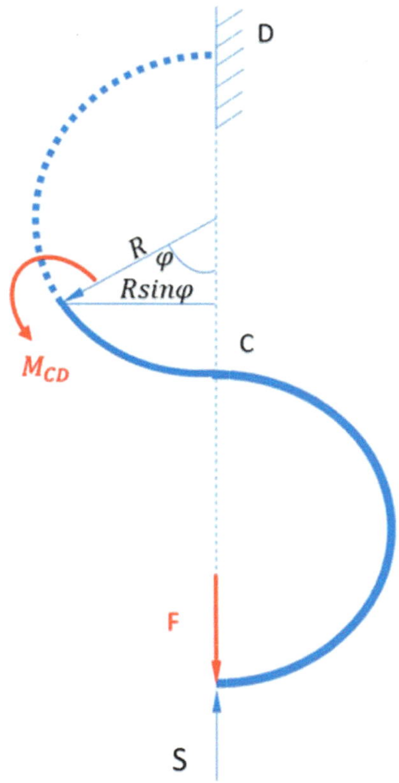

$$M_{CD} = (F - S)R \sin \varphi$$

Taking the derivative of the above equation with respect to S, we get:

$$\frac{\partial M_{CD}}{\partial S} = -R \sin \varphi$$

Then Eq. (1) yields:

$$\frac{2}{EI}\int_0^{\pi} (F - S)R \sin \theta(-R \sin \theta)Rd\theta + \frac{S}{k} = 0$$

or

$$-\frac{2R^3}{EI}(F - S)\int_0^{\pi} \sin^2 \theta \cdot d\theta + \frac{S}{k} = 0$$

The above equation can provide the result for the force S:

$$S = \frac{k\pi R^3}{EI + k\pi R^3} F$$

The displacement on point A is equal to the contraction u_A of the spring. The contraction u_A can be calculated by Hooke's law of the spring:

$$S = k \cdot u_A$$

or

$$u_A = \frac{S}{k} = \frac{\pi R^3}{EI + k\pi R^3} F$$

PROBLEM 10

A cantilever beam of a rectangular cross section is subjected to a vertical load P on its free end G. The cross section of the beam is linearly variable along the beam. Calculate the vertical deflection of the free end G.

Given data:

Force: **P**

Dimensions: **L, a, c, b**

Elasticity modulus: **E**

Usefulformula: $\int (mx + n)^{-3} x^2 dx = \frac{1}{2m^3} \left(\frac{n(3n + 4mx)}{(n + mx)^2} + 2 \log[n + mx] \right)$

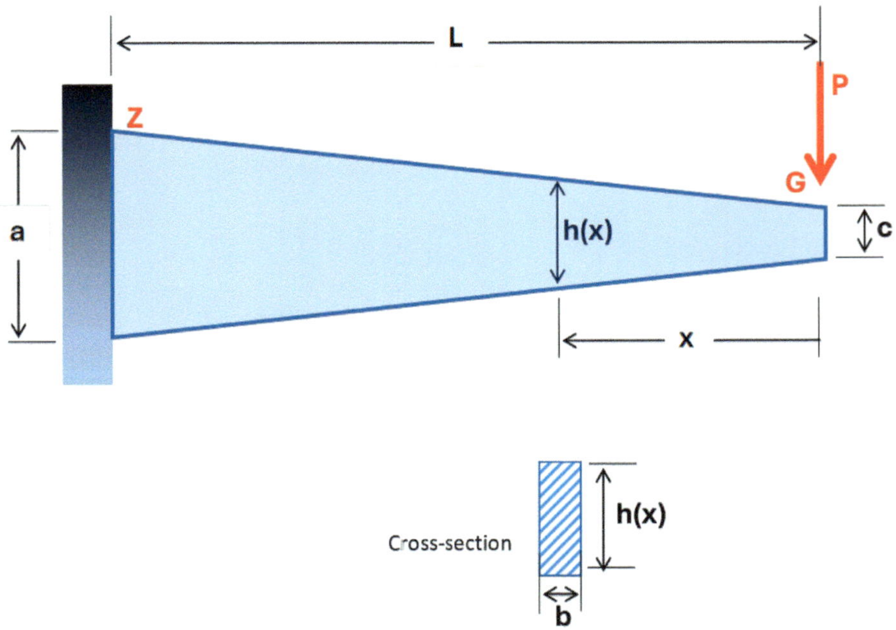

FIG. A.11

SOLUTION

The vertical displacement on G can be calculated with Castigliano's theorem:

$$U = \frac{1}{2E}\int_0^L \frac{M^2}{I(x)}dx$$

The moment of inertia in the above equation varies along x:

$$I(x) = \frac{bh^3(x)}{12}$$

The variation of h(x) is linear:

$$h(x) = Ax + B \tag{1}$$

For $x = 0,\ h(x) = c$
Then Eq. (1) yields:

$$c = A \cdot 0 + B$$

or

$$B = C$$

For $x = L,\ h(x) = a$
Then Eq. (1) yields:

$$a = AL + c$$

or

$$A = \frac{a-c}{L}$$

Having the results for the parameters A and B, the following formula for $h(x)$ can be obtained:

$$h(x) = \tfrac{a-c}{L}x + c$$

Using the above formula, the following equation for the moment of inertia can be obtained:

$$I(x) = \tfrac{b}{12}\left(\tfrac{a-c}{L}x + c\right)^3$$

Then, Castigliano's theorem yields:

$$u_G = \frac{1}{E}\int_0^L \frac{1}{I(x)}M(x)\frac{\partial M(x)}{\partial P}dx \tag{2}$$

Bending moments

Taking the bending moment equilibrium in the following free body diagram

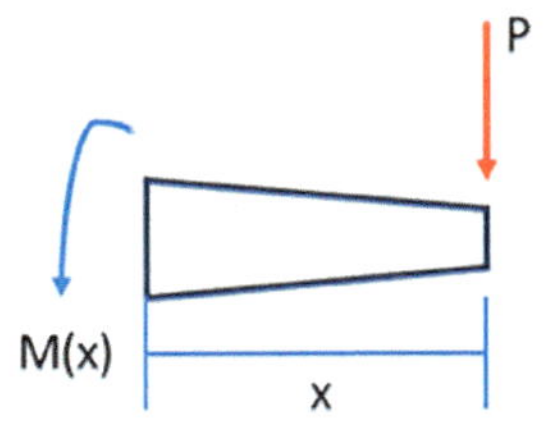

we get:

$$M(x) = Px$$

The derivative of the above equation is:

$$\frac{\partial M(x)}{\partial P} = x$$

Then Eq. (2) yields:

$$u_G = \frac{12P}{Eb} \int_0^L \left(\frac{a - c}{L} x + c \right)^{-3} x^2 dx$$

The above integration can be performed by the following formula:

$$\int_0^L (mx + n)^{-3} x^2 dx = \frac{1}{2m^3} \left[-3 + \frac{n(4Lm + 3n)}{(Lm + n)^2} - 2\ \ln(n) + 2\ \ln(Lm + n) \right]$$

Index

Note: Page numbers followed by *f* indicate figures; *t*, tables, *b*, boxes.

W

Y